Manufacturing Process Management

Manufacturing Process Management

Edited by Michelle Vine

Clanrye International
www.clanryeinternational.com

Clanrye International,
750 Third Avenue, 9th Floor,
New York, NY 10017, USA

ISBN: 978-1-63240-625-5

Cataloging-in-Publication Data

Manufacturing process management / edited by Michelle Vine.
p. cm.
Includes bibliographical references and index.
ISBN 978-1-63240-625-5
1. Manufacturing processes. 2. Manufacturing processes--Management. 3. Production engineering.
I. Vine, Michelle.
TS183 .M36 2017
670--dc23

For information on all Clanrye International publications
visit our website at www.clanryeinternational.com

CLANRYE INTERNATIONAL

Printed in China.

Contents

Preface

This book has been an outcome of determined endeavour from a group of educationists in the field. The primary objective was to involve a broad spectrum of professionals from diverse cultural background involved in the field for developing new researches. The book not only targets students but also scholars pursuing higher research for further enhancement of the theoretical and practical applications of the subject.

This book on manufacturing process management discusses the entire cycle of the manufacturing process. This includes the conversion of raw materials to the finished commodity. Topics included in this book delve into the efficient, effective and improved methods by which goods are produced. It traces the progress of this field and highlights some of its key concepts and applications. It includes some of the vital pieces of work being conducted across the world, on various topics related to manufacturing process and its management. All the processes of the life cycle of a commodity are described in detail. This book aims to equip students and experts with the advanced topics and upcoming concepts in this area. It will be very useful to students and teachers in the fields of industrial engineering, supply chain management and manufacturing engineering.

It was an honour to edit such a profound book and also a challenging task to compile and examine all the relevant data for accuracy and originality. I wish to acknowledge the efforts of the contributors for submitting such brilliant and diverse chapters in the field and for endlessly working for the completion of the book. Last, but not the least; I thank my family for being a constant source of support in all my research endeavours.

Editor

A dichotomous search-based heuristic for the three-dimensional sphere packing problem

Mhand Hifi[1]* and Labib Yousef[1]

*Corresponding author: Mhand Hifi, Laboratory EPROAD – EA 4669 (ROAD Team), UFR des Sciences, University of Picardie Jules Verne, 7 rue du Moulin Neuf, 80000 Amiens, France

E-mail: hifi@u-picardie.fr

Reviewing editor: Zude Zhou, Wuhan University of Technology, China

Abstract: In this paper, the three-dimensional sphere packing problem is solved by using a dichotomous search-based heuristic. An instance of the problem is defined by a set of *n* unequal spheres and an object of fixed width and height and, unlimited length. Each sphere is characterized by its radius and the aim of the problem is to optimize the length of the object containing all spheres without overlapping. The proposed method is based upon beam search, in which three complementary phases are combined: (i) a greedy selection phase which determines a series of eligible search subspace, (ii) a truncated tree search, using a width-beam search, that explores some promising paths, and (iii) a dichotomous search that diversifies the search. The performance of the proposed method is evaluated on benchmark instances taken from the literature where its obtained results are compared to those reached by some recent methods of the literature. The proposed method is competitive and it yields promising results.

Subjects: Combinatorics; Operational Mathematics; Mathematical Modeling

Keywords: beam; dichotomous; heuristic; optimization; packing; production

1. Introduction

Cutting and Packing problems (CP) is a family of natural combinatorial optimization problems, admitted in numerous real-world applications from industrial engineering, logistics, manufacturing, production process, automated planning, etc. One of the more recent papers addressing an optimization with a packing problem is due to Sutou and Dai (2002), where the unequal sphere problem has been used for tackling an application of the automated radiosurgical treatment planning. Wang (1999) has also considered sphere packing problems as an optimization tool for the radiosurgical treatment planning.

ABOUT THE AUTHORS

Mhand Hifi is a full professor of Computer Science and Operations Research at the University of Picardie Jules Verne, France. He is the head of the laboratory EPROAD (Eco-PRocédés, Optimisation et Aide à la Décision) of the University of Picardie Jules Verne. He is area editor for several international journals: COR, IJMOR, AOR, AJOR, etc. His research interest is NP Hard combinatorial optimisation (sequential and parallel approaches) applied to logistic and OR problems.

Labib Yousef is a PhD student at the University of Picardie Jules Verne. He received his BS in Management at the University of Techrine, Syria. He got his MS in Industrial Engineering (Management de l'Innovation Technologique) from Grenoble INP, France.

PUBLIC INTEREST STATEMENT

Cutting and Packing problems (CP) is a family of natural combinatorial optimization problems, admitted in numerous real-world applications from industrial engineering, logistics, manufacturing, production process, automated planning, etc.

This paper tackles a version of the sphere packing problem (belonging to the CP family) with a dichotomous search-based method.

Such a problem can be encountered in the radiosurgical treatment where the unequal sphere problem has been used for determining an automated radiosurgical treatment planning.

Other problems of the CP family have been described and redefined in Wascher, Haussner, and Schumann (2007) where an instance of these problems can be defined by a set of predetermined items to be packed in one or many larger containers (objects) so as to minimize the unused area/space or in some cases to maximize a utility function. Furthermore, the items are bounded by their dimensions (rectangular, circular, or irregular) and the objects can be bounded (rectangular, circular, ...) or unbounded (strips/parallelepipeds, ...).

This paper deals with the problem of packing spheres (called items) in a container (parallelepiped or object) with unlimited length. The problem is known as the *Three-Dimensional Sphere Packing Problem* (3DSPP). An instance of 3DSPP is defined by a set N of n unequal items and an object $\mathcal{P}$ of fixed width W and height H and, unlimited length (for the rest of the paper, the length representing the variation of P's length is denoted L). Moreover, each item $i \in N = \{1, \ldots, n\}$ is characterized by its radius r_i. The goal of the problem is to optimize the length L of the object $\mathcal{P}$ such that all items of N are packed (positioned) in the target object, without overlapping.

The 3DSPP may be formulated as follows:

$$\min \quad L \tag{1}$$

$$(x_i - x_j)^2 + (y_i - y_j)^2 + (z_i - z_j)^2 \geq (r_i + r_j)^2 \quad \forall (i,j) \in N^2, i < j \tag{2}$$

$$r_i \leq x_i \leq L - r_i \qquad \forall i \in N \tag{3}$$

$$r_i \leq y_i \leq H - r_i \qquad \forall i \in N \tag{4}$$

$$r_i \leq z_i \leq W - r_i \qquad \forall i \in N \tag{5}$$

$$\underline{L} \leq L \leq \overline{L} \tag{6}$$

where the objective function (1) minimizes the length of the object $\mathcal{P}$ containing all the items of N, Equation (2) ensures the nonoverlap constraint of any pair of distinct items (i,j) of N; that is, the distance between the centers of both items which must be greater than or equal to the sum of their radii and, Equations (3)–(5) ensure that all items of N belong to the target object $\mathcal{P}$ of dimensions (L,W,H). Note that since the aim of the problem is to find the smallest length of the object containing all items of N, then it is easy to start any method trying to solve the above problem by a trivial value representing the sum of the spheres' area affected to $\underline{L}$ (Equation (6)) and a quick (feasible) solution value with a greedy algorithm affected to $\overline{L}$.

The rest of the paper is organized as follows. Section 2 gives a literature review for the 3DSPP and its variants. The problem representation is discussed in Section 3.1. Section 3.2 exposes the principle of the greedy selection phase which serves to determine a subset of eligible positions employed for packing the predefined items in the target object. Section 3.3 presents a truncated tree search using a width beam search phase that serves to explore some promising paths. Section 3.3.2 exposes the dichotomous search where the first two phases are embedded in an interval search in order to diversify the search process. Section 4 proposes an experimental part in which the performance of the proposed algorithm is given: its provided results are compared to those reached by some best algorithms of the literature. Finally, Section 5 concludes by summarizing the contribution of this paper.

2. Related work

The 3DSPP, belonging to Cutting and Packing (CP) family (Wascher et al. (2007)), has received very little attention. Most published papers focus on packing (or cutting) items in two-dimensions (cf. Birgin & Sobral, 2008; Farr, 2013; Hifi, M'Hallah, and Saadi, 2008; Kubach, Bortfeldt, Tilli, & Gehring, 2009; Stoyan, Yaskow, & Scheithauer, 2003) but very few papers addressing the problem of packing the sphere packing are available.

Among existing papers addressing sphere packing problems, we cite Lochmann, Oger, and Stoyan (2006) who proposed a statistical analysis for packing random spheres with variable radius distribution. Li and Ji (2013) discussed a dynamics-based collective method for random sphere packing and they tried to apply it to the problem of packing sphere into a cylinder container. In this paper, the stability of the method and its convergence were tackled. Farr (2013) studied the problem of random close packing fractions of lognormal distributions of hard spheres. The author tailored a one-directional approach in order to predict a close packing of spheres of lognormal distributions of sphere sizes.

Packing spheres into a container has been addressed by Sutou and Dai (2002) who proposed a global optimization approach. Stoyan et al. (2003) designed a mathematical model in order to pack spheres into an open container, where both height and widths are fixed whereas the length is unfixed. They use a neighboring search based upon extremum points in order to construct and improve a series of solutions. M'Hallah, Alkandari, and Mladenović (2013) proposed a heuristic based on combining VNS with nonlinear programming solver. The method iterates some moves of the current configuration and complete the partial configuration with a solver dedicated for nonlinear programs. M'Hallah and Alkandari (2012) considered the principle used in M'Hallah et al. (2013) to solve the problem of packing identical spheres into the smallest containing cube. Soontrapa and Chen (2013) tackled the problem of packing identical spheres into a smallest containing sphere by using a random search according to Monte Carlo's method. Finally, Birgin and Sobral (2008) proposed twice-differentiable nonlinear programming models for the problem of packing both circles and spheres into different containers where the containers may be circular, rectangular, etc. In order to find a global solution for their proposed models, ALGENCAN solver was used for generating a multiple starts solutions (an extensive efficient models and methods for packing both circular and sphere problems were reviewed in Hifi & M'Hallah, 2009b).

In this paper, we propose a three phase solution procedure that combines the width-beam search and a greedy selection procedure for approximately solving the 3DSPP. The last two phases are embedded in a dichotomous procedure that is used for searching the best length L^* for the object $\mathcal{P}$. Note that the proposed algorithm can be viewed as an adaptation of the method proposed in Hifi et al. (2008) for solving the two-staged two-dimensional cutting problem and later in Akeb and Hifi (2008) for solving another variant of CP family. The novelty in this approach is based upon (i) a new representation used for defining a restricted space search; that is, a structured set based on target eligible positions which serves to limit the search process without degrading the quality of solutions (Sections 3.1 and 3.2) and (ii) a nice dichotomous search (Section 3.4) that tries to avoid premature convergence toward local stagnant solutions whenever the width-beam search is applied (Section 3.3).

3. Tackling 3DSPP with a dichotomous search

Herein, the problem representation and strategies used by the proposed algorithm are discussed. First, Section 3.1 describes how an item can be represented in a space and which eligible positions may be chosen for assigning items of N to the target object $\mathcal{P}$. Second, Section 3.2 describes the principle of the simple constructive procedure which tries to guaranty a feasible packing of all items of N for the 3DSPP according to the set of eligible positions. Such a simple constructive procedure is based on the minimum distance between items and the edges of the objet $\mathcal{P}$ that serves to create a set of eligible positions for the current item to pack. Third and last, Section 3.3 discusses the principle of the proposed algorithm and its main steps.

3.1. Representation of 3DSPP

In this section, we present the principle of the *local strategy* used for filtering the huge number of possible positions when an item will be assigned to the object $\mathcal{P}$. The local strategy is based on the simple Greedy Principle (called GP) where the minimum distance position is favored for packing a predefined item. GP is then used as an evaluation operator for finding a subset of eligible positions of the next item to pack. Furthermore, the GP can also be used either as a complete solution procedure that provides a final feasible solution for the 3DSPP or as a greedy strategy in order to complete a partial solution.

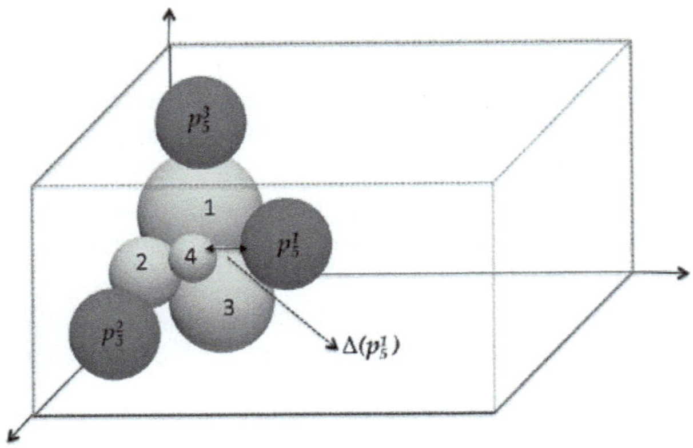

Figure 1. Illustration of the mechanism used for computing eligible positions.

For the rest of the paper, the following notations are introduced:

- The bottom-left-depth corner of $\mathcal{P}$ is positioned at $(0,0,0)$ and $\mathcal{P}$ is characterized by a set formed with six labels (namely faces): $\mathbb{F}=\{\text{left, top, right, bottom, depth, front}\}$. Then, $\mathcal{P}$ is represented in the Euclidian space, as illustrated in Figure 1.
- The center of the ith item belonging to N is positioned at (x_i, y_i, z_i).
- The distance, namely $\delta_{i,j}$, between two different items i and j belonging to N is computed as follows:

$$\delta_{i,j}=\sqrt{(x_i-x_j)^2+(y_i-y_j)^2+(z_i-z_j)^2}-(r_i+r_j), \quad \forall(i,j)\in N^2 \tag{7}$$

Observe that positioning any item i of N into $\mathcal{P}$ induces to define a distance between an item belonging to N and faces of $\mathbb{F}$. Indeed, for instance, assigning an item $i\in N$ to an eligible position of $\mathcal{P}$ while respecting the nonoverlap of the left-face of $\mathcal{P}$ requires to satisfy the following distance: $\delta_{i,\text{left}}=x_i-r_i$. For all faces of $\mathcal{P}$, Table 1 reports the distance to be satisfied whenever a preselected item i of N is assigned to an eligible position of $\mathcal{P}$.

3.2. Defining eligible positions

It is well known that tailored heuristics are mainly based on the strategies which are able to guide well the search process. These strategies may be using some selection criteria in order to provide either partial or final solutions for the problem to solve. Herein, we consider a simple GP which is based on searching the position realizing the minimum distance position between items and feces. In fact, GP is used as a selection criterion for defining a set of eligible positions to assign to the predefined item i (not already positioned) among all eligible positions representing the set P_{I_i}.

In what follows, we assume that the (center of the) first item $i=1$ of N is positioned at the position (r_1, r_1, r_1) and $\forall i\in N$, $i\geq 2$, the following notations are considered:

Table 1. The distance between an item *i* and a face *f*

f	$\delta_{i,f} \mid i\in N, f\in\mathbb{F}$
Left	x_i-r_i
Bottom	y_i-r_i
Depth	z_i-r_i
Right	$L-x_i-r_i$
Top	$H-y_i-r_i$
Front	$W-z_i-r_i$

- I_i denotes the set of items of N already positioned in the current object $\mathcal{P}$.
- $\overline{I}_i$ contains the items of N which are not yet assigned to $\mathcal{P}$.
- P_{I_i} denotes the set of distinct eligible positions for the next item i to pack given the set of packed items I_i.
- An eligible position $p_{i+1} \in P_{I_i}$ (for the item i) is determined by using three elements e_1, e_2 and e_3 where an element is either an item of N already positioned (representing the set I_i) or one of the six faces belonging to $\mathbb{F}$.
- $\mathcal{T}_{p_{i+1}}$ represents the set composed of the three elements e_1, e_2 and e_3.

Figure 1 illustrates the mechanism used by the GP strategy on a small example. We assume that the first four items are already positioned in the object $\mathcal{P}$; then, there are three eligible positions that emerge for the next item 5 to pack. Following the above notations, $I_4 = \{1,2,3,4\}$ and $P_{I_4} = \{p_5^j, j = 1, \ldots, 3\}$. First, the position p_5^1 touches both items 1 and 3 and, the "bottom" face of $\mathcal{P}$. Second, the position p_5^2 is obtained by using the item 2 and both faces "left" and "bottom" of $\mathcal{P}$. Third and last, the position p_5^3 is computed by using the item 1 and both faces "left" and "depth" of $\mathcal{P}$. Finally, it follows that $\mathcal{T}_{p_5^1} = \{1,2,\text{bottom}\}$, $\mathcal{T}_{p_5^2} = \{2,\text{left, bottom}\}$ and $\mathcal{T}_{p_5^3} = \{1,\text{left},\text{depth}\}$.

We can observe that the item 4 is positioned around the three items 1–3 that means $\mathcal{T}_4 = \{1,2,3\}$. For the next item 5 to pack, its coordinates are computed by using both items 1 and 3 and one of the faces (the "bottom" in this case) of $\mathcal{P}$ that gives $\mathcal{T}_5 = \{1,3,\text{bottom}\}$. We recall that the objective of the problem is to minimize the length of the target object $\mathcal{P}$. It means that the right face of $\mathcal{P}$ can be omitted and only the five faces can be considered when optimizing the length of the target $\mathcal{P}$. Hence, all eligible positions may be obtained by using the fifth faces, the positioned items and the current item to pack.

Moreover, GP works as follows: the value corresponding to the $(i+1)$th item to pack, when positioned at the eligible position $p_{i+1}^k \in P_{I_i}$, is computed as follows:

$$\Delta(p_{i+1}^k) = \min_{j \in I_i \cup \mathbb{F}' \setminus \mathcal{T}_{p_{i+1}^k}} \delta_{(i+1,j)} \tag{8}$$

where $\mathbb{F}' = \mathbb{F} \setminus \{\text{ right}\}$.

Finally, when GP is used, then it starts by positioning the first item $i = 1$ at the bottom-left-depth position, i.e. at the position (r_1, r_1, r_1), while the remaining $n-1$ items are successively positioned according to the minimum distance rule (cf. Equation (8)). As illustrated in Figure 1, the item 5 will be placed at position p_5^1 because its corresponding distance realizes the minimum value (of course, this choice may also be modified along the execution of the algorithm).

Of course, one can use GP as a greedy procedure for searching a feasible solution for the 3DSPP. Indeed, by setting the starting length of the target object $\mathcal{P}$ to ∞ (i.e. $L_{\mathcal{P}} = \infty$) and by applying successively GP on the rest of the nonpacked items, GP terminates by building a feasible solution with length L_U. The provided length L_U can also be considered as an upper bound for starting the proposed algorithm.

3.3. A width-beam search heuristic for the 3DSPP

In this section, we first describe the principle of the standard beam search (Section 3.3.1). Second and last, we show how beam search can be adapted for solving the 3DSPP (Section 3.3.2), especially when using a width-beam almost of the standard beam.

3.3.1. A standard beam search

Beam Search (BS) has been first proposed in Ow and Morton (1988) for tackling the scheduling problem and it has since been successfully applied to many other combinatorial optimization problems (some adaptations can be found in Della Croce, Ghirardi, & Tadei, 2004; Hifi & M'Hallah, 2009a;

Hifi et al., 2008; Hifi & Saadi, 2010; Yavuza, 2013). Such an approach can be viewed as a truncated tree search procedure where its objective is to avoid exhaustive search by performing a partial enumeration of the solution space.

At each level of the developed tree, only a subset of nodes (called the *set of elite nodes*) are selected for further branching and the other nodes are discarded, where no backtracking is performed. For each level, the cardinality of the elite nodes to be investigated is fixed to ω that is called the *beam width*. Generally, these selected ω nodes represent those having a high potential to lead the best solutions for the treated problem. Furthermore, each node is assessed via an *evaluation function* whose role is to provide a promising separation mechanism of the nodes of each level of the developed tree.

Algorithm 1 . A standard beam search.

1. **Initialization Step.**
   (a) Let ω be the beam width.
   (b) Set $B = \{B_0\}$ and $B_\omega = \emptyset$, where B is the set of nodes to be investigated, and B_ω the set of nodes branched out of the nodes in B.
   (c) If an initial feasible solution is available, set $z^\star$ to its objective function value; otherwise, set $z^\star = +\infty$ (minimization problem)
2. **Iterative Step.**
   Repeat
   (a) Choose a node $\eta \in B$; branch out η; remove η from B and insert the created nodes into B_ω.
   (b) If a node η of B_ω is a leaf, then
      i. compute its objective function value z_η;
      ii. if $z_\eta > z^\star$, update $z^\star$ and the incumbent solution;
      iii. remove η from B_ω.
   (c) Assess the potential of each node of B_ω using an evaluation function.
   (d) Rank the nodes of B_ω in a decreasing order of their values.
   (e) Insert the $\min\{\omega, |B_\omega|\}$ best nodes (called elite nodes) of B_ω into B; and set $B_\omega = \emptyset$.

   Until $B = \emptyset$.

Algorithm 1 describes the main steps of the standard beam search which is characterized by a beam width ω used for filtering the set of offspring nodes B_ω. A node corresponds to a partial feasible solution and the set B of current nodes is initialized to the root node B_0 whereas B_ω containing the offspring nodes is initialized to the empty set. Each node of B generates a set of offspring nodes, and moves them to B_ω. If a node η of B_ω is a leaf (i.e. no further branching is possible out of η), then its objective function value z_η is computed and compared to $z^\star$. If $z_\eta > z^\star$, then the incumbent solution is set to the leaf node; $z^\star$ is then updated: $z^\star = z_\eta$; and η is removed from B_ω. All nodes belonging to B_ω are assessed using an evaluation function, and then ranked in decreasing order of their values. The first ω nodes of B_ω are then considered as the elite nodes which are appended to the set B, whereas the remaining nodes of B_ω are discarded and B_ω is reset to the empty set. This process is iterated until no further branching is possible, i.e. until $B = \emptyset$.

3.3.2. Adaptation of the beam search for the 3DSPP

Applying BS to 3DSPP requires defining the nodes of the tree and the branching mechanism out of the nodes of B. Herein, a node η_i is represented by the pair of subsets:

(1) The first subset I_i containing all items assigned to the target object $\mathcal{P}$ and,

(2) The complementary subset $\overline{I}_i$ containing the unassigned items.

The first subset represents a *partial solution* realized when assigning all the items to the object $\mathcal{P}$ and the second subset serves to provide a *complementary solution*. Such a complementary solution is computed by applying the MLDP used as greedy procedure (as discussed in Section 3.2).

Moreover, branching out of a node η_i is equivalent to create at most $|P_{I_i}|$ branches emanating out of the current node (related to the eligible positions as described in Section 3.2). Each resulting node corresponds to packing the subset of items I_i and assigning to the current item i a favorite eligible position. Moreover, each of these created nodes will be represented by a pair of two complementary subsets of items of N.

As mentioned above, because of the huge number of feasible positions that a predefined item may generate, we preferred the use of the width-beam search almost of the standard beam search. Indeed, as we shall see in Section 3.4, the proposed algorithm uses a dichotomous search according to a series of values assigned to ω (the beam width). Therefore, in order to try the exploration of a maximum number of promising paths, the beam search applies the width search where all nodes belonging to the same level are simultaneously evaluated following an estimator operator and only the best ones are selected for the rest of the search.

Algorithm 2 . Adaptation of the Beam Search for the 3DSPP: BS

Input. A node η^{ℓ}.

Output. `feasible` // setting equal to `true` whenever a feasible packing is reached, `false` otherwise

1: **Initialization Step.**
2: Let ω be a predefined beam width.
3: Set $B = \{\eta^{\ell}\}$, where η^{ℓ} denotes the input node associated to the ℓ-th level.
4: Set the variable `feasible` to `false` /* no feasible solution at hand */
5: **Iterative Step.**
6: **while** $\big((B \neq \emptyset)$ and (the runtime limit is not performed)$\big)$ **do**
7: Branch from the current level ℓ by selecting the ω eligible positions for each node $\eta_{\ell_i} \in B$;
8: Insert all obtained nodes into B_{ω};
9: Evaluate the potential of each node belonging to B_{ω} using GP for completing the path.
10: If a feasible solution is given by GP, then set `feasible` to `true` and **exit**;
11: Replace B by the best ω nodes of B_{ω} realizing highest densities and, increment the level ℓ.
12: **end while**

The aforementioned process is described by Algorithm 2. The adapted algorithm works according to a given node, namely η^{ℓ}, taken at the level ℓ of the developed tree. Thereafter, the initialization step (Lines 1–3) is applied for starting the set B containing the best provided nodes regarding the starting node η^{ℓ} and, the initialization of the variable feasible to false. This variable controls the (un) feasibility of the series of the solutions built. The main loop (lines 7) starts by choosing the best eligible positions for each node belonging to B. These positions are computed by using GP's selection (cf. Section 3.2). Second, all created nodes are stored in a provisional set B_{ω} where the potential of each of these nodes are evaluated according to the final solution provided by iteratively applying GP as a heuristic (cf. as discussed in the last paragraph of Section 3.2). Thereafter, for each fin al solution (either feasible or unfeasible for the target object $\mathcal{P}$), the potential of a node $\eta \in B_{\omega}$ is represented by the *density of the positioned items* in $\mathcal{P}$. Whenever one of these constructed solutions provides a feasible solution (line 10), i.e. all items are positioned in the target object $\mathcal{P}$, then the algorithm stops with a feasible solution (i.e. setting the variable feasible to true). Otherwise, the set B of the best nodes is updated (line 11) by the ω nodes which realizing the highest densities and the current level of the developed tree is incremented. Finally, the process is iterated until either B is reduced to an empty set or when the fixed runtime limit is performed.

3.4. Using a dichotomous search

The proposed algorithm applies a series of BS for a predefined interval search. First, the interval search starts by $[\underline{L},\overline{L}]$, where $\underline{L}$ denotes a lower bound for the 3DSPP and $\overline{L}$ its upper bound (in the case of a feasible solution exists, its objective value is assigned to $\overline{L}$). Second, for each fixed interval, BS tries to construct the best feasible solution by packing all the items into the current target object; that is, $(L^{\star},W,H)$, where $L^{\star} \in [\underline{L},\overline{L}]$.

Algorithm 3 . A Dichotomous Search Based Heuristic: DSBH

Input. An instance of 3DSPP.

Output. An object $\mathcal{P}$ of dimensions (L_{best}, W, H) and the coordinates of all items of N.

1: **Initialization step**
2: Call an iterative GP on the open strip (∞, W, H) and let $\overline{L}$ be the starting length reached.
3: Set $\underline{L} \leftarrow \frac{4\pi}{3\times W\times H}\sum_{i\in N}(r_i^3)$ and $L^\star = \overline{L}$.
4: Set ω to a predefined minimum value.
5: **Iterative step**
6: **while** $\Big($the runtime limit is not performed$\Big)$ **do**
7:   **repeat**
8:     $L^\star = (\overline{L} + \underline{L})/2$
9:     Generate the starting node η^1 with its three sets I_i, $\overline{I}_i$ and P_I^1.
10:     Set `feasible` $\longleftarrow$ BS(η^1), where BS is called with $(L^\star, W, H)$
11:     **If** $\Big($`feasible=true`$\Big)$ **then** set $\overline{L} = L^\star$; $\underline{L} = L^\star$ otherwise
12:   **until** $(\overline{L} - \underline{L} \geq \alpha)$
13:   Set $\underline{L} \leftarrow \frac{4\pi}{3\times W\times H}\sum_{i\in N}(r_i^3)$ and increment ω.
14: **end while**

The main steps of the dichotomous search are summarized in Algorithm 3. First, its starts by defining the initial interval $[\underline{L},\overline{L}]$ where the upper bound $\overline{L}$ is obtained by applying GP as a heuristic on the open object, i.e. (∞,W,H). The main loop repeat (cf. lines 7–12) of the dichotomous procedure serves to explore a series of neighborhoods depending on the values of ω. At line 8, a new target upper bound is computed, namely $L^\star = (\overline{L}+\underline{L})/2$. Line 9 generates the initial node positioned at the bottom-left-depth corner (in the position (r_1,r_1,r_1)) and creates its corresponding sets I_i, $\overline{I}_i$ and P_I^1 (as discussed in Section 3.2). At line 10, BS is called with the target value of the object $(L^\star,W,H)$ and the created sets reached at the next step. Line 11 serves to update the interval search where its upper bound is updated whenever a feasible solution i s obtained, the lower bound is updated otherwise. Thereafter, the process is iterated till the gap between both lower and upper bounds becomes closest to a certain tolerance, namely α. Finally, the aforementioned process is iterated a certain number of times following the values of ω (line 13) and according to the runtime limit fixed.

4. Computational results

In this section, we investigate the effectiveness of the width beam search-based heuristic (DSBH) on two sets of benchmark instances: Set1 and Set2.

The first set "Set1" contains six instances (SYS1,...,SYS6) extracted from Stoyan et al. (2003), where the number of the predefined items varies from 25 to 60. These instances have been already tested using Stoyan et al.'s (2003), Birgin and Sobral's (2008) and Kubach, Bortfeldt, Tilli, and Gehring (2011) approaches.

The second set "Set2" contains six instances (KBTG1, KBTG2, KBTG3, KBTG7, KBTG8, and KBTG9) taken from Kubach et al. (2011). For each instance, both dimensions W and H of the object are fixed to 10 whereas the number of the predefined items is fixed to 30 (resp. 50) for the first (resp. last) three instances. Moreover, these six instances have been already tested in Kubach et al. (2011) where they represent the six instances with unequal spheres.

The rest of the experimental part is organized as follows. First, the behavior of the proposed algorithm DSBH is evaluated on the first set of instances Set1 (Section 4.1). The results reached by DSBH are thereafter compared to those reached by four heuristics: Stoyan et al.'s (2003) method-based heuristic (noted SYS), Birgin and Sobral's (2008) method-based heuristic and both sequential and parallel heuristics proposed by Kubach et al.'s (2009, 2011) (noted $KBTG_s$ and $KBTG_p$, respectively).

Second, for the instances of Set2 (Section 4.2), the results provided by DSBH are compared to those reached by Kubach et al. (2011). Note also that the proposed algorithm was coded in C++ and tested on an Intel Core 2 Duo (2.53 GHz and with 4 GB of RAM) and the runtime limit was fixed to one hour.

4.1. Performance of DSBH vs. three heuristics: Set1

Generally, when using approximate algorithms to solve optimization problems, it is well known that different parameter settings for the approach lead to results of variable quality. As discussed in Section 3.3.2, DSBH uses two parameters: the beam width ω and the maximum runtime limit to fix. Our computational study was conducted by varying ω in the discrete interval $\{5,6,7,\ldots\}$ and the maximum runtime limit was fixed to 3600 s (which can be considered as a standard runtime limit considered by algorithms of the literature). Of course, the upper value of ω depends on the limited runtime and the size of the instance. In order to show the effect of these parameters, we discuss the quality of the solutions obtained by DSBH when the beam width ω increases.

First, we do it on an instance extracted from Stoyan et al. (2003) by varying the value of ω from 1 (the minimum value assigned to ω in Algorithm 3) to the upper value (incrementing ω with one unit) till attaining the fixed runtime limit. Figure 2(a) shows the behavior of DSBH when varying the beam width ω for the instance SYS5. From the curve labeled Sol., one can observe that the value of the length $L^\star$ of the target object $\mathcal{P}$ oscillates but globally its average solution value decreases and converges toward one of the best solutions (generation of a series of local optima). This is the reason why we proposed to reiterate the GP rule in order to get the final solution realizing the best local optimum. On the other hand, according to the second curve labeled Best, one can observe that the behavior of DSBH provides a series of decreased solution values; in this case, only the best solution was retained for the whole values of ω. For SYS5, the best local optimum reached by DSBH realizes the value closest to 10.94 and its solution configuration is illustrated by Figure 2(b). In fact, the provided solution is equal to 10.9359 (as shown later in the second part of the experimental part) and it dominates both values 11.2170 and 10.9359 reached by the sequential KBTG algorithm (Kubach et al., 2011) and its parallel version (Kubach et al., 2009) representing the best solution values of the literature. Note that $\omega = 12$ seems to give the best solution value that balances between the quality of the solution and the fixed runtime since for higher values of ω the solution remains unchanged.

Second, for the instances of Set1, Table 3 compares the results of DSBH to those reached by four algorithms: SYS (cf. Stoyan et al., 2003), BSA (Birgin & Sobral, 2008), KBTG_s (Kubach et al., 2011) and its parallel version noted KBTG_s (proposed in Kubach et al., 2009), where the known solutions of the literature are taken from Kubach et al., 2009, 2011).

Columns from 1 to 4 show the instance label, problem size n and both height H and width W of $\mathcal{P}$. Column 5 displays the solution value $L^\star_{\text{SYS}}$ reached by STS whereas column 6 displays BSA's solution values (noted $L^\star_{\text{BSA}}$). Columns 7 and 8 report the best solutions (noted $L^\star_{\text{KBTG}}$) provided by KBTG algorithm for both fixed runtimes: 300 and 3600 seconds. Column 8 shows the best solutions reached by

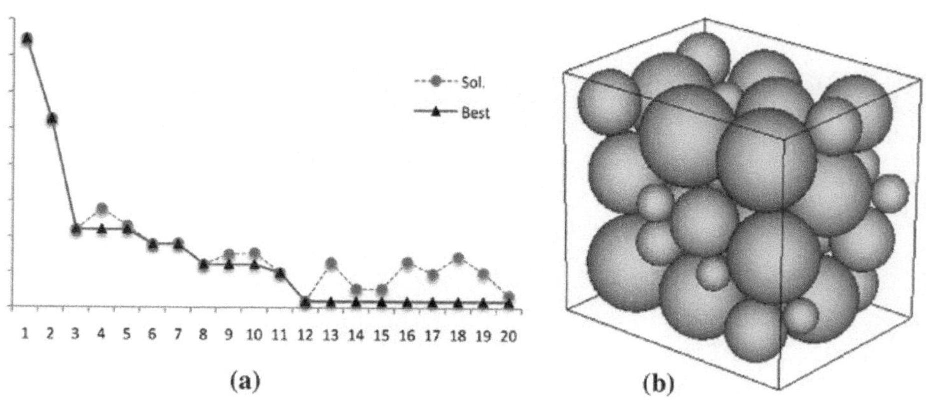

Figure 2. Illustration of DSBH's behavior on SYS5 instance: (a) variation of solution's quality and (b) the best solution's structure reached by DSBH.

the parallel KBTG version (noted $L^{\star}_{KBTG_p}$) without fixing the runtime limit. Columns 9 and 10 display the solution (noted $L^{\star}$) realized by the proposed algorithm DSBH for both fixed runtimes (300 and 3600 s, respectively). Finally, column 11 reports the best value of ω for which its best solution for the second runtime limit is performed.

All results of Table 2 are summarized in Table 3, where it tallies the percentage improvement (when it happens) yielded by DSBH when compared to the results reached by the four considered algorithms. Table 3 is composed of two parts: the first part contains the informations about the instance and the second part shows the percentage improvement realized by DSBH over the four heuristics tested (noted %SYS, %BSA, %$KBTG_s$ and %$KBTG_p$ according to the heuristics SYS, BSA, $KBTG_s$ and $KBTG_p$, respectively) especially for the second runtime limit for DSBH.

The analysis of the results of both Tables 2 and 3 follows:

(1) First, DSBH outperforms the three algorithms SYS, BSA, and $KBTG_s$ whenever the runtime limit was fixed to 300 s. Indeed, it is able to reach the best solution for all instances of Set1.

(2) Second, DSBH outperforms SYS, BSA and both $KBTG_s$ and $KBTG_p$ algorithms since for the instances of Set1 it is able to provide better results.

(3) Third, when comparing DSBH's results to SYS's ones, one can observe that the percentage of the improvement varies from 6.49% (instance SYS5) to 8.82% (instance SYS3). This percentage improvement remains interesting when comparing DSBH's results to those reached by BSA: in this case, such improvement varies between 5.96% (instance SYS1) and 7.65% (instance SYS6). Moreover, it stills interesting when comparing DSBH's results to those provided by KBTG algorithm. Indeed, the improvement varies from 0.48% (instance SYS1) to 6.26% (SYS4).

Table 2. Performance of BSBH versus the four heuristics of the literature on instances of Set1. The symbol " " (resp. "⋄") means that the value for this instance is not available (resp. corresponds to the best solution)

				Kubach et al.'s heuristics							
#Inst.				**SYS**	**BSA**	**$KBTG_s$**		**$KBTG_p$**	**DSBH**		
Label	*n*	*H*	*W*	$L^{\star}_{SYS}$	$L^{\star}_{BSA}$	$L^{\star}_{KBTG_s}$	$L^{\star}_{KBTG_s}$	$L^{\star}_{KBTG_p}$	$L^{\star}$	$L^{\star}$	ω^*
SYS1	25	5.5	6.9	9.912	9.7942	9.5397	9.2874	9.2656	9.2635	9.2431⋄	23
SYS2	35	6.5	7.9	9.623	-	9.2608	9.1280	8.9301	9.1585	8.9164⋄	21
SYS3	40	5.5	6.9	9.473	9.3090	91794	8.9850	8.7178	8.9085	8.7055⋄	24
SYS4	45	8.5	9.9	11.086	11.0962	10.9991	10.8760	10.4042	10.7224	10.2357⋄	23
SYS5	50	8.5	9.9	11.646	11.6211	11.4877	11.3494	10.9865	11.1772	10.9359⋄	29
SYS6	60	8.5	9.9	12.842	12.7215	12.7110	12.3745	11.8399	12.4149	11.8178⋄	28
Average											

Table 3. Percentage improvements between all tested heuristics: BSBH, SYS, BSA and both $KBTG_s$ and KBTG*p* on instances of Set1

#Inst.				**DSBH vs. all heuristics (% Improvement)**			
Label	*n*	*H*	*W*	**%SYS**	**%BSA**	**%$KBTG_s$**	**%KBTG**
SYS1	25	5.5	6.9	7.24	5.96	0.48	0.24
SYS2	35	6.5	7.9	7.92	-	2.37	0.15
SYS3	40	5.5	6.9	8.82	6.93	3.21	0.14
SYS4	45	8.5	9.9	8.31	8.41	6.26	1.65
SYS5	50	8.5	9.9	6.49	6.27	3.78	0.46
SYS6	60	8.5	9.9	8.67	7.65	4.71	0.19
Average				7.91	7.04	3.47	0.47

(4) Fourth and last, DSBH's results remain better than those reached by the parallel algorithm $KBTG_p$ (we recall that the parallel algorithm ran without runtime limit). Indeed, all the solution values for the instances of Set1 are improved. In this case, the percentage of improvement varies from 0.14% (instance SYS3) to 1.65% (instance SYS4), which remains very interesting for a sequential algorithm.

Figure 3 shows the behavior of DSBH on the instances of Set1 where each curve represents the variation of the improvements realized according to the algorithm SYS, BSA and both $KBTG_s$ and $KBTG_p$, respectively. Figure 4(a) illustrates GP's configuration (when applied as a heuristic for providing a starting solution for DSBG) and Figure 4(b) displays DSBH's configuration. From the starting configuration (cf. Figure 4(a)), one can observe the failure of GP when used as a heuristic: in this case, all items are concentrated on the bottom-left-depth position and all positions located along the length of $\mathcal{P}$ to the top are omitted. However, the configuration performed by DSBH shows the correction made when GP's process is repeated.

4.2. Performance of DSBH versus KBTG heuristic: Set2

This section compares the results reached by DSBH to those reached by $KBTG_s$ (note that, for this type of instances, $KBTG_s$ realizes the best solution values of the literature). This comparison is performed on the instances of the second group Set2 extracted from Kubach et al. (2011). For this type of instances, instead of determining the minimum length $L^\star$ of the target container $\mathcal{P}$, Kubach et al. (2011) computed the density of all packed items in the final object $\mathcal{P}$. Therefore, in addition to the density, the best length $L^\star$ of the final object $\mathcal{P}$ corresponding to that density is reported in the same table.

The results realized by the two tested methods (DSBH and $KBTG_s$, respectively) are reported in Table 4. Columns 1–4 display all the characteristics related to the instance. Column 5 reports the solution value (expressed in term of density) realized by $KBTG_s$'s algorithm (extracted from Kubach et al. (2009, 2011)). The next three columns report the best length $L^\star$ reached by DSBH, its corresponding density and the best value of ω for which the best solution is reached. Finally, the last column displays the percentage improvement realized by DSBH according to the solution values reached by $KBTG_s$. The analysis of the results of Table 4 follows.

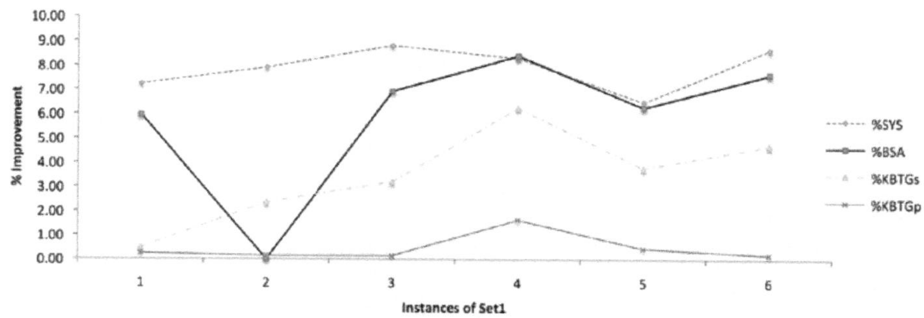

Figure 3. Variation of the percentage improvement realized by DSBH when compared to the results of the four algorithms (SYS, BSA, and both $KBTG_s$ and $KBTG_p$) on the instances of Set1.

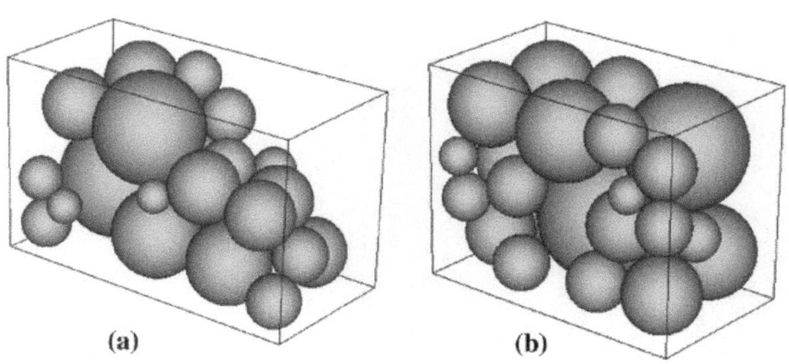

Figure 4. Illustration of DSBH's behavior on SYS1 instance (Set1: (a) the starting solution with a length equal to 11.5051 and (b) the best solution's structure and its length (9.2431) reached by DSBH.

Table 4. Performance of BSBH vs. KBTG$_s$ on instances of Set2

#Inst.					Dichotomous search-based heuristic			
Label	n	H	W	$d^{\star}_{KBTG_s}$	$L^{\star}$	$d^{\star}$	ω^{*}	%$KBTG_s$
KBTG1	30	10	10	54.096	10.9031	56.0092°	18	3.42
KBTG2	30	10	10	30.071°	1.9900	30.071°	23	0.00
KBTG3	30	10	10	51.387	18.2415	53.6243°	21	4.17
KBTG7	50	10	10	55.372	13.0997	57.5662°	17	3.81
KBTG8	50	10	10	45.060	2.5825	47.004°	15	4.14
KBTG9	50	10	10	52.732	27.8033	55.3203°	13	4.68
Average				48.120		49.932		3.63

(1) DSBH remains competitive since it improves most solutions reached by KBTG$_s$. Indeed, it is able to improve five out of six best solutions while it matches the other solution (instance KBTG2) when compared to the results reached by KBTG$_s$.

(2) For the improved solutions, DSBH realizes an improvement varying from 3.42% (instance KBTG1) to 4.68% (instance KBTG8). Globally, the average improvement over all instances is equal to 3.63%.

(3) In term of density, DSBH realizes a density of 49.932% which is much better than those realized by KBTG$_s$ (48.120%), as illustrated in Figure 5.

Finally, Figure 6(a) illustrates the starting solution reached by iterating GP till packing all the items, whereas Figure 6(b) shows the final solution when DSBH is applied. We can observe that the starting configuration leaves also an important unoccupied area along the length of the object $\mathcal{P}$ to the depth. However, the configuration leads a better packing when GP rule is repeated on different eligible positions of the search space.

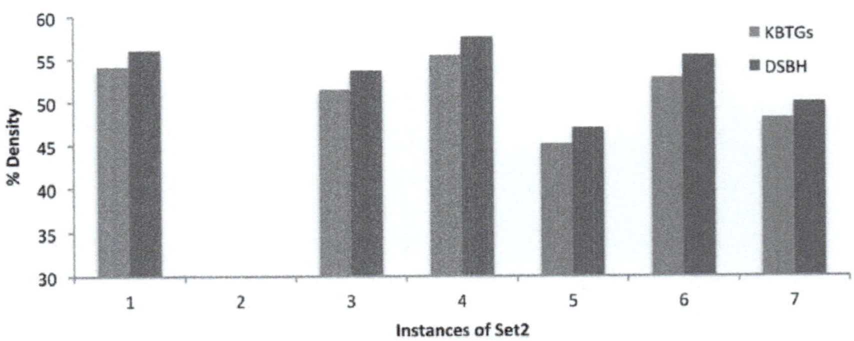

Figure 5. Variation of the percentage of the density realized by both DSBH and KBTG$_s$ on the instances of Set2.

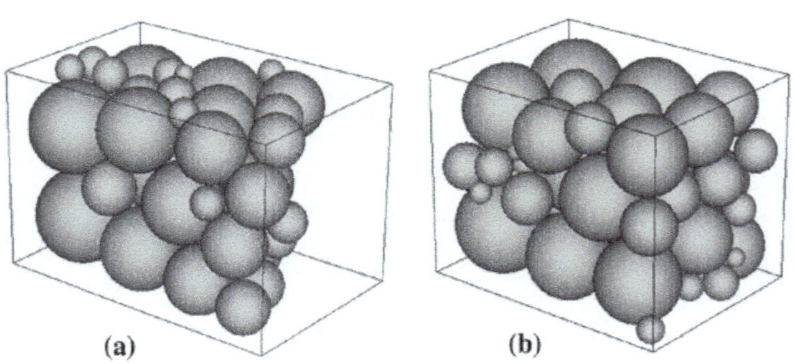

Figure 6. Illustration of DSBH's behavior on KBTG7 instance (Set2): (a) the starting solution with a length equal to 15.3817 and (b) the best solution's structure and its length (13.0997) reached by DSBH.

5. Conclusion

In this paper, the three-dimensional sphere packing problem was solved with a dichotomous search-based heuristic. The proposed method is an adaptation of the beam search which combines the following components: (i) a greedy selection phase for performing a local restricted space search that contains eligible positions, (ii) a standard width-beam search phase for exploring some promising paths, and (iii) a dichotomous search for diversifying the search space. Through extensive testing on different instances of the literature, the proposed method is able to improve a significant number of the solutions provided by recent methods available in the literature.

Acknowledgements
The authors thank the anonymous referees for their helpful comments and suggestions which contributed to the improvement of the contents of this paper.

Funding
The authors received no direct funding for this research.

Author details
Mhand Hifi[1]
E-mail: hifi@u-picardie.fr
Labib Yousef[1]
E-mail: labib.yousef@etud.u-picardie.fr
[1] Laboratory EPROAD - EA 4669 (ROAD Team), UFR des Sciences, University of Picardie Jules Verne, 7 rue du Moulin Neuf, 80000 Amiens, France.

References

Akeb, H., & Hifi, M. (2008). Algorithms for the circular two-dimensional open dimension problem. *International Transactions in Operational Research, 15*, 685–704.

Birgin, E. G., & Sobral, F. N. C. (2008). Minimizing the object dimensions in circle and sphere packing problems. *Computers & Operations Research, 35*, 2357–2375.

Della Croce, F., Ghirardi, M., & Tadei, R. (2004). Recovering beam search approach for combinatorial optimization problems. *Journal of Heuristics, 10*, 89–104.

Farr, R. S. (2013). Random close packing fractions of lognormal distributions of hard spheres. *Powder Technology, 245*, 28–34.

Hifi, M., & M'Hallah, R. (2009a). Beam search and non-linear programming tools for the circular packing problem. *International Journal of Mathematics in Operational Research, 1*, 476–503.

Hifi, M., & M'Hallah, R. (2009b). A literature review on circle and sphere packing problems: models and methodologies. *Advances in Operations Research*, 22 pp. doi:10.1155/2009/150624. Article ID 150624.

Hifi, M., M'Hallah, R., & Saadi, T. (2008). Algorithms for the constrained two-staged two-dimensional cutting problem. *INFORMS, Journal on Computing, 20*, 212–221.

Hifi, M., & Saadi, T., (2010). A cooperative algorithm for constrained two-staged two-dimensional cutting problems. *International Journal of Mathematics in Operational Research, 9*, 104-124.

Kubach, T., Bortfeldt, A., Tilli, T. & Gehring, H. (2009). *Parallel greedy algorithms for packing unequal spheres into a cuboidal strip or a cuboid.* (Working Paper No 440). Hagen: Department of Management Science, University of Magdeburg, (Diskussionsbeitrag der Fakultät für Wirtschaftswissenschaft der FernUniversität in Hagen).

Kubach, T., Bortfeldt, A., Tilli, T., & Gehring, H. (2011). Greedy algorithms for packing unequal sphere into a cuboidal strip or a cuboid. *Asia-Pacific Journal of Operational Research, 28*, 739–753.

Li, Y., & Ji, W., (2013). Stability and convergence analysis of a dynamics-based collective method for random sphere packing. *Journal of Computational Physics, 250*, 373-387.

Lochmann, K., Oger, L., & Stoyan, D. (2006). Statistical analysis of random sphere packings with variable radius distribution. *Solid State Sciences, 8*, 1397–1413.

M'Hallah, R., & Alkandari, A. (2012). Packing unit spheres into a cube using VNS. *Electronic Notes in Discrete Mathematics, 39*, 201–208.

M'Hallah, R., Alkandari, A., & Mladenović, N. (2013). Packing unit spheres into the smallest sphere using VNS and NLP. *Computers & Operations Research, 40*, 603–615.

Ow, P. S., & Morton, T. E. (1988). Filtered beam search in scheduling. *International Journal of Production Research, 26*, 297–307.

Soontrapa, K., & Chen, Y. (2013). Mono-sized sphere packing algorithm development using optimized Monte Carlo technique. *Advanced Powder Technology, 24*, 955–961.

Stoyan, Y., Yaskow, G., & Scheithauer, G. (2003). Packing of various radii solid spheres into a parallelepiped. *Central European Journal of Operational Research, 11*, 389–407.

Sutou, S., & Dai, Y. (2002). Global optimization approach to unequal sphere packing problems in 3D. *Journal of Optimization Theory and Applications, 114*, 671–694.

Wang, J. (1999). Packing of unequal spheres and automated radiosurgical treatment planning. *Journal of Combinatorial Optimization, 3*, 453–463.

Wascher, G., Haussner, H., & Schumann, H. (2007). An improved typology of cutting and packing problems. *European Journal of Operational Research, 183*, 1109–1130.

Yavuza, M. (2013). Iterated beam search for the combined car sequencing and level scheduling problem. *International Journal of Production Research, 51*, 3698–3718.

2

An integrated multi-criteria decision-making methodology for conveyor system selection

Pairat Jiamruangjarus[1] and Thanakorn Naenna[1*]

*Corresponding author: Thanakorn Naenna, Faculty of Engineering, Department of Industrial Engineering, Mahidol University, Nakhonpathom 73170, Thailand

E-mail: thanakorn.nae@mahidol.ac.th

Reviewing editor: Zude Zhou, Wuhan University of Technology, China

Abstract: Material handling equipment (MHE) is important for every industry because it has an effect on the productivity of manufacturing. Conveyor systems are presently one popular type of MHE. This paper presents an integration of the analytic network process (ANP) with the benefits, opportunities, costs and risk (BOCR) model in order to select the best conveyor system. The proposed model established a network with four merits, six strategies criteria, and twenty six sub-criteria with four alternatives (present, roller conveyor, chain conveyor, and monorail). The ANP is to determine the relative weights of an evaluative criteria and decision alternatives. Therefore, the final ranking of the alternatives are calculated by synthesizing the score of each alternative under BOCR. The results showed that the best alternative under all five methods is the chain conveyor. These research results can be easily applied, adapted and used to improve performance of selecting the conveyer system in small and medium enterprises through large industries.

Subjects: Industrial Design; Manufacturing & Processing; Operations Research; Production Systems

Keywords: material handling equipment; analytic network process (ANP); BOCR

1. Introduction

Although we are not particularly interested in the practice of material handling, in everyday life material handling equipment (MHE) can be found almost anywhere, especially in trade and industry. In fact, in modern life, it is not rare for materials handling to be used for such purposes as lifting, moving, storage and other activities. The study on metal processing industries has been a

ABOUT THE AUTHOR

Thanakorn Naenna received his MS in Manufacturing and Systems Engineering in 1998 and MEng in Operation Research and Statistics from Rensselaer Polytechnic Institute, USA in 2002, the PhD in Engineering Science from Rensselaer Polytechnic Institute, USA in 2003. He is currently an assistant professor in Mahidol University. His research interests include production and manufacturing.

PUBLIC INTEREST STATEMENT

In this paper, we propose a multi-criteria decision-making model for conveyor system selection. Selection of the appropriate conveyor system is a very difficult task for manufacturing company. There are many conflicting quantitative and qualitative factors influencing the conveyor system selection decision. Multi-criteria decision-making model has been found to be useful approach to analyze these factors. This paper proposes an integrated multi-criteria decision-making model with benefits, opportunities, costs and risks (BOCR) to support decision-making in the problem of conveyor system selection. Analytic network process (ANP) with BOCR is utilized for assigning weights of criteria and is designed to select the most proper conveyor system alternative using the criteria weights attained by ANP.

discontinuous process; it was found that only five percent of total production time is spent working on a machine, and the other ninety-five percent is spent on waiting and movement (Srisom & Sriuthai, 2004). It was also found that in each industry, the cost of material handling will account for approximately 30–70% or more of the total cost of production, depending on the type of industry (Dongre & Mohite, 2015). The initial cost of manufacturing operation can be reduced by 15–30% by efficient management of material handling (Sule, 1994; Kulak, 2005; Sujono & Lashkari, 2007; Tuzkaya, Gülsün, Kahraman, & Özgen, 2010).

An investment analysis in choosing a material handling system (MHS) is extremely complex, and there are multiple solutions for particular situations (Swaminathan, Matson, & Mellichamp, 1992). The material handling selection problem is important (Chan, Ip, & Lau, 2001), and there are many factors concerned that should be considered. It may be significantly affected by constraints or other factors such as product size, the characteristics of the product that is to be handled, space and time required, etc. (Marcello, Gabbrielli, & Miconi, 2001). These main factors consist of MHE, manufacturing type, working area, product appearance, environment, functional equipment, material handling methods, and other factors. Many times, investors will only consider the benefits of investment and investment cost, which makes the mistake of considering the major two factors: the opportunities and risks arising from the investment.

This research focuses on the investment conveyor system. The complexity of conveyor equipment selection is a problem for many manufacturers (Fonseca, Uppal, & Greene, 2004). There are several factors and limitations involved in conveyor equipment selection. A conveyor system is a part of the mechanical handling equipment that is used to move materials from one location to another (Tompkins, White, Bozer, & Tanchoco, 2002). The types of conveyors can be categorized in several ways. For example, a belt conveyor can be used for bulk and unit loads, so it can be located overhead or on the floor. Bulk materials such as grain, dry chemicals and saw dust can be conveyed using a chute, belt, bucket or vibrating conveyors. Unit materials such as castings, machined parts, and materials can be placed on pallets and cartons or tote boxes, and can be conveyed using chute, belt, roller, wheel, or tow conveyors. Materials can be conveyed on belt, roller, wheel, vibrating, pneumatic or tow conveyors.

This paper describes a tool to support decision-making in the problem of conveyor system selection. For this problem, this paper used the analytic network process (ANP) with the benefits, opportunities, costs and risk (BOCR) as a development tool. The ANP, as one of widely used multiple criteria decision-making (MCDM) method, is often implemented in BOCR analysis to improve the performance of decision analysis (Chen, Lee, & Kang, 2010; Cho, Kim, & Heo, 2015; Erdogmus, Kapanoglu, & Koc, 2005; Jaafari, Najafi, & Melónc, 2015; Krishna Mohan, Reformat, & Pedrycz, 2013; Lee, Chang, & Lin, 2009; Malmir, Hamzehi, & Farsijani, 2013; Mili, 2014; Saatya & Sagir, 2015; Sakthivela, Ilangkumaranb, & Gaikwada, 2015; Tornjanski, Marinković, & Lalić, 2014; Ustun & Demirtas, 2008; Wang, Lee, Peng, & Wud, 2013; Wiratanaya, Darmawan, Kolopaking, & Windia, 2015; Wijnmalen, 2007; Yazgan, Boran, & Goztepe, 2010). The rest of this paper is organized as follows. Section 2 describes the MHS, the ANP and presents the five methods of BOCR. Section 3 describes the methodology and algorithm for the ANP model with BOCR. In Section 4, selected examples are presented to apply the model in real cases. The final section provides concluding remarks.

2. Literature review

2.1. Material handling system

Material handling is defined by American Society of Mechanical Engineers as the art and science involving the moving, packaging, and storing of substances in a form. Although the best solution to the problem of materials handling is no handling, or the simplest solution to the handling of materials being no movement no cost. Both solutions are hardly possible for a complete manufacturing process. A MHS is a system for improving the performance of a manufacturing system, such as reducing work-in-process (WIP) by delivering the right amount of materials, to the right place, at the right time, and at the lowest cost (Kulak, 2005).

In a manufacturing system, MHE is the most important part, and it plays an increasingly important role in the productivity of manufacturing. The selection of an MHE system is complex, and there is a considerable amount of capital investment required. As handling activities account for 30–40% of production costs (Tompkins & White, 1984), an appropriate MHE should be selected by aiming to reduce production costs and increase profit. For these reasons, researchers have to find solutions by using various methods such as expert systems, mathematical models, MCDM method, etc. For this study, the researcher has placed an emphasis on MCDM methods. Types of MHE have been divided into seven main groups: conveyors, overhead conveyors, cranes, industrial trucks, automated guided vehicles (AGVs), robots, and storage/retrieval systems.

The complexity of conveyor equipment selection is a problem for many manufacturers (Fonseca et al., 2004). There are several factors and limitations involved in conveyor equipment selection. A conveyor system is a part of mechanical handling equipment that is used to move materials from one location to another (Tompkins et al., 2002).

2.2. Analytic network process

The ANP was introduced by T. L. Saaty, as a generalization of the analytic hierarchy process (AHP) (Saaty, 2004a). The ANP is an improved model of the AHP. The AHP was proposed in 1980 by Thomas L. Saaty as a decision-making method. The ANP permits mutual dependence and feedback among criteria, therefore the ANP is different from the AHP (Liang & Li, 2008).

The AHP is designed for solving the independence problem among alternatives or criteria problems, while the ANP is designed for solving the dependence problem among alternatives or criteria problems (Lee & Kim, 2000). Therefore, the AHP would not be appropriate for complex relationships, because the structure is linear from top-to-bottom. The ANP allows for complex interrelationships among clusters or elements, by replacing hierarchies with networks as shown in Figure 1.

- Step 1: Model construction and problem structuring: The problem should begin distinctly and decomposed into a rational system as a network. This network structure can be obtained through the opinion of decision-makers, through brainstorming or other appropriate methods (Chung, Lee, & Pearn, 2005).
- Step 2: Pairwise comparison matrices and priorities: At each cluster, all pairs of the decision elements are compared with respect to the importance of the elements toward their control criteria. The clusters are also compared pairwise themselves, with respect to their contribution to the purpose. An expert who acts as a decision-maker is asked to determine the relative importance of each criterion on a scale of 1 to 9 (Wijnmalen, 2007).

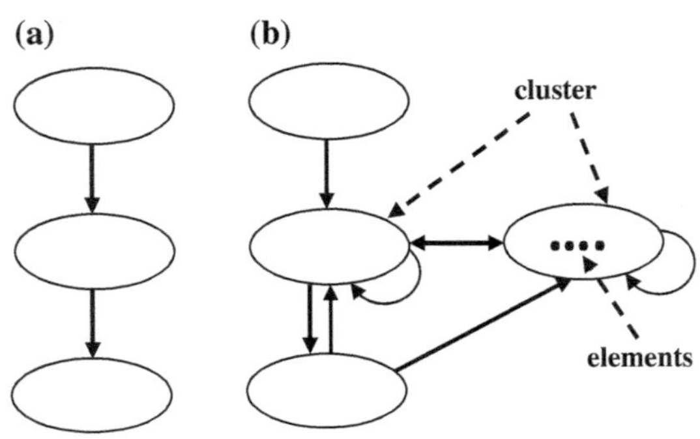

Figure 1. Structural difference between (a) hierarchy and (b) network.

Source: Chung et al. (2005).

Reversing the comparison between already compared elements, a reciprocal value is assigned to the reverse comparison; that is $a_{ij} = 1/a_{ij}$ (Lee et al., 2009) and then, a pairwise comparison matrix was developed and solved by the following equation (1):

$$A \times w = \lambda_{max} \times w, \tag{1}$$

where A is the matrix of pairwise comparison, w is the eigenvector, and λ_{max} is the largest eigenvalue according to an approximation of w from several algorithms by Saaty (Lee et al., 2009).

- Step 3: Supermatrix formation: The supermatrix concept resembles the Markov chain process (Yazgan et al., 2010). The local priority vectors are entered into the appropriate column of a matrix, known as a supermatrix, by aiming to obtain global priorities in a system with interdependent influences. As a result, a supermatrix is actually a partitioned matrix, where each matrix segment represents a relationship between two nodes in a system (Wijnmalen, 2007). Let the clusters of a decision system be C_k, $k = 1, 2,..., N$, and each cluster k has n_k elements, denoted by $e_{k1}, e_{k2}, \cdots, e_{kn_k}$. The value obtained in the previous steps are clustered and placed in the appropriate positions in a supermatrix, based on the influence flow from one cluster to another, or from a cluster to itself, as in a loop. A standard form for a supermatrix is as in formulate (2) (Yazgan et al., 2010):

$$W = \begin{array}{cc} & \begin{array}{c} e_{11}\ e_{12} \cdots e_{1m1} \cdots e_{k1} e_{k2} \cdots e_{kmk} \cdots e_{n1} e_{n2} \cdots e_{nmn} \end{array} \\ \begin{array}{cc} & e_{11} \\ C_1 & e_{12} \\ \vdots & \vdots \\ & e_{1m1} \\ & \vdots \\ & e_{k1} \\ & e_{k2} \\ C_k & \vdots \\ & e_{kmk} \\ \vdots & \vdots \\ & e_{n1} \\ C_n & e_{n2} \\ & \vdots \\ & e_{nmn} \end{array} & \begin{bmatrix} W_{11} & \cdots & W_{1k} & \cdots & W_{1n} \\ \vdots & & \vdots & \vdots & \vdots \\ W_{k1} & \cdots & W_{kk} & \cdots & W_{kn} \\ \vdots & & & \vdots & \\ W_{n1} & \cdots & W_{nk} & \cdots & W_{nn} \end{bmatrix} \end{array} \tag{2}$$

As an example, the representation of the supermatrix for a hierarchy with three levels (Yazgan et al., 2010):

$$W = \begin{bmatrix} 0 & 0 & 0 \\ W_{21} & 0 & 0 \\ 0 & W_{32} & I \end{bmatrix}. \tag{3}$$

where W_{21} is a vector that represents the effect of the objective on the criteria, W_{32} is a matrix that represents the effect of the criteria on each of the alternatives, I is the identity matrix, and all of the zeros correspond to those elements that have no influence. For the previous example, if the criteria are interrelated among themselves, the hierarchy is replaced by a network as shown in Equation (2). The presence of the matrix element W_{22} of the supermatrix W_n can be represented by the interdependency that the supermatrix would present (Yazgan et al., 2010):

$$W_n = \begin{bmatrix} 0 & 0 & 0 \\ W_{21} & W_{22} & 0 \\ 0 & W_{32} & I \end{bmatrix}. \tag{4}$$

If there is an interrelationship of elements within a cluster or between two clusters, then any zero value in the supermatrix can be replaced by a matrix. As there is usually interdependence among clusters in a network, the columns of a supermatrix may sum to more than one. However, the supermatrix must be modified so that each column of the matrix sums to a unified value. Saaty (Yazgan et al., 2010) has recommended the approach of determining the relative importance of the cluster as the controlling component (Wijnmalen, 2007). Raising a matrix to powers gives the long-term influences of the elements relative to each other. To obtain a convergence on the importance of weights, the weighted supermatrix is raised to the power of $2k + 1$, where k is an arbitrarily large number, and forms a new matrix that is called the limit supermatrix (Yazgan et al., 2010). The limit supermatrix form is similar to the weighted supermatrix, except that all of the columns of the limit supermatrix are the same. The normalizing of each cluster of this supermatrix can be reached with the final priorities of all elements in the matrix.

- Step 4: Selection of the best alternatives: The final priorities can be obtained by normalizing each alternative's column in the limit supermatrix.

The ANP is a technique which is similar to the AHP. The characteristic between criteria and sub-criteria of the ANP is a network, while AHP is a hierarchy. In addition, the ANP is also similar to other techniques as well. For example, the multi-attribute utility technique (MUAT), cross-impact analysis and cost-consequences analysis (CCA), these are categorized as alternative selection analyses, as well as the AHP or the ANP. Saaty has proposed one of the general theories of the ANP (Saaty, 2004a), which is the BOCR or benefits, opportunities, costs, and risks in a decision-making process (Chen et al., 2010). The BOCR is a combination of the score of each alternative by five methods.

1. Additive: $P_i = bB_i + oO_i + c[(1/C_i)_{Normalized}] + r[(1/R_i)_{Normalized}]$ (5)

2. Probabilistic additive: $P_i = bB_i + oO_i + c(1 - C_i) + r(1 - R_i)$ (6)

3. Subtractive: $P_i = bB_i + oO_i - cC_i - rR_i$ (7)

4. Multiplicative priority powers: $P_i = B_i^b O_i^o [(1/C_i)_{Normalized}]^c [(1/R_i)_{Normalized}]^r$ (8)

5. Multiplicative: $P_i = B_i O_i / C_i R_i$ (9)

An example of recent research using BOCR is the Disney decision. Examining the construction of a new theme park in greater China, this research uses an integration of BOCRs models. It has an objective of searching for new market areas. An area for a new Disney park will require a minimal investment, mostly on returns through royalties, licensing and income streams. The final result for this project is Hong Kong as the first site to get into China, although Shanghai is a more costly option with a higher potential future market than Hong Kong. Another research paper on the topic is about the model of the buyer–supplier relationship. It has integrated the ANP and BOCR concepts. The result of the research provides advice to select the most suitable form of relationship between supplier and manufacturer.

3. Methodology and algorithm

The adoption of the BOCR concept, the ANP model with BOCR is suggested in this segment to select a solution to the conveyor system problem. The steps are shown as follows:

Step 1. Define the problem, set criteria, and set definition criteria by experts through brainstorming or other appropriate methods.

Step 2. Set a control network for the problem by specifying strategic criteria and the four merits, benefits, opportunities, costs, and risks (Saaty, 2004b, 2004c) as shown in Figure 2. The four merits in the control network are used to rate the B, O, C, and R.

Step 3. Create a questionnaire with a nine-point scale as outlined by Saaty, as a comparison in determining the pairwise strategic criteria.

Step 4. Use a five-step scale that can evaluate the importance of BOCRs for each strategic criterion. Values and wording of five-scale 0.42, 0.26, 0.16, 0.10, and 0.06, of very high, high, medium, low, and very low, respectively. (Erdogmus et al., 2005; Erdogmus, Aras, & Koc, 2006; Saaty, 2004b, 2004c). The panelist opinions are aggregated by the geometric mean method.

Step 5. Calculate the score of each strategic criterion from Step 4, and the respective strategic criteria from Step 3. The value of B, O, C, and R came from the normalized priorities of BOCRs, respectively.

Step 6. Set criteria and sub criteria according to the merit of each and organization for all four merits. Based on brainstorming or other appropriate methods, a network is constructed in the form as shown in Figure 3. The goal must be connected with the four merits, benefits (B), opportunities (O), costs (C), risks (R).

Step 7. Prepare a questionnaire established on a pairwise comparison of elements at each level. The questionnaire used in this section is the nine-point-scale questionnaire.

Step 8. After all experts have filled out the questionnaire, create an aggregate of opinions by the geometric mean method. After that, calculate the unweighted supermatrix, weighted supermatrix, and limit supermatrix for each merit.

Step 9. Calculate each alternative by using the five ways, additive, probabilistic additive, subtractive, multiplicative priority powers, and multiplicative.

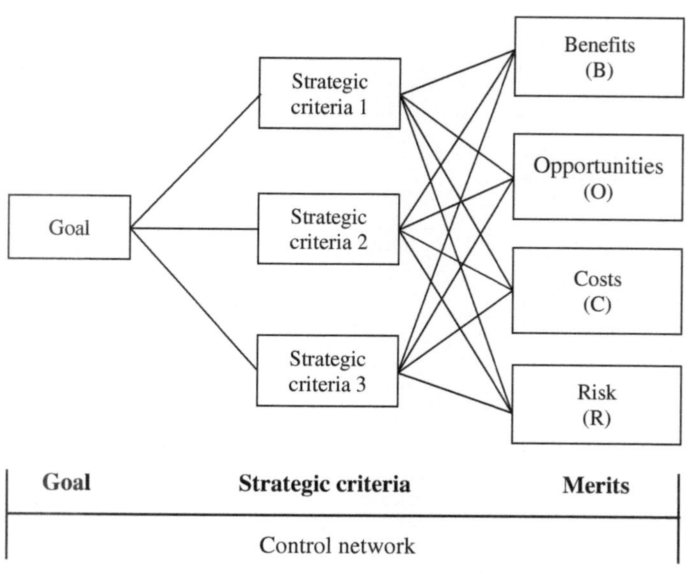

Figure 2. The network of the decision problem (control network).

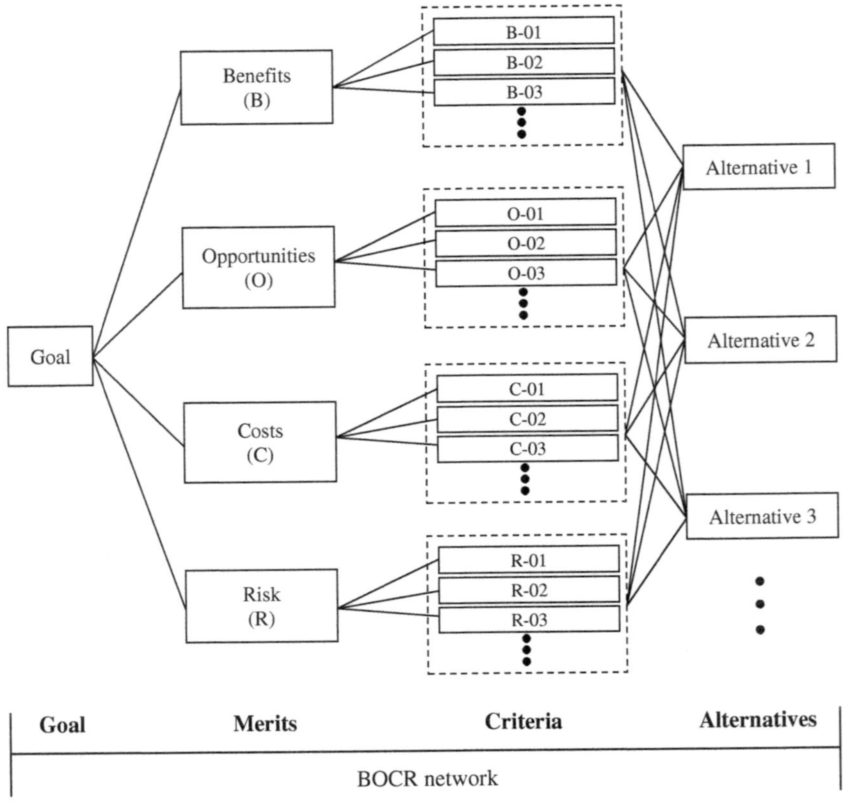

Figure 3. The network of the decision problem (BOCR network).

4. Experimental results

A committee should be formed composed of three experts, including a manager, senior manager from the company logistics department, and a manager from the design department from the conveyor system company.

In Figure 4, the strategic criteria for the conveyor selection system is shown. The company wants to select a conveyor system, and the strategic criteria are flexibility, manufacturing, future plans, productivity, safety, and quality. Flexibility is related with the quality of being adaptable or variable. Manufacturing is related with reputation and relationships. Future plans are related with capacity plans and process plans. Productivity is related with the quality of being productive or having the power to produce. Safety is related with safety device design at an ergonomics design level. Quality is related with fulfilling the customer's requirements and expectations, at all times. In the third level, there are four merits: benefits (B), opportunities (O), costs (C), risks (R). For example, an expert's opinion of a questionnaire with a nine-point scale is shown in Table 1. A proportion of 5:1 between flexibility and manufacturing implies that flexibility is five times more important than manufacturing.

The eigenvalue method is used for calculating eigenvector and eigenvalue. Calculating the consistency index (CI) and consistency ratio (CR) value is used for checking consistency of comparison (Saaty, 1980).

$$Ws1 = \begin{matrix} \text{flexibility} \\ \text{manufacture} \\ \text{future plan} \\ \text{productivity} \\ \text{safety} \\ \text{quality} \end{matrix} \begin{bmatrix} 0.3169 \\ 0.0975 \\ 0.0723 \\ 0.1739 \\ 0.0973 \\ 0.2419 \end{bmatrix} \quad \text{and } \lambda_{max} = 6.3827 \tag{10}$$

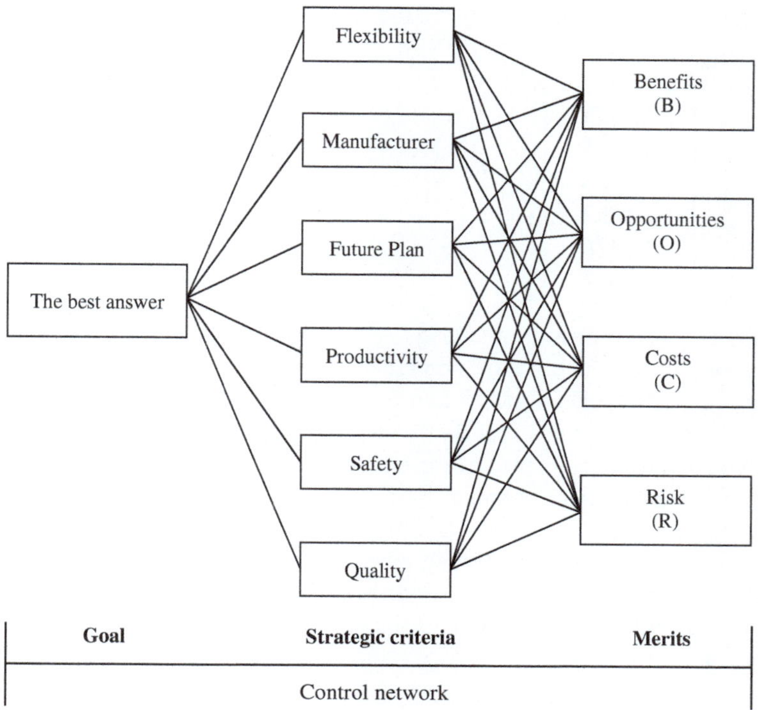

Figure 4. Control network.

Table 1. Scale of relative importance

Intensity of importance	Definition	Explanation
1	Equal importance	Two activities contribute equally to the objective
3	Moderate importance	Experience and judgment slightly favor one activity over another
5	Strong importance	Experience and judgment strongly favor one activity over another
7	Very strong or demonstrated importance	An activity is favored very strongly over another; its dominance demonstrated in practice
9	Extreme importance	Sometimes one needs to interpolate a compromise judgment numerically because there is no good word to describe it
2, 4, 6, 8	For compromise between the above values	When compromise is needed

Note: Saaty, 1980.

$$CI = \frac{\lambda_{max} - n}{n-1} = \frac{6.3827 - 6}{6-1} = 0.0765 \tag{11}$$

$$CR = \frac{CI}{RI} = \frac{0.0765}{1.25} = 0.0612 \tag{12}$$

If the value of CR is less than 0.1, the mean comparison is consistent (Ergu, Kou, Shi, & Shi, 2014). A combination of opinions from all experts is used as a geometric mean method. For example, the pairwise comparison between flexibility and manufacturing from all respondents are (5:1), (3:1), and (5:1). Therefore, putting the value in the geometric mean method is $(5 \times 3 \times 5)^{1/3} = 4.2172$. For the six strategic criteria calculated shown in Table 2, the synthesized priorities of strategic criteria are as follows:

$$Ws = \begin{matrix} \text{flexibility} \\ \text{manufacture} \\ \text{future plan} \\ \text{productivity} \\ \text{safety} \\ \text{quality} \end{matrix} \begin{bmatrix} 0.2934 \\ 0.1415 \\ 0.1025 \\ 0.1662 \\ 0.0928 \\ 0.2037 \end{bmatrix} \tag{13}$$

For BOCR, each expert will be asked to answer according to a five-step scale. According to Step 4 in the methodology and algorithm section, the values according to the five-scale were 0.42, 0.26, 0.16, 0.10, and 0.06, of very high, high, medium, low, and very low, respectively. The final value after brainstorming all the expert opinions is shown in Tables 3 and 4. The final criteria in the BOCR network came from experts through the Delphi method as shown in Figure 5. For example, calculations will show the network of benefits/used subnet as shown in Figure 6.

After calculating all of the comparisons, the unweighted supermatrix and the weighted supermatrix for benefits/used are formed as presented in Tables 5 and 6. And after that, they are transformed to the limit supermatrix. The final decision for used/benefits subnet is shown in the limit supermatrix, and the highest final score is alternative 4.

Table 2. Pairwise comparison

	Absolute		**Very strong**		**Strong**		**Weak**		**Equal**	**Weak**		**Strong**		**Very strong**		**Absolute**		
	9:1	**8:1**	**7:1**	**6:1**	**5:1**	**4:1**	**3:1**	**2:1**	**1:1**	**1:2**	**1:3**	**1:4**	**1:5**	**1:6**	**1:7**	**1:8**	**1:9**	
In order to have the best selection for conveyor system.																		
Flexibility					x													Manufacture
Flexibility							x											Future plan
Flexibility							x											Productivity
Flexibility							x											Safety
Flexibility									x									Quality
Manufacture							x											Future plan
Manufacture											x							Productivity
Manufacture									x									Safety
Manufacture											x							Quality
Future plan											x							Productivity
Future plan									x									Safety
Future plan											x							Quality
Productivity									x									Safety
Productivity									x									Quality
Safety											x							Quality

Table 3. Comparison matrix for strategic criteria

	Flexibility	Manufacturer	Future plan	Productivity	Safety	Quality	Synthesized priorities
Flexibility	1	4.2172	2.0801	2.4662	2.0801	1.0000	0.2934
Manufacturer	0.2371	1	2.0801	1.0000	1.0000	1.0000	0.1415
Future plan	0.4807	0.4807	1	0.4807	1.4422	0.5503	0.1025
Productivity	0.4055	1.0000	2.0801	1	2.0801	1.0000	0.1662
Safety	0.4807	1.0000	0.6934	0.4807	1	0.3333	0.0928
Quality	1.0000	1.0000	1.8171	1.0000	3.0000	1	0.2037

Table 4. Priorities of BOCRs

	Flexibility (0.2934)	Manufacturer (0.1415)	Future plan (0.1025)	Productivity (0.1662)	Safety (0.0928)	Quality (0.2037)	Priorities for BOCR merits	Normalized priorities
Benefit	High	High	High	Very high	High	High	0.2866	0.2977 (b)
Opportunity	High	High	Medium	High	High	Very high	0.2823	0.2932 (o)
Cost	Low	High	High	Medium	Low	Low	0.1490	0.1547 (c)
Risk	High	High	Very high	Medium	Low	High	0.2449	0.2544 (r)

The alternative's final ranking is a combination of scores under the B, O, C, and R of each alternative by using the five methods for combining scores. The final ranking of the alternatives are as shown in Table 6. For example, alternative 1 (As-Is: Present) is calculated by the five methods.

Additive:

$$P_1 = bB_1 + oO_1 + c\left[(1/C_1)_{Normalized}\right] + r\left[(1/R_1)_{Normalized}\right]$$
$$= 0.2977 \times 0.1018 + 0.2932 \times 0.0999 + 0.1547 \times 0.5480 + 0.2544 \times 0.5618$$
$$= 0.2873$$

Probabilistic additive:

$$P_1 = bB_1 + oO_1 + c(1 - C_1) + r(1 - R_1)$$
$$= 0.2977 \times 0.1018 + 0.2932 \times 0.0999 + 0.1547 \times (1{-}0.0792) + 0.2544 \times (1{-}0.0747)$$
$$= 0.4375$$

Subtractive:

$$P_1 = bB_1 + oO_1 - cC_1 - rR_1$$
$$= 0.2977 \times 0.1018 + 0.2932 \times 0.0999 - 0.1547 \times 0.0792 - 0.2544 \times 0.0747$$
$$= 0.0283$$

Multiplicative priority powers:

$$P1 = B_1^b O_1^o [(1/C1)_{Normalized}]^c [(1/R1)_{Normalized}]^r$$
$$= (0.1018^{0.2977})(0.0999^{0.2932})(0.5480^{0.1547})(0.5618^{0.2932})$$
$$= 0.2029$$

Multiplicative:

$$P_1 = B_1 O_1 / C_1 R_1$$
$$= 0.1018 \times 0.0999 / 0.0792 \times 0.0747$$
$$= 1.7194$$

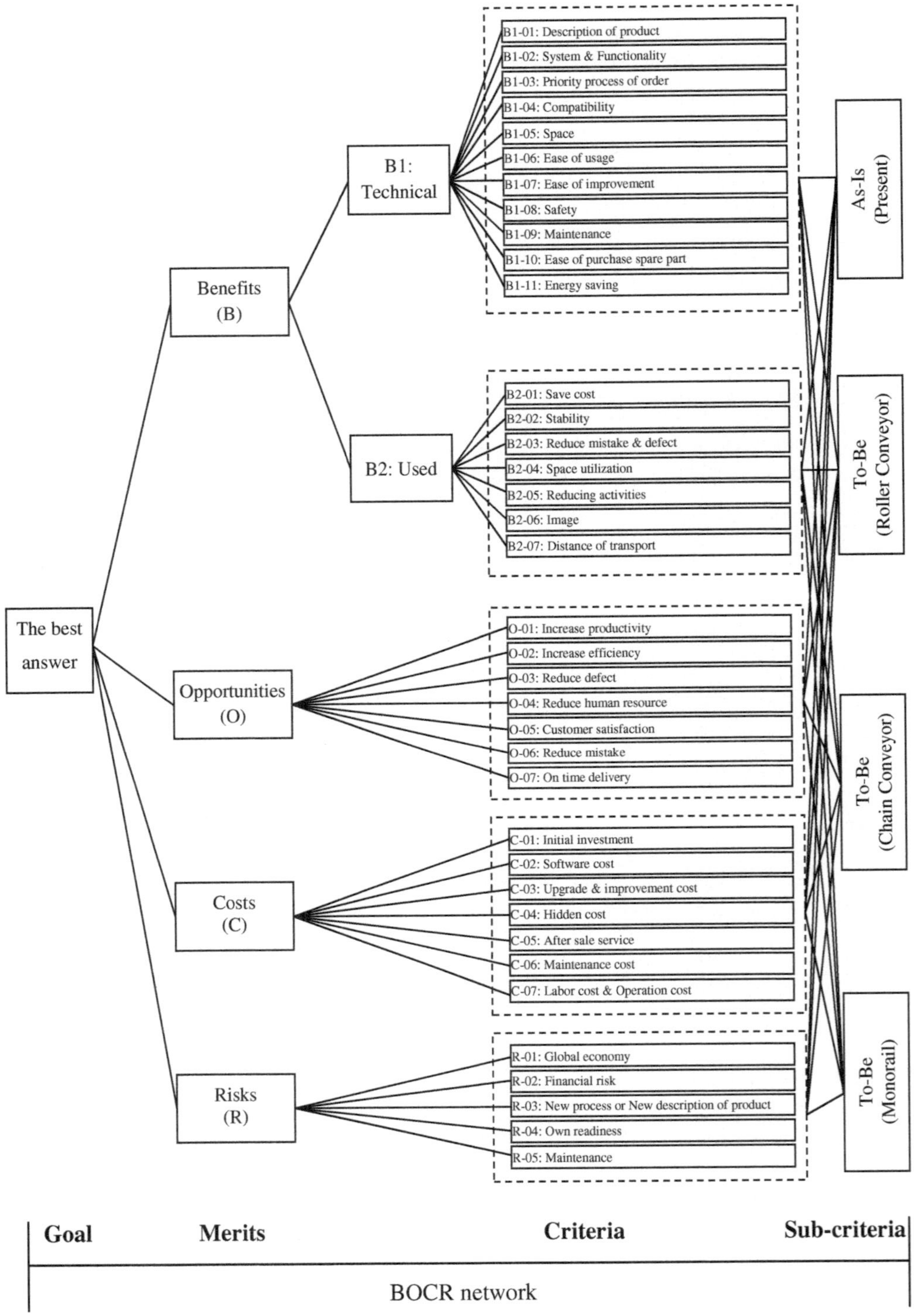

Figure 5. BOCR network.

Under the benefits and opportunities merits in Table 7, to-be: Monorail is the best with 0.3654, 0.3918, respectively. Nevertheless, under the costs and risks merits, as-is: Present has the best cost and the least risky alternative, with normalized reciprocals of 0.5480 and 0.5618, respectively.

The final calculation of alternatives uses the five methods under B, O, C, and R. The results of ranking the alternatives are shown in Table 8. From final result, first priority of the best alternative under all five methods is the chain conveyor. While roller conveyor always stay, respectively, as the third and the last

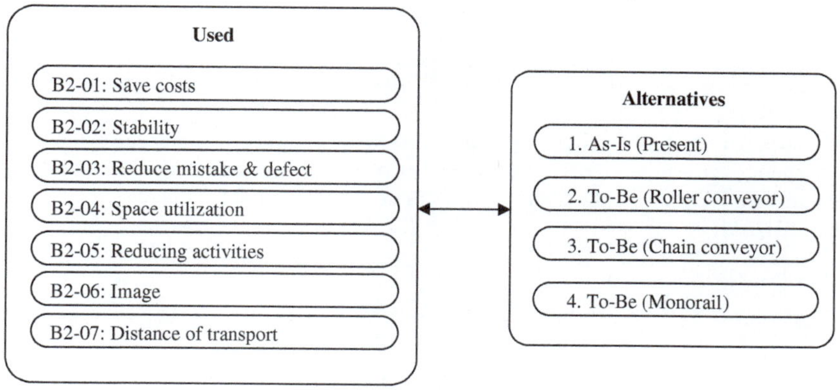

Figure 6. Clusters with element under used benefits.

Table 5. Weighted supermatrix under benefits/used subnet

	Alternatives				Benefits/Used						
	As-Is: present (I)	To-be: roller Conveyor (II)	To-be: chain Conveyor (III)	To-be: monorail (IV)	Save costs (B2–01)	Stability (B2–02)	Reduce mistake & defect (B2–03)	Space utilization (B2–04)	Reducing activities (B2–05)	Image (B2–06)	Distance of transport (B2–07)
As-Is: present (I)	0	0	0	0	0.0935	0.0786	0.0844	0.0562	0.0973	0.0810	0.1081
To-be: roller conveyor (II)	0	0	0	0	0.2151	0.1876	0.2711	0.4412	0.0820	0.1540	0.3206
To-be: chain conveyor (III)	0	0	0	0	0.2149	0.3389	0.4063	0.2435	0.3266	0.3885	0.3104
To-be: monorail (IV)	0	0	0	0	0.4765	0.3949	0.2382	0.2591	0.4942	0.3765	0.2609
Save costs (B2–01)	0.2054	0.3318	0.2498	0.3039	0	0	0	0	0	0	0
Stability (B2–02)	0.1561	0.2205	0.2164	0.1738	0	0	0	0	0	0	0
Reduce mistake & defect (B2–03)	0.1224	0.0877	0.1399	0.1149	0	0	0	0	0	0	0
Space utilization (B2–04)	0.1390	0.1051	0.1580	0.1817	0	0	0	0	0	0	0
Reducing activities (B2–05)	0.0559	0.0539	0.0957	0.0561	0	0	0	0	0	0	0
Image (B2–06)	0.0853	0.0360	0.0611	0.0679	0	0	0	0	0	0	0
Distance of transport (B2–07)	0.2360	0.1651	0.0792	0.1018	0	0	0	0	0	0	0

system. Under probabilistic additive (0.4771), subtractive (0.0680), and the multiplicative priority powers (0.2326), chain conveyor is the first best, and monorail is the second. In addition, the ranking under multiplicative method is not the same as other, major reason is that the method does not take into account the priorities B, O, C, and R. Nevertheless, with the objective of selecting two systems, the first and second select chain conveyor and monorail, respectively, under all five methods.

Table 6. The Limit supermatrix under benefits/used subnet

	Alternatives				Benefits/Used						
	As-Is: present (I)	**To-be: roller Conveyor (II)**	**To-be: chain Conveyor (III)**	**To-be: monorail (IV)**	**Save costs (B2–01)**	**Stability (B2–02)**	**Reduce mistake & defect (B2–03)**	**Space utilization (B2–04)**	**Reducing activities (B2–05)**	**Image (B2–06)**	**Distance of transport (B2–07)**
As-Is: present (I)	0.0426	0.0426	0.0426	0.0426	0.0426	0.0426	0.0426	0.0426	0.0426	0.0426	0.0426
To-be: roller conveyor (II)	0.1254	0.1254	0.1254	0.1254	0.1254	0.1254	0.1254	0.1254	0.1254	0.1254	0.1254
To-be: chain conveyor (III)	0.1477	0.1477	0.1477	0.1477	0.1477	0.1477	0.1477	0.1477	0.1477	0.1477	0.1477
To-be: mono-rail (IV)	0.1843	0.1843	0.1843	0.1843	0.1843	0.1843	0.1843	0.1843	0.1843	0.1843	0.1843
Save costs (B2–01)	0.1433	0.1433	0.1433	0.1433	0.1433	0.1433	0.1433	0.1433	0.1433	0.1433	0.1433
Stability (B2–02)	0.0983	0.0983	0.0983	0.0983	0.0983	0.0983	0.0983	0.0983	0.0983	0.0983	0.0983
Reduce mis-take & defect (B2–03)	0.0581	0.0581	0.0581	0.0581	0.0581	0.0581	0.0581	0.0581	0.0581	0.0581	0.0581
Space utiliza-tion (B2–04)	0.0759	0.0759	0.0759	0.0759	0.0759	0.0759	0.0759	0.0759	0.0759	0.0759	0.0759
Reducing activities (B2–05)	0.0336	0.0336	0.0336	0.0336	0.0336	0.0336	0.0336	0.0336	0.0336	0.0336	0.0336
Image (B2–06)	0.0297	0.0297	0.0297	0.0297	0.0297	0.0297	0.0297	0.0297	0.0297	0.0297	0.0297
Distance of transport (B2–07)	0.0612	0.0612	0.0612	0.0612	0.0612	0.0612	0.0612	0.0612	0.0612	0.0612	0.0612

Table 7. Priorities of alternatives under four merits

Merits	**Benefits (0.2977)**			**Opportunities (0.2932)**		
			Normalized			Normalized
Alternative						
As-Is: present			0.1018			0.0999
To-be: roller conveyor			0.2482			0.2258
To-be: chain conveyor			0.2846			0.2824
To-be: monorail			0.3654			0.3918
	Costs (0.1547)			Risks (0.2544)		
	Normalized	Reciprocal	Normalized recipro-cal	Normalized	Reciprocal	Normalized reciprocal
Alternative						
As-Is: PRESENT	0.0792	12.6228	0.5480	0.0747	13.3890	0.5618
To-be: roller conveyor	0.2440	4.0985	0.1779	0.2549	3.9230	0.1646
To-be: chain conveyor	0.2528	3.9555	0.1717	0.2375	4.2112	0.1767
To-be: monorail	0.4240	2.3586	0.1024	0.4329	2.3098	0.0969

Table 8. Final synthesis of priorities of alternatives

Alternatives	Synthesizing methods									
	Additive		Probabilistic additive		Subtractive		Multiplication priority powers		Multiplicative	
	Priority	Rank	Priority	Rank	Priority	Rank	Priority	Rank	Priority	Rank
As-Is: present	0.2873	1	0.4375	4	0.0283	4	0.2029	4	1.7194	1
To-be: roller conveyor	0.2095	4	0.4466	3	0.0375	3	0.2065	3	0.9010	3
To-be: chain conveyor	0.2390	3	0.4771	1	0.0680	1	0.2326	1	1.3389	2
To-be: monorail	0.2642	2	0.4570	2	0.0479	2	0.2186	2	0.7801	4

5. Conclusions

In the industrialized world from past to present, one very important aspect to every industry is materials handling. For an investment analysis in choosing a MHS, the process is extremely complex and there are multiple solutions for any particular situation. Therefore, this paper differs from previous studies of the conveyor selection system problem by using decision-making techniques to select the best conveyor system for a company.

In this research, a model, which performs an analysis using the ANP with benefits-opportunities-costs-risks (BOCR), is used to evaluate conveyor system selection. The model in this paper can help perform a stable evaluation of the various types of conveyor systems. The multi-criteria decision-making techniques used by the ANP are the same as the AHP, but the ANP has been featured for its relationship between alternatives on the criteria. The suggested model not only considers the cost and benefit factors, similar to other decision-making investment models, but this paper also takes into account opportunity and risk factors. Therefore, the proposed model can be used to properly evaluate any conveyor system in any industry, to help select the best form of conveyor system.

This research suggests key factors which used for a decision to help consider and investment in material handling conveyor system. Moreover, this research provides the conveyor selection model that aimed to support decision-making of executive or plant manager who are interested for conveyor system investment analysis while this model is also useful as to increase reliability for manufacturers of conveyor and to encourage clients to be more participated in conveyor system selection.

The criteria in the model are not fixed, but may differ across the type of situation. Therefore, the criteria should be removed or added conditional upon the development of the model. This model can also be profitable for use in future research.

Funding
The authors received no direct funding for this research.

Author details
Pairat Jiamruangjarus[1]
E-mail: paijiam@hotmail.com
Thanakorn Naenna[1]
E-mail: thanakorn.nae@mahidol.ac.th
[1] Faculty of Engineering, Department of Industrial Engineering, Mahidol University, Nakhonpathom 73170, Thailand.

References

Chan, F. T. S., Ip, R. W. L., & Lau, H. (2001). Integration of expert system with analytic hierarchy process for the design of material handling equipment selection system. *Journal of material processing technology, 116*, 137–145. doi:10.1016/S0924-0136(01)01038-X

Chen, H. H., Lee, A. H. I., & Kang, H. Y. (2010). A model for strategic selection of feeder management systems: A case study. *International journal of Electrical power and energy systems, 32*, 421–427. doi:10.1016/j.ijepes.2009.09.023

Cho, S., Kim, J., & Heo, H. (2015). Application of fuzzy analytic hierarchy process to select the optimal heating facility for Korean horticulture and stockbreeding sectors. *Renewable and Sustainable Energy Reviews, 49*, 1075–1083. doi:10.1016/j.rser.2015.04.105

Chung, S. H., Lee, A. H. I., & Pearn, W. L. (2005). Analytic network process (ANP) approach for product mix

planning in semiconductor fabricator. *International journal of production economics, 96*, 15–36. doi:10.1016/j.ijpe.2004.02.006
Dongre, A., & Mohite, N. Y. (2015). Significance of selection of material handling system design in industry—A review. *International Journal of Engineering Research and General Science., 3*, 76–79.
Erdogmus, S., Kapanoglu, M., & Koc, E. (2005). Evaluating high-tech alternatives by using analytic network process with BOCR and multiactors. *Evaluation and Program Planning, 28*, 391–399. doi:10.1016/j.evalprogplan.2005.07.003
Erdogmus, S., Aras, H., & Koc, E. (2006). Evaluation of alternative fuels for residential heating in Turkey using analytic network process (ANP) with group decision making. *Renewable and Sustainable Energy Reviews, 10*, 269–279. doi:10.1016/j.rser.2004.09.003
Ergu, D., Kou, G., Shi, Y., & Shi, Y. (2014). Analytic network process in risk assessment and decision analysis. *Computers & Operations Research, 42*, 58–74. doi:10.1016/j.cor.2011.03.005
Fonseca, D. J., Uppal, G., & Greene, T. J. (2004). A knowledge-based system for conveyor equipment selection. *Expert systems with applications, 26*, 615–623. doi:10.1016/j.eswa.2003.12.011
Jaafari, A., Najafi, A., & Melónc, M. G. (2015). Decision-making for the selection of a best wood extraction method: An analytic network process approach. *Forest Policy and Economics, 50*, 200–209. doi:10.1016/j.forpol.2014.09.010
Krishna Mohan, K., Reformat, M. Z. & Pedrycz, W. (2013, June). *Interval-based analysis of BOCR (benefits, opportunities, costs and risks) models evaluated by multiple experts*. IEEE Conference Publications. doi:10.1109/IFSA-NAFIPS.2013.6608407
Kulak, O. (2005). A decision support system for fuzzy multi-attribute selection of material handling equipments. *Expert systems with applications, 29*, 310–319. doi:10.1016/j.eswa.2005.04.004
Lee, J. W., & Kim, S. H. (2000). Using analytic network process and goal programming for interdependent information system project selection. *Computer and operations research, 27*, 367–382. doi:10.1016/S0305-0548(99)00057-X
Lee, A. H. I., Chang, H.-J., & Lin, C.-Y. (2009). An evaluation model of buyer-supplier relationships in high-tech industry—The case of an electronic components manufacturer in Taiwan. *Computers & Industrial Engineering, 57*, 1417–1430. doi:10.1016/j.cie.2009.07.012
Liang, C., & Li, Q. (2008). Enterprise information system project selection with regard to BOCR. *International Journal of Project Management, 26*, 810–820. doi:10.1016/j.ijproman.2007.11.001
Malmir, R., Hamzehi, E., & Farsijani, H. (2013). A Multi stage decision making model to evaluate suppliers by using MOLP and ANP in a strategic approach. *International Journal of Application or Innovation in Engineering & Management, 2*, 563–577.
Marcello, B., Gabbrielli, R., & Miconi, D. (2001). Material handling device selection in cellular manufacturing. *Journal of multi-criteria decision analysis, 10*, 303–315. doi:10.1002/mcda.310
Mili, K. (2014). Six sigma approach for the straddle carrier routing problem. EWGT2013—16th Meeting of the EURO Working Group on Transportation. *Procedia—Social and Behavioral Sciences, 111*, 1195–1205. doi:10.1016/j.sbspro.2014.01.154
Saaty, T. L. (1980). *The analytical hierarchy process: Planning. Priority setting, resource allocation* (1st ed.). New York, NY: McGraw-Hill.
Saaty, T. L. (2004a). Decision making—The analytic hierarchy and network process (AHP/ANP). *Journal of Systems Science and Systems Engineering, 13*(1), 1–35. doi:10.1007/s11518-006-0151-5
Saaty, T. L. (2004b). Fundamentals of the analytic network process—Dependence and feedback in decision-making with a single network. *Journal of Systems Science and Systems Engineering, 13*, 129–157. doi:10.1007/s11518-006-0158-y
Saaty, T. L. (2004). The analytic network process dependence and feedback in decision making (Part 1) theory and validation. In *MCDM 2004*. Whistler, BC.
Saatya, T. L., & Sagir, M. (2015). Choosing the best city of the future. *Journal of Urban Management, 4*, 3–23. doi:10.1016/j.jum.2015.06.003
Sakthivela, G., Ilangkumaranb, M., & Gaikwada, A. (2015). A hybrid multi-criteria decision modeling approach for the best biodiesel blend selection based on ANP-TOPSIS analysis. *Ain Shams Engineering Journal, 6*, 239–256. doi:10.1016/j.asej.2014.08.003
Srisom, Y., & Sriuthai, P. (2004). Increasing the efficient of material handling. *Mechanical Electrical Industrial, 21*, 1–22.
Sujono, S., & Lashkari, R. S. (2007). A multi-objective model of operation allocation and material handling system selection in FMS design. *International Journal of Production Economics, 105*, 116–133. doi:10.1016/j.ijpe.2005.07.007
Sule, D. R. (1994). *Manufacturing facilities: Location, planning and design* (2nd ed.). Boston, MA: PWS.
Swaminathan, S. R., Matson, J. O., & Mellichamp, J. M. (1992). EXCITE: Expert consultant for in-plant transportation equipment. *International Journal of Production Research, 30*, 1969–1983. doi:10.1080/00207549208948133
Tompkins, J. A., & White, J. A. (1984). *Facilities planning*. New York, NY: Wiley.
Tompkins, J. A., White, J. A., Bozer, Y. A., & Tanchoco, J. M. A. (2002). *Facilities planning*. New York, NY: Wiley.
Tornjanski, V., Marinković, S. & Lalić, N. (2014, June). Application of ANP method based on a BOCR model for decision-making in banking. In *XIV International Symposium. New Business Models and sustainable Competitiveness Symposium Proceeding, SYMORG*. Zlatibor.
Tuzkaya, G., Gülsün, B., Kahraman, C., & Özgen, D. (2010). An integrated fuzzy multi-criteria decision making methodology for material handling equipment selection problem and an application. *Expert systems with applications, 37*, 2853–2863. doi:10.1016/j.eswa.2009.09.004
Ustun, O., & Demirtas, E. A. (2008). Multi-period lot-sizing with supplier selection using achievement scalarizing functions. *Computers & Industrial Engineering, 54*, 918–931. doi:10.1016/j.cie.2007.10.021
Wang, W. M., Lee, A. H. I., Peng, L. P., & Wud, Z. L. (2013). An integrated decision making model for district revitalization and regeneration project selection. *Decision Support Systems, 54*, 1092–1103. doi:10.1016/j.dss.2012.10.035.
Wijnmalen, D. J. D. (2007). Analysis of benefits, opportunities, costs, and risks (BOCR) with the AHP-ANP: A critical validation. *Mathematical and computer modelling, 46*, 892–905. doi:10.1016/j.mcm.2007.03.020
Wiratanaya, G. N., Darmawan, D. P., Kolopaking, L. M., & Windia, W. (2015). Selection of beef production systems in Bali: An analytical network with BOCR approach. *Journal of Economics and Sustainable Development, 6*, 45–59.
Yazgan, H. R., Boran, S., & Goztepe, K. (2010). Selection of dispatch rules in FMS: ANP model based on BOCR with choquet integral. *The International Journal of Advanced Manufacturing Technology, 49*, 785–801. doi:10.1007/s00170-009-2416-x

Effect of roasting on physicochemical and functional properties of flaxseed flour

Azhar Khan[1]* and C.S. Saini[1]

*Corresponding author: Azhar Khan, Department of Food Engineering and Technology, Sant Longowal Institute of Engineering & Technology, Longowal 148 106, Punjab, India

E-mail: azharkhanmuslim@gmail.com

Reviewing editor: Shashi Dubey, Hindustan College of Engineering, India

Abstract: An investigation was carried out on the physical, physicochemical, and functional properties of flaxseed. Physical properties viz. seed shape and size, geometric and arithmetic mean diameter, sphericity, aspect ratio, bulk and true density, porosity, angle of repose, and static friction coefficient were determined. Geometric and arithmetic mean diameter were 2.19 and 3.51 mm while average sphericity and aspect ratio were 40.34 and 62.58%. The average true density, bulk density, and porosity were 1.34 g/cm^3, .66 g/cm^3, and 51.56%. Angle of repose was 19.40° and coefficient of static friction obtained on glass, stainless steel, plywood perpendicular, and plywood parallel was .32, .36, .33, and .33, respectively. Flaxseeds were roasted and compositional and functional properties like water absorption capacity (WAC), oil absorption capacity (OAC), foaming capacity, foaming stability, sedimentation value, and least gelation concentration of roasted and unroasted flaxseed flour were performed. Foaming capacity (9.23%) and foaming stability (54.43%) were significantly higher for unroasted flaxseed than roasted flaxseed flour (7.82 and 48.60%). Roasted flour was observed to have highest values of WAC, bulk density, WSI, ash, fiber, carbohydrate, and lowest values of moisture, protein, fat, OAC, tap density, porosity, angle of repose, WAI, and sediment value as compared to unroasted flour.

Subjects: Engineering & Technology; Food Engineering; Food Science & Technology

Keywords: flaxseed; roasting; physical; functional; physicochemical; Environmental sustainability engineering

ABOUT THE AUTHORS

Azhar Khan did his B.Tech. (Food Technology) from Marathwada Agriculture University (MAU) Parbhani, Maharashtra and M.Tech. (Food Engineering & Technology) from Sant Longowal Institute of Engineering & Technology. He has 1.5 years of experience in consultancy/advisory services in Agriculture and Food processing sector and is presently working as production supervisor in Food industry.

Charanjiv Singh is an associate professor at the Department of Food Engineering and Technology, Sant Longowal Institute of Engineering & Technology.

PUBLIC INTEREST STATEMENT

The physical properties of flaxseeds were evaluated to understand structural and functional behaviour of the flaxseed flour. The results of the experimental investigation may help to optimize engineering parameters of separation and processing equipments like in the processing of flaxseed flour and extraction of flaxseed oil. Roasting of flaxseed enhanced the nutty flavour and flaxseed flour sample being a rich source of good quality protein, minerals and fat can be used for utilization in bakery industry as a composite flour to improve the nutritional quality of products. Great potential, therefore, exists for incorporating flaxseed into human food products like cookies, cake, buns etc., as a fat supplement and a functional agent in a variety of formulated foods. Flaxseed products will boost the economy and status of bakery industry by providing essential nutrients in addition to phytochemical constituents for malnutrition, arthritis, cardiovascular and other diseases. Hence, greater use of the flaxseed as food is recommended.

1. Introduction

Flaxseed belongs to the family *Linaceae* and is an economically important multipurpose oilseed crop that yields ample amount of tiny smooth and flat seeds with a hard shiny shell that is light to reddish brown in color. The seeds possess wide range of health benefits due to the presence of omega-3 fatty acid especially alpha-linolenic acid, which prevents heart disease, inflammatory bowel disease, arthritis etc. and are thus consumed in whole, milled, and oil form. The seeds have been consumed from ancient times for its medicinal purposes to relieve abdominal pains and also as energy source (Oplinger, Oelke, Doll, Brundy, & Schuler, 1989). The oil percentage in Flaxseed is in the range of 28–42% depending on the variety and cultivation condition.

Roasting helps in the formation of desired flavor, and the quality of roasted nuts and oilseeds highly depends on roasting conditions as it improves the flavor, brown color, texture, and overall acceptability of the product (Özdemir & Devres, 2000a; Pittia, Dalla Rosa, & Lerici, 2001). However, the developed roasted flavor and aroma depends upon the temperature and time of roasting besides the type of nuts and techniques applied. Moreover, roasting results in lipid damage due to oxidation reactions, but the damage is less due to the antioxidant nature of phytochemicals present like tocopherols and polyphenols which play an important role for protection of nuts and oilseeds against fat deterioration (Chun, Lee, & Eitenmiller, 2005). Roasting also results in the formation of melanoidins due to Maillard's reaction, that are known to possess pronounced antioxidant properties, thereby improving the oxidative stability of nuts and seeds (Rizzi, 2003). Maillard reaction also plays an important role in the formation of typical color and flavor during roasting (Ziegleder, 1991). The roasting treatment has also resulted in loosening the shell of beans roasted (Krysiak & Motyl-Patelska, 2006).

The physical properties of agricultural produce are important in designing and constructing equipment and structures for handling, transportation, processing, storage, and also for assessing the product quality (Kashaninejad, Mortazavi, Safekordi, & Tabil, 2006; Sirisomboon, Pornchaloempong, & Romphophak, 2007). Physical and mechanical properties of fruit, nut, seed, and kernel are important to design equipment for dehulling, nut shelling, drying, oil extraction, and other processes like transportation and storage (Sirisomboon, Kitchaiya, Pholpho, & Mahuttanyavanitch, 2007). Physical properties such as size, shape, sphericity, aspect ratio, true density, bulk density, and porosity, and mechanical properties such as coefficient of friction, angle of repose as well as fracture resistance are very important in the design of processing machines for major agricultural crops (Owolarafe, Olabige, & Faborode, 2007). Many studies have been reported on the chemical and physical–mechanical properties of fruits and kernels, such as apricot kernel, berries, cherry laurel, cornelian cherry, fresh okra fruit, orange, rose fruit, wild plum etc. The objective of this study was to determine the physical properties of flaxseed and effect of roasting on the physical, physicochemical, and functional properties of flaxseed flour in comparison to wheat flour.

2. Material and methods

2.1. Materials

Flaxseeds (*Linum usitatissimum*) were purchased from local market of Parbhani, Maharashtra, India. The wheat flour (Brand name Rajdhani) was purchased from local market of Sangrur, Punjab, India. The seeds were cleaned and used as basic raw material for the study. For roasting, 50 g seeds were placed in a single layer on Petri dishes and were roasted in convection oven (Memmert, UNB 500) at 180°C for 10 min. After roasting, the seeds were allowed to cool at room temperature. After cooling (30 min), roasted samples were immediately ground. The raw and roasted samples were grinded using laboratory blender (Waring, USA) at low speed.

2.2. Proximate composition

Moisture content, ash, fat, fiber, and protein were determined using the method of Association of Official Analytical Chemists (2000) and total carbohydrate was determined by difference method of Raghuramulu, Nair, and Kalyanasundaram (1983). The energy value of the sample was calculated applying factors 4, 9, and 4 for each gram of protein, lipid, and carbohydrate, respectively (Shrestha & Noomhorm, 2002)

$$\text{Energy value(Calories)} = 4(\text{protein}) + 9(\text{fat}) + 4(\text{carbohydrate})$$

2.3. Determination of physical properties

2.3.1. Color characteristics

Color analysis was done by standard method using Hunter Lab Color Spectrophotometer (Gretag Macthbeth, I-5, USA)

2.4. Determination of geometrical properties

2.4.1. Determination of shape

The shape was expressed in terms of its sphericity index and aspect ratio. The higher the sphericity value, the closer is the shape to a sphere. For the sphericity index (ϕ), the dimensions obtained for the 10 selected seeds used to compute the sphericity index based on the equation (Karababa, 2006) as:

$$\phi(\%) = \frac{(X \cdot Y \cdot Z)^{\frac{1}{3}}}{X}$$

where, X = length (L) in mm; Y = width (mm); and Z = thickness.

For the aspect ratio same seeds were selected for conducting the experiment. The aspect ratio (R_a) was calculated as recommended by Maduako and Faborode (1990):

$$R_a(\%) = \frac{Y}{X} \times 100$$

2.4.2. Arithmetic and geometric mean diameter

The arithmetic mean diameter (D_a) and geometric mean diameter (D_g) of seeds were calculated from the geometrical dimensions by the formula given by Goyal, Kingsly, Kumar, and Walia (2007). Both arithmetic and geometric mean are expressed in mm and calculated by the following equation:

$$D_a = \frac{X + Y + Z}{3}$$

$$D_g = (X \cdot Y \cdot Z)^{\frac{1}{3}}$$

2.4.3. Surface area

Surface area (S) was estimated by the formula corresponding to the geometrical shape. The surface area of seed was determined by using the following equation (Goyal et al., 2007):

$$S\ (\text{mm}^2) = \pi D_g^2$$

where, D_g = geometrical mean diameter (mm).

2.4.4. Thousand kernel weight

Thousand kernel weight was determined by counting 100 seeds randomly and weighing them in an electronic balance (.001) and then multiplied by 10 to give the mass of 1,000 kernels.

Thousand kernel weight was expressed in grams (g).

2.5. Determination of gravimetric properties

2.5.1. Determination of bulk and true density

The bulk density is the ratio of the mass of the sample to its container volume occupied. For bulk density measurement, an empty cylindrical container was filled with seeds to a known volume. Tapping during the filling was done to obtain uniform packaging and to minimize the wall effect, if any. The filled sample was weighed and the bulk density was calculated using the below equation Mohsenin (1980);

$$\text{Bulk density}\left(\frac{\text{g}}{\text{cm}^3}\right) = \frac{M}{V}$$

where, M = mass of the sample (g); V = volume of the filled sample (cm^3).

The true density is defined as the ratio of mass of the sample to its true volume. It was determined by the toluene displacement method in order to avoid absorption of water during experiment. Five grams of sample was weighed and immersed into a 100 ml measuring cylinder containing 50 ml of toluene (Mwithiga & Sifuna, 2006). It was ensured that the seeds were submerged during immersion. The net volumetric displacement was recorded from the graduated scale of the cylinder. The true density was then calculated using equation below:

$$\text{True Density}(\text{g/cm}^3) = \frac{\text{Weight of seed (g)}}{\text{Rise in tolune level (cm}^3)}$$

2.5.2. Determination of tap density

The tap density of a material can be used to predict the flow properties and compressibility of a particular material. The tap density of a sample was determined by the method of Deshpande and Poshadri (2011) as;

$$\text{Tap density}\left(\frac{\text{g}}{\text{cm}^3}\right) = \frac{\text{weight of sample}}{\text{volume of tapped sample}}$$

2.5.3. Determination of porosity of seed

Porosity, ε (%) indicates the amount of pores in the bulk material and was calculated as per Mohsenin (1980). The porosity of the seed was calculated from the average values of bulk density and true density using the relationship.

$$\varepsilon(\%) = \left[1 - \frac{\rho_b}{\rho_t}\right] \times 100$$

where, ρ_b = bulk density (g/cm^3); ρ_t = true density (g/cm^3)

2.6. Determination of frictional properties

2.6.1. Angle of repose

The angle of repose is the angle with the horizontal at which the material will stand when piled. The cylinder was placed over a plain surface and flaxseeds were filled in. Tapping during filling was done to obtain uniform packing and to minimize the wall effect if any. The tube was slowly raised above the floor so that whole material could slide and form a natural slope. The height of heap above the floor and the diameter of the heap at its base were measured and the angle of repose (Φ) was calculated by following equation;

$$\Phi = \frac{h}{R}$$

or

$$\Phi = \tan^{-1}\frac{2h}{D}$$

where, Φ = angle of repose (o); h = height of the pile (cm); and D = diameter of the pile (cm).

2.6.2. Coefficient of static friction

The coefficient of static friction (μ) was determined from three structural materials namely glass, stainless steel, and plywood. A plastic cylinder of 30 mm diameter and 35 mm height was placed on an adjustable tilting flat plate faced with the test surface and filled with the sample. The cylinder was raised slightly without touching the surface. The structural surface with the cylinder resting on it was inclined gradually, until the cylinder starts to slide down. The angle of tilt was noted from a graduated scale (Dutta, Nema, & Bhardwaj, 1988) with the following equation;

$$\mu = \tan\theta$$

2.7. Determination of functional properties of flour

2.7.1. Color characteristics

A Hunter Lab Color Spectrophotometer (Gretag Macthbeth, I-5, USA) was employed to measure the color of the flour samples. A white and black standard tile was used for calibration. The color values were expressed as L^* (whiteness/darkness), a^* (redness/greenness), and b^* (yellowness/blueness). For the color analysis, the flour samples were packed in a transparent zip bag and placed against the light source to measure the color. Three measurements were analyzed for each substitution level and were analyzed at three different locations.

The color difference (ΔE) was calculated by applying the equation;

$$\Delta E = \left[\left(L^* - L_S\right)^2 + \left(a^* - a_S\right)^2 + \left(b^* - b_S\right)^2 \right]^{1/2}$$

The whiteness index values that combine lightness and yellow-blue into a single term was calculated as described by Hsu, Chen, Weng, and Tseng (2003) as follows:

$$\text{WI} = 100 - [(100 - L)^2 + a^2 + b^2]^{1/2}$$

2.7.2. Water absorption index (WAI) and water solubility index (WSI)

WAI and WSI of flour were determined as described by Singh, Sandhu, and Kaur (2005). Flour sample (2.5 g) was dispersed into 30 ml of distilled water, using a glass rod, and heated at 90°C for 15 min in a water bath. The cooked paste was cooled to room temperature and transferred to tarred centrifuge tube, and centrifuged at 3,000 × g for 10 min. The supernatant was transferred into a tarred evaporating dish for determination of dry solid content by evaporating the supernatant overnight at 110°C while the sediment was weighed. Triplicate determinations were carried out. WSI and WAI were calculated by the equations;

$$\text{WAI}(\text{g/g}) = \frac{\text{Weight of sediment}}{\text{Weight of flour sample}}$$

$$WSI(\%) = \frac{\text{Weight of dissolved solids in supernatant}}{\text{Weight of flour sample}} \times 100$$

2.7.3. Water absorption capacity

WAC of flour was measured by the centrifugation method of Sosulski (1962). Sample (3.0 g) was dispersed in 25 ml of distilled water and placed in pre-weighed centrifuge tube. The dispersions were stirred occasionally, held for 30 min, followed by centrifugation for 25 min at 3,000 × g. The supernatant was poured in pre-weighed Petri dish and dried in hot air oven for 25 min at 50°C and the solids remained after drying were weighed. Triplicate determinations were carried out and the water absorption capacities were expressed as gram of water bound per gram of the sample on a dry basis.

$$\mathrm{WAC}\left(\frac{g}{g}\right)=\frac{(\text{wt of tube with sample after drying}-\text{wt of tube})-\text{Sample weight}}{\text{Sample weight}}$$

2.7.4. Oil absorption capacity

For the determination of oil absorption capacity (OAC), the method of Lin, Humbert, and Sosulski (1974) was used. Sample (.5 g) was mixed with 6 ml of corn oil in pre weighed centrifuge tube. The contents were stirred for 1 min with a thin brass wire to disperse the sample in the oil. After a holding period of 30 min, the tubes were centrifuged for 25 min at 3000 × *g*. The separated oil was then removed with a pipette and the tubes were inverted for 25 min to drain the oil prior to weighing. Triplicate determinations were carried out and the oil absorption capacities were expressed as gram of oil bound per gram of the sample on a dry basis.

$$\mathrm{OAC}\left(\frac{g}{g}\right)=\frac{(\text{weight of tube with sample after removing oil}-(\text{tube weight }+\text{Sample weight})}{\text{Sample weight}}$$

2.7.5. Least Gelation Concentration

Gelation properties were studied in triplicates by employing the method of Sathe, Deshphande, and Salunkhe (1981). Test tubes, containing suspensions of 2, 4, 6, 8, 10, 12, 14, 16, 18, and 20% (w/v) of material in 5 ml distilled water were heated for 1 h in boiling water, followed by rapid cooling in ice cold water. The tubes were further cooled at 4°C for 2 h. The least gelation concentration (LGC) was determined when the sample in the inverted test tube did not fall down or slip. The results were expressed as no gelation, gel, firm gel, and very firm gel.

2.7.6. Foaming properties

Foaming properties were determined according to the method of Okaka and Potter (1977). One gram of flour was dispersed in 50 ml of distilled water, in a capped test tube, by shaking vigorously for 5 min followed by immediate pouring into a 250-ml graduated cylinder. The volume of the foam formed was then recorded as the foam capacity (%). A final observation was made after 60 min for recording the foam stability (%).

$$\text{Foaming capacity}(\%)=\frac{\text{Volume after whipping}-\text{Volume before whipping}}{\text{Volume before whipping}}$$

$$\text{Foaming stability }(\%)=\frac{\text{Foam Volume after 1hour}}{\text{Initial foam volume}}\times 100$$

2.7.7. SDS-sedimentation volume test

Sodium dodecyl sulfate (SDS) sedimentation volume of flour samples was estimated according to the procedure of Axford, Mcdermott, and Redman (1978). Flour (5 g, 14% moisture basis) was added to water (50 ml) in a cylinder, a stopclock was started, and the material was dispersed by rapid shaking for 15 s. The contents were re-shaken for 15 s after 2 and 4 min. After shaking, SDS-lactic acid reagent (50 ml) was added and mixed by inverting the cylinder four times before re-starting the clock. The SDS-lactic acid reagent was prepared by dissolving SDS (20 g) in distilled water (1 L) and then adding a stock diluted lactic acid solution (20 ml; 1 part lactic acid plus 8 parts distilled water by volume). Inversion (four times) was repeated at 2, 4, 5, and 6 min before finally starting the clock once again. The contents of the cylinder were allowed to settle for 40 min before reading sedimentation volume.

2.8. Statistical analysis

The reported data in Table 1 are an average of 10 observations and in rest of the tables are an average of 3 observations The significant differences were obtained by a one-way analysis of variance followed by Duncan's multiple range test ($p < .05$).

Table 1. Physical properties of flaxseed	
Parameters	**Flaxseed**
Moisture content	
Moisture content (%w.b.)	7.66 ± .52
Moisture content (%d.b.)	8.29 ± .62
Color values	
L^*	46.63 ± .25
a^*	8.21 ± .37
b^*	8.96 ± .01
Geometrical properties	
Length X(mm)	5.43 ± .25
Width Y(mm)	3.39 ± .15
Thickness Z (mm)	1.69 ± .24
Geometric mean diameter D_g (mm)	2.19 ± .03
Arithmetic mean diameter D_a (mm)	3.51 ± .17
Sphericity ϕ (%)	40.34 ± 1.38
Surface area S (mm^2)	15.08 ± .49
Gravimetric properties	
Bulk density, ρ_b (g/cm^3)	.66 ± .08
True density, ρ_s (g/cm^3)	1.34 ± .09
Porosity, ε (%)	51.56 ± 4.72
Aspect ratio, R_a (%)	62.58 ± 3.25
Thousand kernel weight (g)	7.14 ± .26
Friction properties	
Angle of repose (°)	19.40 ± 1.27
Static coefficient of friction	
Glass	.32 ± .01
Stainless steel	.36 ± .02
Plywood perpendicular	.33 ± .01
Plywood parallel	.33 ± .01

Note: Results are expressed as mean ± SD of triplicate observations.

3. Results and discussion

3.1. Physical properties

The results of the determined physical parameters of the flaxseed are shown in Table 1. Moisture content of flaxseed was observed as 7.66% on wet basis and 8.29% on dry basis (Table 1). Flaxseed with low moisture content 7.66% is suitable for long-term storage as higher moisture content could cause decomposition of fatty acids by microbial action.

Color parameters including L^*, a^*, and b^* were observed to be 46.63, 8.21, and 8.96, respectively, as shown in Table 1. Based upon color, flaxseed is commercially divided into two parts viz. golden yellow flaxseed and brown color flaxseed. Golden flaxseed is having less omega-3 content as compared to brown color flaxseed. However, brown color flaxseed is preferable for bakery products like cookies, bun, bread etc., and is also utilized in oil processing industry due to high omega-3 content.

3.2. Geometrical properties of flaxseed

Basic geometric characteristics of the flaxseed in terms of length, width, thickness, geometrical mean diameter, arithmetic mean diameter, sphericity, and surface area are presented in Table 1,

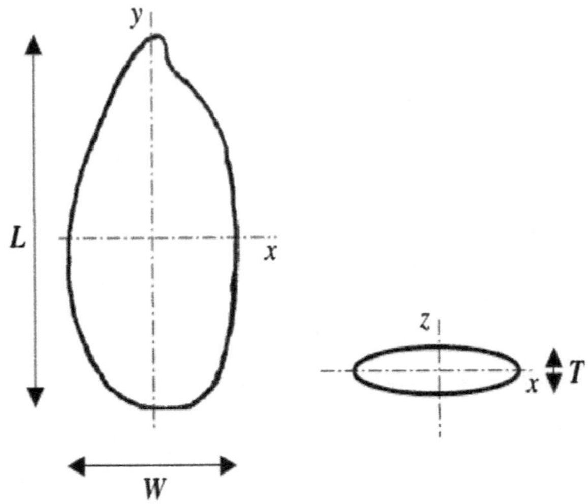

Figure 1. Dimensional representation of flaxseed.

and graphical representation is shown in Figure 1. These parameters depend upon the size of the seed, with bigger size seeds having higher values (Owolarafe et al., 2007). Selvi, Pınar, and Yeşiloğlu (2006) reported the length of linseed in the range of 4.58–4.7 mm, and no change in the width and thickness for the moisture range from 8.25 to 11.75% dry basis (d.b.). The sphericity of the flaxseed was 40.34%, which indicates that the shape of the seeds is flat and thus makes it difficult to roll on surface as flat seeds slide easier than spherical seeds, which roll on structural surfaces. Thus, the flat shape of the seeds enables them to slide, and this property is important in the development of hopper and dehuller designs for flaxseed. This is very important in design of hoppers for machines (Omobuwajo, Sanni, & Balami, 2000). The geometric mean diameter of flaxseed was found to be lower than pumpkin, sunflower, and guna seeds, and higher than that of sesame seeds (Gupta & Das, 1997; Joshi, Das, & Mukherjee, 1993).

3.3. Gravimetric properties of flaxseed

Bulk density of flaxseed is presented in Table 1, and the value of bulk density was found to be .66 g/cm^3. Coşkuner and Karababa (2007) has reported the bulk density values of 726.6–555.6 kg/m^3 for the moisture range of 6.09–16.81% (d.b.), whereas Selvi et al. (2006) observed the bulk density of 690.5–545.0 kg/m^3 for the moisture range of 6.09–16.81% (d.b.) for the commercial variety of linseed. The decrease in bulk density of flaxseed may be due to the increase in seed size with moisture content which gives rise to decrease in quantity of seeds occupying the same bulk volume. The true density of flaxseed was observed to be 1.34 g/cm^3 and was found to be higher than sunflower kernel (Gupta & Das, 1997) and guna seed (Aviara, Gwandzang, & Haque, 1999). Porosity depends on size, shape, and boldness of seeds, and includes the air spaces in between the seeds during storage. The flaxseed was having porosity value of 51.56% (Table 1). According to Hettiarachchy, Hareland, and Ostenson (1990), presence of more voids in between the seeds leads to accumulation of moisture thereby decreasing the storage stability of flaxseeds. Porosity of the mass of seeds determines the resistance to airflow during aeration and drying procedures. The aspect ratio (R_a) was found to be 62.58%. The 1000 kernel weight was found to be higher as 7.14 gm than reported by Selvi et al. (2006), which was 6.0 g at 8.25% (d.b.) moisture content for linseed.

3.4. Friction properties of flaxseed

The angle of repose indicates the cohesion among the individual units of a material. The higher the cohesion, higher is the angle of repose. Coefficient of static friction of seeds was higher on stainless steel and lowest on glass surface, i.e. .36 ± .02 and .32 ± .01, respectively. Values of the coefficient of static friction of the flaxseed grains on four surfaces are given in Table 1. The results showed higher coefficients for the rough surface such as stainless steel and plywood than that of the smooth surface such as glass. Change in moisture content has a pronounced effect on the friction coefficient of flaxseed. The angle of repose of flaxseed was 19.40°. The angle of repose of flaxseed is influenced by the moisture content present in the seeds, and seeds with higher moisture content have higher values for angle of repose. The angle of repose of flaxseed was found higher than fenugreek seeds (Altuntaş, Özgöz, & Taşer, 2005) and faba bean grains (Altuntaş & Yıldız, 2007).

3.5. Physical properties of wheat flour and flaxseed samples

Physical properties of wheat flour and flaxseed samples are presented in Table 2. The bulk density of flaxseed is important in relation to its packaging as it determines the capacity of storage, packaging, and transport system (James, 2005). The maximum bulk density value of .49 g/cm^3 was found in WF followed by .48 g/cm^3 in RFF and .47 g/cm^3 in URFF, respectively. It is revealed from the results that roasting significantly increased the bulk density. The lower values of bulk density could be due to low volumes of flaxseeds. The variation in density of flaxseed may be due to the effect of random harvesting of seeds at different maturity stages. A decreased bulk density would be an advantage in infant food formulation. Highest tap density of .81 g/cm^3 was found in URFF followed by .77 g/cm^3 in RFF and .60 g/cm^3 in wheat flour, respectively. It was concluded from the data that true density was significantly affected by treatment and maximum true density was observed in URFF i.e. 1.77 g/cm^3 followed by 1.58 g/cm^3 in RFF while minimum value was found in WF (1.40 g/cm^3). Porosity is a measure of the voids between the solid particles in a material. Pore space can be filled with fluids including gas and/or water. Air filled porosity allows gases to move within the material. URFF was having the maximum porosity of 73.01% while minimum value was found in WF (64.44%). Angle of repose is directly affected by moisture content or fat content of the sample. Higher the moisture content, higher is the angle of repose and was found to be maximum in URFF i.e. 39.71° followed by 36.56° in RFF and 33.60° in wheat flour, respectively. A significant difference in WSI and WAI was observed between wheat flour and different flaxseed samples with a maximum value of WAI i.e. 5.96 g/g in URFF while minimum value was found in WF (4.89 g/g). WAI was found to be higher in URFF samples as compared to RFF samples. Water solubility index is measure of soluble starch content in flour. It was observed that RFF was having highest WSI i.e. 10.83% while minimum value of 4.24% was found in WF.

Table 2. Physical and functional properties of wheat and flaxseed flour

Parameters	WF	URFF	RFF
Bulk density (g/cm^3)	.49 ± .01^a	.47 ± .01^a	.48 ± .01^a
Tap density (g/cm^3)	.60 ± .03^b	.81 ± .01^a	.77 ± .06^a
True density (g/cm^3)	1.40 ± .13^b	1.77 ± .19^a	1.58 ± .13^{ab}
Porosity (%)	64.44 ± 3.72^b	73.01 ± 2.80^a	69.09 ± 2.99^{ab}
Angle of repose (°)	33.60 ± 1.25^c	39.71 ± .52^a	36.56 ± .26^b
WAI (g/g)	4.89 ± .18^b	5.96 ± .05^a	5.33 ± .57^{ab}
WSI (%)	4.24 ± .26^c	6.61 ± .20^b	10.83 ± .74^a
Functional properties			
WAC (g/g)	.63 ± .04^b	1.47 ± .17^a	1.61 ± .27^a
OAC (g/g)	1.14 ± .03^a	1.08 ± .19^a	.96 ± .20^a
Foaming capacity (%)	30.40 ± .76^a	9.23 ± .23^b	7.82 ± .02^c
Foaming stability (%)	56.05 ± 5.56^a	54.43 ± 1.07^{ab}	48.60 ± .51^b
Sedimentation value (ml)	23.56 ± 1.25^a	24.93 ± .58^a	23.60 ± .50^a

Notes: Mean ± SD with different superscripts in a row differ significantly ($p < .05$) ($n = 3$). WF = wheat flour. URFF = unroasted flaxseed flour; RFF = roasted flaxseed flour. WAI: water absorption index. WSI: water solubility index; WAC: water absorption capacity; OAC: oil absorption capacity.

Table 3. Physicochemical properties and color of wheat and flaxseed flour

Parameters	WF	URFF	RFF
Moisture (%)	13.55 ± .25^a	4.71 ± .23^b	4.13 ± .58^b
Protein (%)	11.46 ± 1.25^b	23.78 ± 2.19^a	22.83 ± 2.40^a
Crude Fat (%)	1.30 ± .43^b	32.27 ± .39^a	31.05 ± 1.05^a
Ash (%)	.78 ± .05^b	3.16 ± .05^a	3.29 ± .085^a
Crude fiber (%)	.57 ± .19^b	9.34 ± .22^a	9.63 ± .29^a
Carbohydrate (%)	71.47^a	26.72^b	29.06^b
Energy value (Cal/100 g)	346.84^b	492.51^a	487.04^a
Color			
L^*	92.65 ± .28^a	45.40 ± .31^b	34.43 ± .11^c
a^*	2.13 ± .07^c	7.61 ± .15^a	6.95 ± .05^b
b^*	9.89 ± .15^b	10.72 ± .07^a	9.27 ± .06^c
ΔE	-	47.57 ± .30^b	58.42 ± .10^a
Whiteness index	87.48^a	43.84^b	33.41^c

Notes: Mean ± SD with different superscripts in a row differ significantly ($p < .05$) ($n = 3$). WF = wheat flour. URFF = unroasted flaxseed flour; RFF = roasted flaxseed flour.

3.6. Physicochemical properties of wheat flour and flaxseed samples

3.6.1. Color characteristics

Color is an important quality factor that typically relates to the acceptability, marketability, and wholesomeness of foods (Berrois, Wood, Whitehand, & Pan, 2004). Color characteristics of wheat flour and flaxseed samples are presented in Table 3. The L^* value of 92.65 was found for wheat flour, 45.40 for URFF, and 34.43 for RFF, respectively. Wheat flour was having highest L^* value, lowest a^* value, while lowest L^* value was found in RFF. ΔE value for URFF and RFF was found to be 47.57 and 58.42, respectively. Whiteness index value was found to be highest in wheat flour i.e. 87.48 followed by 43.84 in URFF and 33.41 in RFF, respectively.

3.6.2. Chemical composition

The wheat flour, roasted and unroasted flaxseed flour samples were analyzed for physicochemical properties. The proximate composition of the samples is presented in Table 3. As observed from the data, the moisture content was minimum in RFF and maximum in URFF samples, but both the samples were having less moisture content as compared to wheat flour. The protein content was found to be highest in flaxseed than in wheat flour. The protein content of URFF was observed to be 23.78% highest than 22.83% in RFF. The results for decreased protein content may be due to the fact that roasting may have destroyed some of the protein content. These results are in accordance with those described by Flax council of Canada (2004) and Broihier (1999). The crude fat content of URFF was higher than RFF samples as shown in table 3. The values for crude fat content were 32.27% in URFF and 31.05% in RFF and the decreased fat content is due to the destruction of fat during the treatment process. The ash content of wheat flour was found to be minimum i.e. .78% as compared to other flaxseed samples. The highest ash content was observed in RFF (3.29%) followed by URFF (3.16%) and no significant difference was seen between the ash content of URFF and RFF. The crude fiber content of wheat flour (.57%) was significantly different from both the flaxseed samples and no significant difference was observed in the crude fiber content of URFF and RFF. The carbohydrate content of wheat flour was observed to be maximum (71.47%) and significantly different from flaxseed flour samples. Among the flaxseed samples, maximum carbohydrate content was found in RFF i.e. 29.06% followed by URFF 26.72%, respectively. The decrease in the carbohydrate content of URFF is due to increased fat, protein, and ash content of the sample. The energy value was calculated on the basis of protein, fat, and carbohydrate content of the sample and the energy value of URFF was found to be maximum, i.e. 492.51 Cal/100g, followed by the energy value of RFF and wheat flour, respectively.

3.7. Functional properties of flour

The functional properties of wheat, roasted and unroasted flaxseed flour samples are presented in Table 2. It was revealed from the results that RFF was having the highest water absorption capacity (WAC) of 1.61 g/g, followed by URFF and WF i.e. 1.47 g/g and .63 g/g, respectively. Flour with high water-binding capacity is associated with more hydrophilic constituents like polysaccharides, non-starch components mainly mucilage (Aboubakar, Njintang, Scher, & Mbofung, 2008). The WAC of wheat flour was found significantly different from flaxseed flour samples. The higher WAC of Flaxseed samples may be due to the presence of hull mucilage polysaccharides which adsorb water and swell (Dev & Quensel, 1986) or due to the existing differences in the conformational characteristics of its proteins. Moreover, gelation of carbohydrates and swelling of crude fibers may also influence the water adsorption of oilseed samples (Narayana & Narasinga Rao, 1982). The OAC is a prominent factor in food formulations as it improves flavor and increases the mouth feel of foods. The OAC was observed to be highest in WF i.e. 1.14 g/g followed by URFF (1.08 g/g) while RFF was having lowest value of .96 g/g. Oil binding capacity of food component is important for various applications because it relies mainly on this capacity to physically entrap oil by a complex capillary attraction process and this property of flour leads to better flavor retention, a consistency trait and an increase in mouth-feel (Khattab & Arntfield, 2009). The foaming capacity of wheat flour (30.40%) was found to be highest and significantly different from flaxseed flour samples. Foaming properties depend on the proteins and carbohydrates present in the flour (Sreerama, Sashikala, Pratape, & Singh, 2012). Among the flaxseed samples, maximum foaming capacity was observed in URFF i.e. 9.23% followed by RFF 7.82%, respectively. Foaming stability of WF was found to be maximum (56.05%), and minimum value was observed in RFF i.e. 48.60%. The higher foam capacity and stability of wheat flour suggests the presence of higher amount of solubilized native protein in the sample. Owing to a large increase in the surface area in the liquid/air interphase, proteins denature and aggregate during whipping and are important for flour to be used in leavened food products such as baked goods, cakes, and biscuits (Sreerama, Sasikala, & Pratape, 2008). Foaming properties are important in the maintenance of the texture and structure of different food products like bakery and ice cream products. The foamability of the flour depends on the presence of the flexible protein molecules which may decrease the surface tension of water (Sathe & Salunkhe, 1982). It was observed from the results that the foam capacity and stability of RFF was less as compared to URFF and may be due to loss of surface proteins during roasting process. Sedimentation value is the indirect measurement of quality and composition of gluten proteins. The sedimentation value was observed to be highest (24.93 ml) in URFF.

Table 4. Least gelation concentration of wheat flour and different flaxseed samples

Flour conc. (%)	WF	URFF	RFF
2	No gel	No gel	No gel
4	No gel	No gel	No gel
6	No gel	No gel	Gel
8	No gel	Gel	Gel
10	No gel	Firm gel	Firm gel
12	Gel	Firm gel	Firm gel
14	Gel	Very firm gel	Firm gel
16	Firm gel	Very firm gel	Very firm gel
18	Firm gel	Very firm gel	Very firm gel
20	Very firm gel	Very firm gel	Very firm gel

Notes: WF = wheat flour; URFF = unroasted flaxseed flour; RFF = roasted flaxseed flour.

3.7.1. Least gelation concentration

LGC of wheat flour and flaxseed samples is presented in Table 4. The ability of proteins to form gels is measured by LGC. No gel formation was observed in WF up to the concentration of 10% while gel was formed in URFF at 8% and in RFF at 6%.

4. Conclusion

The physical properties including seed shape, size, geometric mean diameter, sphericity, bulk density, porosity, surface area, static coefficient of friction and angle of repose, physicochemical and functional properties were investigated. The study reveals that engineering properties including physical and mechanical properties of flaxseed may be useful in designing equipment for postharvest handling and processing operations like deshelling of seed, oil extraction, and pulverization of seed kernel into flour etc. These properties are necessary for the design of equipment for harvesting, separating, processing, packing, and transportation.

This study also provides insight on the physicochemical and functional properties of roasted and un-roasted flaxseed flour in comparison to wheat flour. Statistical analysis showed significant difference between the properties of flaxseed and wheat flour studied. The findings showed that flaxseed is rich in protein and fat, and possess good physicochemical properties that could be exploited for nutrition and food formulation. Flaxseed also possesses good functional properties that may be useful in food systems where they can play many functional roles. The water absorption and OAC of flaxseed flour make it useful for various products that require water and oil retention for their textural integrity like oil retention capability helps retain flavor and provides good mouth feel. Hence, flaxseed can be incorporated in commercial flours which are low in protein as composite flour that can be utilized in bakery products like cookies, muffins, biscuits, and buns. Consumption of foods formulated or fortified with flaxseed flours would be an important step toward relieving protein malnutrition in the poor countries of the world.

Nomenclature

X	Length (mm)
Y	Width (mm)
Z	Thickness (mm)
D_g	Geometrical mean diameter (mm)
R_a	Aspect ratio (%)
D_r	Density ratio (%)
ϕ	Sphericity (%)
D_a	Arithmetic mean diameter (mm)
ρ_t	True density (g/cm^3)
ρ_b	Bulk density(g/cm^3)
ε	Porosity (%)
Φ	Angle of repose (o)
h	Height of the pile (cm)
D	Diameter of the pile (cm)
μ	Coefficient of static friction
S	Surface area (mm^2)

Funding
The authors received no direct funding for this research.

Author details
Azhar Khan[1]
E-mail: azharkhanmuslim@gmail.com
C.S. Saini[1]
E-mail: charanjiv_cjs@yahoo.co.in
[1] Department of Food Engineering and Technology, Sant Longowal Institute of Engineering & Technology, Longowal 148106, Punjab, India.

References

Aboubakar, Njintang.Y. N.., Scher, J., & Mbofung, C. M. F., (2008). Physicochemical, thermal properties and microstructure of six varieties of taro (*Colocasia esculenta* L. Schott) flours and starches. *Journal of Food Engineering, 86*, 294–305. http://dx.doi.org/10.1016/j.jfoodeng.2007.10.006

Altuntaş, E., & Yıldız, M. (2007). Effect of moisture content on some physical and mechanical properties of faba bean (*Vicia faba* L.) grains. *Journal of Food Engineering, 78*, 174–183. http://dx.doi.org/10.1016/j.jfoodeng.2005.09.013

Altuntaş, E., Özgöz, E., & Taşer, Ö. F. (2005). Some physical properties of fenugreek (*Trigonella foenum-graceum* L.) seeds. *Journal of Food Engineering, 71*, 37–43. http://dx.doi.org/10.1016/j.jfoodeng.2004.10.015

Association of Official Analytical Chemists. (2000). Official methods of analysis (14th ed.) Arlington, VA. Retrieved from http://www.fspublishers.org/published_papers/41283_pdf

Aviara, N. A., Gwandzang, M. I., & Haque, M. A. (1999). Physical properties of guna seeds. *Journal of Agricultural Engineering Research, 73*, 105–111. http://dx.doi.org/10.1006/jaer.1998.0374

Axford, D. W. E., Mcdermott, E. E., & Redman, D. G. (1978). Small-scale tests of bread making quality. *Milling Feed and Fertilizer, 161*, 18–20.

Berrois, J. D. J., Wood, D. F., Whitehand, L., & Pan, J. (2004). Sodium bicarbonate and the microstructure, expansion and color of extruded black beans. *Journal of Food Processing and Preservation, 28*, 321–335.

Broihier, K. (1999). Just the flax. Retrieved August 2, 2004, from http://www.egginnovations.compdf/smartch.pdf updated on 2004 January 6, 2004

Chun, J., Lee, J., & Eitenmiller, R. R. (2005). Vitamin E and oxidative stability during storage of raw and dry roasted peanuts packaged under air and vacuum. *Journal of Food Science, 70*, 292–297.

Coşkuner, Y., & Karababa, E. (2007). Some physical properties of flaxseed (*Linum usitatissimum* L.)*Journal of Food Engineering, 78*, 1067–1073. http://dx.doi.org/10.1016/j.jfoodeng.2005.12.017

Deshpande, H. W., & Poshadri, A. (2011). Physical and sensory characteristics of extruded snacks prepared from Foxtail millet based composite flours. *International Journal of Food Research, 18*, 751–756.

Dev, D. K., & Quensel, E. (1986). Functional properties and microstructural characteristics of linseed flour and protein isolate. *LWT-Food Science and Technology, 19*, 311–317.

Dutta, S. K., Nema, V. K., & Bhardwaj, R. K. (1988). Physical properties of gram. *Journal of Agricultural Engineering Research, 39*, 259–268. http://dx.doi.org/10.1016/0021-8634(88)90147-3

Flax council of Canada. (2004). Flax nutrition. Retrieved August 11, 2004, from http://www.flaxcouncil.ca/pdf/smartch updated on February 2, 2004

Goyal, R. K., Kingsly, A. R. P., Kumar, P., & Walia, H. (2007). Physical and mechanical properties of aonla fruits. *Journal of Food Engineering, 82*, 595–599. http://dx.doi.org/10.1016/j.jfoodeng.2007.03.019

Gupta, R. K., & Das, S. K. (1997). Physical properties of sunflower seeds. *Journal of Agricultural Engineering Research, 66*, 1–8. http://dx.doi.org/10.1006/jaer.1996.0111

Hettiarachchy, N. S., Hareland, G. A., & Ostenson, A. (1990). Chemical composition of 11 flaxseed varieties grown in North Dakota. *Proc. Flax Institute, 53*, 36–39.

Hsu, C. L., Chen, W., Weng, Y. M., & Tseng, C. Y. (2003). Chemical composition, physical properties, and antioxidant activities of yam flours as affected by different drying methods. *Food Chemistry, 83*, 85–92. http://dx.doi.org/10.1016/S0308-8146(03)00053-0

James, G. B. (Ed.). (2005). *Food processing handbook* (p. 27). Weinheim: Wiley.

Joshi, D. C., Das, S. K., & Mukherjee, R. K. (1993). Physical properties of pumpkin seeds. *Journal of Agricultural Engineering Research, 54*, 219–229. http://dx.doi.org/10.1006/jaer.1993.1016

Karababa, E. (2006). Physical properties of popcorn kernels. *Journal of Food Engineering, 72*, 100–107. http://dx.doi.org/10.1016/j.jfoodeng.2004.11.028

Kashaninejad, M., Mortazavi, A., Safekordi, A., & Tabil, L. G. (2006). Some physical properties of Pistachio (*Pistacia vera* L.) nut and its kernel. *Journal of Food Engineering, 72*, 30–38. http://dx.doi.org/10.1016/j.jfoodeng.2004.11.016

Khattab, R. Y., & Arntfield, S. D. (2009). Functional properties of raw and processed canola meal. *LWT - Food Science and Technology, 42*, 1119–1124.

Krysiak, W., & Motyl-Patelska, L. (2006). Effects of air parameters on changes in temperature inside roasted cocoa beans. *Acta Agrophysica, 7*, 113–127.

Lin, M. J. Y., Humbert, E. S., & Sosulski, F. W. (1974). Certain functional properties of sunflower meal products. *Journal of Food Science, 39*, 368–370. http://dx.doi.org/10.1111/jfds.1974.39.issue-2

Maduako, J. N., & Faborode, M. O. (1990). Some physical properties of cocoa pods in relation to primary processing. *Journal of Technology, 2*, 1–7.

Mohsenin, N. N. (1980). *Physical properties of plant and animal materials* (p. 407, 2nd ed.). New York, NY: Gordon and Breach Science.

Mwithiga, G., & Sifuna, M. M. (2006). Effect of moisture content on the physical properties of three varieties of sorghum seeds. *Journal of Food Engineering, 75*, 480–486. http://dx.doi.org/10.1016/j.jfoodeng.2005.04.053

Narayana, K., & Narasinga Rao, M. S. (1982). Functional properties of raw and heat processed winged bean (*Psophocarpus tetragonolobus*) flour. *Journal of Food Science, 47*, 1534–1538. http://dx.doi.org/10.1111/jfds.1982.47.issue-5

Okaka, J. C., & Potter, N. N. (1977). Functional and storage properties of cowpea powder-wheat flour blends in breadmaking. *Journal of Food Science, 42*, 828–833. http://dx.doi.org/10.1111/jfds.1977.42.issue-3

Omobuwajo, T. O., Sanni, L. A., & Balami, Y. A. (2000). Physical properties of sorrel (*Hibiscus sabdariffa*) seeds. *Journal of Food Engineering, 45*, 37–41. http://dx.doi.org/10.1016/S0260-8774(00)00039-X

Oplinger, E., Oelke, E., Doll, J., Brundy, L., & Schuler, R. (1989). *Flax: alternative field manual* (pp. 114–126). St. Paul, MN: University of Winsconsin, Corporative Extension.

Owolarafe, O. K., Olabige, M. T., & Faborode, M. O. (2007). Physical and mechanical properties of two varieties of fresh oil palm fruit. *Journal of Food Engineering, 78*, 1228–1232. http://dx.doi.org/10.1016/j.jfoodeng.2005.12.049

Özdemir, M., & Devres, O. (2000a). Kinetics of color changes of hazelnuts during roasting. *Journal of Food Engineering, 44*, 31–38. http://dx.doi.org/10.1016/S0260-8774(99)00162-4

Pittia, P., Dalla Rosa, M., & Lerici, C. R. (2001). Textural changes of coffee beans as affected by roasting conditions. *LWT – Food Science and Technology, 34*, 168–175. http://dx.doi.org/10.1006/fstl.2000.0749

Raghuramulu, N., Nair, K. M., & Kalyanasundaram, S. (1983). Food analysis. *A Manual of Laboratory Techniques* (pp. 31–136). Hyderabad: National Institute of Nutrition, ICMR.

Rizzi, G. P. (2003). Free radicals in the Maillard reaction. *Food Reviews International, 19*, 375–395. http://dx.doi.org/10.1081/FRI-120025481

Sathe, A. K., Deshphande, S. S., & Salunkhe, D. K. (1981). Functional properties of lupin seed protein and protein concentrates. *Journal of Food Science, 42*, 491–492.

Sathe, S. K., & Salunkhe, D. K. (1982). Functional properties of the great northern bean (*Phaseolus vulgaris* L.) proteins, emulsion, foaming, viscosity and gelation properties. *Journal of Food Science, 46*, 71–81.

Selvi, K. C., Pınar, Y., & Yeşiloğlu, E. (2006). Some physical properties of linseed. *Biosystems Engineering, 95*, 607–612. http://dx.doi.org/10.1016/j.biosystemseng.2006.08.008

Shrestha, A. K., & Noomhorm, A. (2002). Comparison of physico-chemical properties of biscuits supplemented with soy and kinema flour. *International Journal of Food Science and Nutrition, 3*, 361–368.

Singh, N., Sandhu, K. S., & Kaur, M. (2005). Physicochemical properties including granular, morphology, amylose content, swelling and solubility, thermal and pasting properties of starches from normal, waxy, high amylose and sugary corn. *Progress in Food Biopolymer Research Journal, 1*, 44–54.

Sirisomboon, P., Pornchaloempong, P., & Romphophak, T. (2007). Physical properties of green soybean: Criteria for sorting. *Journal of Food Engineering, 79*, 18–22. http://dx.doi.org/10.1016/j.jfoodeng.2006.01.022

Sirisomboon, P., Kitchaiya, P., Pholpho, T., & Mahuttanyavanitch, W. (2007). Physical and mechanical properties of Jatropha curcas L. fruits, nuts and kernels. *Biosystems Engineering, 97*, 201–207. http://dx.doi.org/10.1016/j.biosystemseng.2007.02.011

Sosulski, F. W. (1962). The centrifuge method for determining flour absorption in hard red spring wheat. *Cereal Chemistry, 39*, 344–350.

Sreerama, Y. N., Sasikala, V. B., & Pratape, V. M. (2008). Nutritional implications and flour functionality of popped/expanded horse gram. *Food Chemistry, 108*, 891–899. http://dx.doi.org/10.1016/j.foodchem.2007.11.055

Sreerama, Y. N., Sashikala, V. B., Pratape, V. M., & Singh, V. (2012). Nutrients and antinutrients in cowpea and horse gram flours in comparison to chickpea flour: Evaluation of their flour functionality. *Food Chemistry, 131*, 462–468. http://dx.doi.org/10.1016/j.foodchem.2011.09.008

Ziegleder, G. (1991). Composition of flavor extracts of raw and roasted cocoas. *Journal of Food Control and Research A, 192*, 521–525.

4

An integer linear programming for a comprehensive reverse supply chain

Hoda Mahmoudi[1] and Hamed Fazlollahtabar[1*]

*Corresponding author: Hamed Fazlollahtabar, Faculty of Industrial Engineering, Iran University of Science and Technology, Tehran, Iran

E-mail: hfazl@iust.ac.ir

Reviewing editor: Zude Zhou, Wuhan University of Technology, China

Abstract: Reverse supply chain is a cycle of recovery for the products and materials used by the customers but can be returned to the chain performing some operations. Due to significance of reverse supply chain in the content of environmental and economical aspects, we formulate a mathematical model of reverse multi-layer multi-product supply chain for minimizing the total costs including returning, disassembly, processing, recycling, remanufacturing, and distribution centers. The presented model is an integer linear programming model being solved using Lingo 9 software. Numerical experiments are conducted to gain insight into the proposed model. The solutions provide a decision aid stream strengthening the concept of reverse supply network design and analysis for profit-making organization.

Keywords: reverse supply chain, mathematical modeling, cost analysis

1. Introduction

Companies focus on setting up a reverse supply chain either because of environmental regulations or to reduce their operating costs by reusing products or components. IBM, for instance, has profited from its programs to receive end-of-use products, promote second hand items internet auctions, and dismantle equipment as a source of spare parts (Fleischmann, Van Nunen, & Gräve, 2003).

ABOUT THE AUTHOR

Hamed Fazlollahtabar has MSc in Industrial Engineering at Mazandaran University of Science and Technology, Babol, Iran. He received Doctorate award from the Gulf University of Science and Technology in Quantitative Approaches in Electronic Systems. Currently, he is studying PhD in Industrial Engineering at Iran University of Science and Technology, Tehran, Iran. He is in the editorial board of WASET (World Academy of Science Engineering Technology) Scientific and Technical Committee on Natural and Applied Sciences, reviewer committee of International Conference on Industrial and Computer Engineering (CIE), and member of the International Institute of Informatics and Systemics (IIIS). He is a member of Iran Elite Council. His research interests are mathematical modeling and optimization in knowledge-based systems and manufacturing systems. He has published over 130 research papers in international book chapters, journals, and conferences.

PUBLIC INTEREST STATEMENT

Due to significance of reverse supply chain in the content of environmental and economical aspects, we formulate a mathematical model of reverse multi-layer multi-product supply chain for minimizing the total costs including returning, disassembly, processing, recycling, remanufacturing, and distribution centers.

A reverse logistics system comprises a series of activities, which form a continuous process to treat return-products until they are properly recovered or disposed. These activities include collection, cleaning, disassembly, test and sorting, storage, transport, and recovery operations. The latter can also be represented as one or a combination of several main recovery options, like reuse, repair, refurbishing, remanufacturing, cannibalization, and recycling (Dekker & Van Der Laan, 1999; Guide, Harrison, & Van Wassenhove, 2003; Thierry, Salomon, Van Nunen, & Van Wassenhove, 1993). In reuse, the returned product can be used more than once in the same form after cleaning or reprocessing. Remanufacturing is an industrial process in which worn-out products are restored to like-new condition. Recycling denotes material recovery without conserving any product structure.

Reverse supply chain refers to the movement of goods from customer to vendor. This is the reverse of the traditional supply chain movement of goods from vendor to customer. Reverse logistics is the process of planning, implementing and controlling the efficient and effective inbound flow, and storage of secondary goods and related information for the purpose of recovering value or proper disposal. Typical examples of reverse supply chain include:

- Product returns and management of their deposition.
- Remanufacturing and refurbishing activities.
- Management and sale of surplus, as well as returned equipment and machines from the hardware leasing business.

In these cases, the resource goes at least one step back in the supply chain. For instance, products move from customer to distributor or manufacturer. Other instances of products reversing direction in the supply chain are manufacturing returns, commercial returns (B2B and B2C), product recalls, warranty returns, service returns, end-of-use returns, and end-of-life returns. There are various types of reverse supply chains, and they arise at different stages of the product cycle; however, most return supply chains are organized to carry out five key processes:

- Product acquisition: Obtaining the used product from the user by the reseller or manufacturer.
- Reverse logistics: Transporting products to a facility for inspecting, sorting, and disposition.
- Inspection and disposition: Assessing the condition of the return and making the most profitable decision for reuse.
- Remanufacturing or refurbishing: Returning the product to its original specifications.
- Marketing: Creating secondary markets for the recovered products.

Kim, Hong, and Goyal (2009) developed a fair and equitable mechanism of sharing the profits achieved due to cooperation in a supply chain between a single manufacturer and a single retailer. Damghani and Taghavifard (2012) considered a generic process in which just-in-time practices and changed them into agility indices, and agility indices were converted into performance measurement in supply chain. Fazlollahtabar, Mahdavi, and Mohajeri (2013) proposed a fuzzy mathematical programming model for a supply chain which considered multiple depots, multiple vehicles, multiple products, multiple customers, and different time periods.

Concerning reverse logistics, a lot of researches have been made on various fields and subjects such as reuse, recycling, remanufacturing logistics, etc. In this paper, we propose a framework and a mathematical model for costs in a multi-layer multi-product in reverse supply chain system.

This paper is organized as follows. In next section, we introduce some key literatures relevant to this study. In Section 3, a general framework and problem definition for reverse supply chain are proposed. Section 4 proposes the mathematical model of the reverse supply chain. In Section 5, numerical experiments are presented. Finally, conclusions and further researches are addressed in the last section.

2. Related literature

Supply chain redesign involves complex competencies such as integration with suppliers and customers to create value as well as the acquisition of external information and its transformation into practices capable of producing marketable outputs (Dobrzykowski, Tran, & Hong, 2011). Supply chain management (SCM) is regarded as a means to maximize the overall value generated. SCM includes all functions involved in receiving and filling a customer request by forging close relationships with customers and suppliers. Strong relationships and mutual trust between buyer and suppliers are essential for the successful implementation of SCM (Kumar, Singh, & Singh, 2011). In a supply chain, selection of right suppliers is a strategic consideration. Supplier selection is the process of finding suitable suppliers from the set of alternative suppliers who can be strategic partners to the organization and provide the right quality of product or services at the right price at the right time. Such a problem is highly complex one because it involves a number of qualitative and quantitative factors having interdependencies among themselves (Sreekumar & Mahapatra, 2011).

Increasing inclination of firms towards supply chain collaboration as a means of developing or enhancing core competency has gained noticeable research attention. Successfully implementing collaboration is a major concern for both industries and academia. Building from the literature base, the study investigates effect of collaborative culture on all collaborative activities, and effect of preparedness that moderates the relationship of collaborative activities to supply chain performance (Kumar & Banerjee, 2012). The rapid advancement in communication technology, coupled with spiraling customer expectations, are increasingly forcing business enterprises to seek inter-organizational coordination for smooth material and informational flow across the total supply chain justifiably. Therefore, given the inherent uncertainties pervading the operational environment within real-world supply chains, it becomes imperative for each partnering echelon to focus on individual information requirements from a viewpoint of global optimization of overall supply chain performance (Dev & Shankar, 2012). Currently, a number of factors encountered in global competition force companies to pursue the more rapid development and introduction of new and innovative products and product generations. Comparatively, little academic research has been conducted on product change projects in the context of SCM (Slamanig & Winkler, 2012).

For the last decade, increasing concerns over environmental degradation and increased opportunities for cost savings or revenues from returned products prompted some researchers to formulate more effective reverse logistics strategies. Kroon and Vrijens (1995) reported a case study concerning the design of a logistics system for reusable transportation packages. The authors proposed a mixed integer programming (MIP), closely related to a classical un-capacitated warehouse location model. Jayaraman, Patterson, and Rolland (2003) proposed a general MIP model and solution procedure for a reverse distribution problem focused on the strategic level. The model decided whether each remanufacturing facility is open considering the product return flow.

Jayaraman, Guide, and Srivastava (1999) presented a MIP to determine the optimal number and locations of remanufacturing facilities for the electronic equipment. They developed heuristic concentration procedures combined with heuristic expansion components to handle relatively large problems. Kim, Song, and Jeong (2006) discussed a notion of remanufacturing system in reverse logistics environment. They proposed a general framework in view of supply planning and developed a mathematical model to optimize the supply planning function. The model determined the quantity of products parts processed in the remanufacturing facilities subcontractors and the amount of parts purchased from the external suppliers while maximizing the total remanufacturing cost saving.

Barros, Dekker, and Scholten (1998) proposed a MIP model considered two-echelon location problems with capacity constraints based on a multi-level capacitated warehouse location problem. Pati, Vrat, and Kumar (2008) developed an approach based on a mixed integer goal programming model to solve the problem recycling in supply chain. The model studied the inter-relationship between multiple objectives of a recycled paper distribution network. The objectives considered were reduction in reverse logistics cost.

Listes and Dekker (2005) suggested a scenario-based stochastic mixed integer programming model to maximize the total profit in a sand recycling network. Lee and Dong (2008) developed an mixed integer linear programming (MILP) model for integrated logistics network design for end-of-lease computer products. They considered a simple network with a single production center and a given number of hybrid distribution-collection facilities to be opened which have been solved using tabu search.

Du and Evans (2008) minimized tardiness and total costs for location and capacity decisions in a closed-loop logistics network operated by third party logistics (3PL) providers. To solve the bi-objective MILP model, a hybrid scatter search method was developed. Salema, Barbosa-Povoa, and Novais (2007) studied the design of a reverse distribution network and found that most of the proposed models on the subject are case based and, for that reason, they lack generality. The model contemplates the design of a generic reverse logistics network where capacity limits, multi-product management, and uncertainty on product demands and returns are considered. A mixed integer formulation was developed. This formulation allowed for any number of products, establishing a network for each product while guaranteeing total capacities for each facility at a minimum cost. But the inventory was not taken into consideration.

Lee, Rhee, and Gen (2007) proposed the reverse logistics network problem minimizing total reverse logistics various shipping costs. However, several researchers studied and analyzed cost problem in reverse logistics, our study focuses on a general framework and state total cost in reverse supply chain.

This paper propose a multi-layer, multi-product reverse supply chain problem which consist of returning center, disassembly center, processing center, manufacturing center, recycling center, material center, and distribution center costs while minimizing of total costs in reverse supply chain for returned products.

3. Problem definition

The reverse supply chain under study is a multi-layer and multi-product one, where the chain includes several products in a same category (e.g. electronic devices in media category, TV, and Radio). Also, several layers are considered in the proposed supply chain namely, supplier, manufacturer, distributor, and consumer. In general, the current reverse logistic systems aim at determining the economic value of reusable items. Recent models focused on the vehicle routing problem and second hand product sale and services. In the proposed model, the returned products after collecting and inspecting divides into two groups of disassemblable and not disassemblable products. The products which can be taken parted to the parts will be sent to the disassembling centers and there, they will be converted to the parts. There they are divided into reusable and not reusable parts. The not reusable parts will rebut safely and the reusable parts will be sent to the processing center. Some of the products that don't need to be disassembling; according to their variety, will be transmitted to the processing center right after collecting centers, then considering to the variety of product and the request of manufacturing centers, they will be sent to them accordingly. In the remanufacturing process, according to the production center's demand, the parts which can be used again, after processing center will be sent to the remanufacturing center and after compounding with the other parts will be changed into new products and can return to the distribution layer in the proposed supply chain. In the recycling process according to the recycling center's demand, the disassembled parts (which can recover again) right after disassembling centers will be sent to the recycling centers. The configuration is shown in Figure 1.

Several tasks in the process of reverse supply chain motivated us to focus on cost analysis over the elements required for the reverse process. In the previous studies, researches considered some of the cost elements and formulated a mathematical model for decision-making. But, a more comprehensive study considering most of the cost elements exist in a reverse supply chain is developed here.

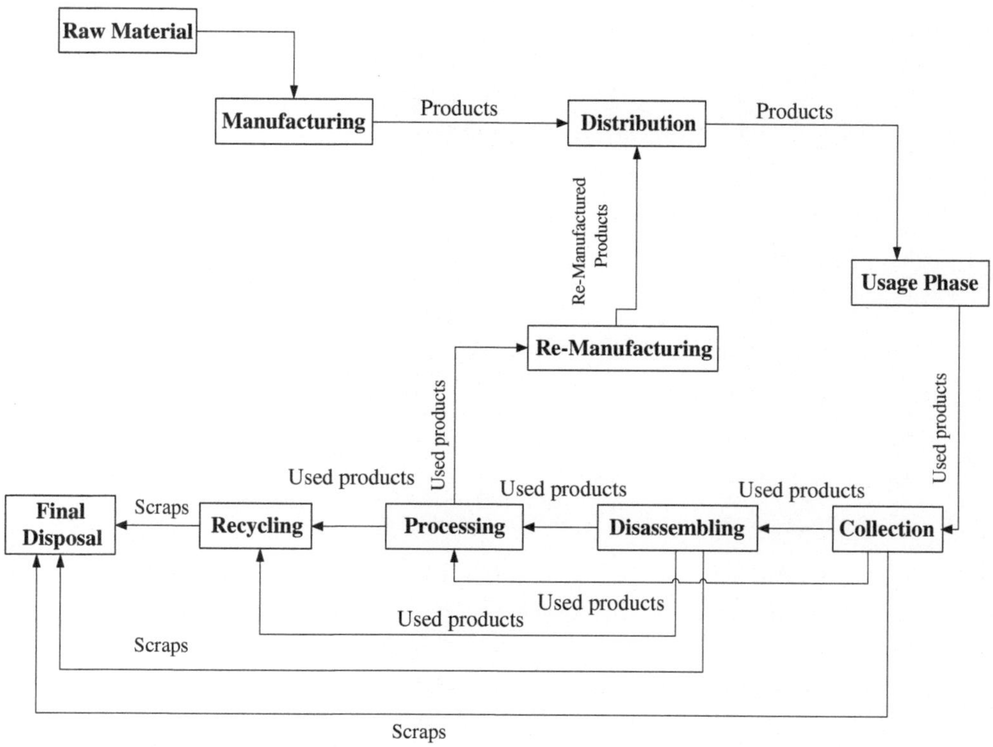

Figure 1. Framework for the proposed reverse supply chain.

3.1. Purpose

In this paper, the reverse supply chain model has been considered for returned products with the purpose of minimizing the reverse supply chain costs.

3.2. Assumptions

- The quantity of return, disassembly, processing, manufacturing, recycling, material, and distribution centers are determined.
- Some products will be transported straightly from return centers to the processing centers.
- Some parts will be transported straightly from disassembly centers to the recycling centers.

3.3. Indices, parameters, and decision variables

3.3.1. Indices

i: index of returning centers

j: index of disassembly centers

k: index of processing center

f: index of manufacturing center

r: index of recycling center

p: index of products

m: index of parts

l: index of distribution centers

c: index of customers

3.3.2. Parameters

a_{ip}: the capacity of returning center *i* for product *p*

b_{jm}: the capacity of disassembly center *j* for parts *m*

u_{km}: the capacity of processing center *k* for part *m*

h_{fm}: the capacity of production center *f* for parts *m*

E_{lm}: the capacity of distribution center *l* for part *m*

DM_{fm}: the manufacturing center's demand *f* for part *m*

$DRCP_{rp}$: the recycling center's demand *r* for product *p*

$DRCM_{rm}$: the recycling center's demand *r* for part *m*

DD_{lm}: the distribution center's demand *l* for part *m*

DC_{cm}: the client's demand *c* for part *m*

n_{mp}: The produced part's amount *m* from disassembling one product *p*

$CSRD_{ijp}$: unit cost of transportation from returning center *i* to disassembly center *j* for product *p*

$CSRP_{ikp}$: unit cost of transportation from returning center *i* into the processing center *k* for product *p*

$CSDP_{jkm}$: unit cost of transportation from disassembly center *j* into processing center *k* for part *m*

$CSDRC_{jrm}$: unit cost of transportation from disassembly center *j* into the recycling center *r* for part *m*

$CSPM_{kfm}$: unit cost of transportation from processing center *k* into the manufacturing center *f* for part *m*

$CSPRC_{krm}$: unit cost of transportation from processing center *k* into the recycling center *r* for part *m*

$CSPDC_{flm}$: unit cost of transportation from manufacturing center *f* into the distribution center *l* for part *m*

$CSDC_{lcm}$: unit cost of transportation from distribution center *l* into the clients *c* for part *m*

$FOCD_{jm}$: the fixed opening cost for disassembly center *j* for part *m*

$FOCP_{km}$: the fixed opening cost for processing centers *k* for part *m*

$FOCR_{ip}$: the fixed opening cost for returning centers *i* for product *p*

$FOCRC_{rm}$: the fixed opening cost for recycling centers *r* for part *m*

RMC_{fm}: unit cost of remanufacturing in manufacturing center *f* for part *m*

IC_{ip}: unit cost of maintaining in returning center *i* for product *p*

OCD_{jm}: unit cost of operations in disassembly center *j* for part *m*

OCP_{km}: unit cost of operations in processing center *k* part *m*

$OCRC_{rm}$: unit cost of operations in recycling center *r* part *m*

NRS_{min}: the minimum amount of returning center for opening and operations

NRS_{max}: the maximum amount of returning centers for operations and opening

NDS_{min}: the minimum amount of disassembling centers for opening and operations

NDS_{max}: the maximum quantity of disassembling centers for opening and operations

NPS_{min}: the minimum amount of processing centers for opening and operations

NPS_{max}: the maximum amount of processing centers for opening and operations

$NRCS_{min}$: the minimum amount of recycling centers for opening and operations

$NRCS_{max}$: the maximum amount of recycling centers for opening and operations

3.3.3. Decision variables

Φ_{ijp}: amount shipped from returning center *i* to disassembling center *j* for product *p*

δ_{ikp}: amount shipped from returning center *i* into the processing center *k* for product *p*

G_{jkm}: amount shipped from disassembly center *j* into the processing center *k* for part *m*

O_{jrm}: amount shipped from disassembly center *j* into the recycling center *r* for part *m*

Q_{kfm}: amount shipped from processing center *k* into the manufacturing center *f* for part *m*

S_{krm}: amount shipped from processing center *k* into the recycling center *r* for part *m*

T_{flm}: amount shipped from manufacturing center *f* into the distribution center *l* for part *m*

V_{lcm}: amount shipped from distribution center *l* into the clients *c* for part *m*

α_{jm}: if the disassembly center *j* is open for part *m*, 1 or otherwise 0

β_{km}: if processing center *k* is open for part *m*, 1 or otherwise 0

γ_{ip}: if the returning center *i* is open for product *p*, 1 or otherwise 0

λ_{rm}: if recycling center *r* is open for part *m*, 1 or otherwise 0

μ_{fm}: the part's flow amount *m* in manufacturing center *f*

X_{ip}: the product's flow amount *p* in returning center *i*

Y_{jm}: the part's flow amount *m* in disassembly center *j*

θ_{km}: the part's flow amount *m* in processing center *k*

τ_{rm}: the part's flow amount *m* in recycling center *r*

4. Mathematical formulation

Considering the purpose and the accompanying assumptions, we formulate the problem in an integer linear mathematical program. The advantages of such a program are: obtaining exact solutions, no essence for linearization causing more computational efforts, capability of the program in large-sized problems and simplicity of implementation in optimization software. The formulation of the mathematical model is given below:

$$\begin{aligned}
\text{Min } Z = & \sum_{i=1}^{I}\sum_{j=1}^{J}\sum_{p=1}^{P} CSRD_{ijp}\Phi_{ijp} + \sum_{i=1}^{I}\sum_{k=1}^{K}\sum_{p=1}^{P} CSRP_{ikp}\delta_{ikp} + \sum_{j=1}^{J}\sum_{k=1}^{K}\sum_{m=1}^{M} CSDP_{jkm}G_{jkm} \\
& + \sum_{j=1}^{J}\sum_{r=1}^{R}\sum_{m=1}^{M} CSDRC_{jrm}O_{jrm} + \sum_{k=1}^{K}\sum_{f=1}^{F}\sum_{m=1}^{M} CSRM_{kfm}Q_{kfm} + \sum_{k=1}^{K}\sum_{r=1}^{R}\sum_{m=1}^{M} CSPRC_{krm}S_{krm} \\
& + \sum_{f=1}^{F}\sum_{l=1}^{L}\sum_{m=1}^{M} CSPDC_{flm}T_{flm} + \sum_{l=1}^{L}\sum_{c=1}^{C}\sum_{m=1}^{M} CSDC_{lcm}V_{lcm} + \sum_{j=1}^{J}\sum_{m=1}^{M} FOCD_{jm}\alpha_{jm} \\
& + \sum_{k=1}^{K}\sum_{m=1}^{M} FOCP_{km}\beta_{km} + \sum_{i=1}^{I}\sum_{p=1}^{P} FOCR_{ip}\gamma_{ip} + \sum_{r=1}^{R}\sum_{m=1}^{M} FOCRC_{rm}\lambda_{rm} + \sum_{f=1}^{F}\sum_{m=1}^{M} RCM_{fm}\mu_{fm} \\
& + \sum_{i=1}^{I}\sum_{p=1}^{P} IC_{ip}X_{ip} + \sum_{j=1}^{J}\sum_{m=1}^{M} OCD_{jm}Y_{jm} + \sum_{k=1}^{K}\sum_{m=1}^{M} OCP_{km}\theta_{km} + \sum_{r=1}^{R}\sum_{m=1}^{M} OCRC_{rm}\tau_{rm}
\end{aligned} \tag{1}$$

s.t.

$$\sum_{j=1}^{J} \Phi_{ijp} \le a_{ip}\gamma_{ip} \quad \forall i,p \tag{2}$$

$$\sum_{k=1}^{K} \delta_{ikp} \le a_{ip}\gamma_{ip} \quad \forall i,p \tag{3}$$

$$X_{ip} \le a_{ip}\gamma_{ip} \quad \forall i,p \tag{4}$$

$$\sum_{k=1}^{K} G_{jkm} \le b_{jm}\alpha_{jm} \quad \forall j,m \tag{5}$$

$$\sum_{r=1}^{R} O_{jrm} \le b_{jm}\alpha_{jm} \quad \forall j,m \tag{6}$$

$$Y_{jm} \le b_{jm}\alpha_{jm} \quad \forall j,m \tag{7}$$

$$\sum_{f=1}^{F} Q_{kfm} \le u_{km}\beta_{km} \quad \forall k,m \tag{8}$$

$$\sum_{r=1}^{R} S_{krm} \le u_{km}\beta_{km} \quad \forall k,m \tag{9}$$

$$\theta_{km} \le u_{km}\beta_{km} \quad \forall k,m \tag{10}$$

$$\sum_{l=1}^{L} T_{flm} \le h_{fm} \quad \forall f,m \tag{11}$$

$$\mu_{fm} \le h_{fm} \quad \forall f,m \tag{12}$$

$$\sum_{c=1}^{C} V_{lcm} \le e_{lm} \quad \forall l,m \tag{13}$$

$$\sum_{k=1}^{K} Q_{kfm} \ge DM_{fm} \quad \forall f,m \tag{14}$$

$$\mu_{fm} \ge DM_{fm} \quad \forall f,m \tag{15}$$

$$\sum_{f=1}^{F} T_{flm} \ge DD_{lm} \quad \forall l,m \tag{16}$$

$$\sum_{l=1}^{L} V_{lcm} \ge DC_{cm} \quad \forall c,m \tag{17}$$

$$\sum_{j=1}^{J} O_{jrm} + \sum_{k=1}^{K} S_{krm} \geq DRCM_{rm} \quad \forall r, m \tag{18}$$

$$\tau_{rm} \geq DRCM_{rm} \quad \forall r, m \tag{19}$$

$$\sum_{i=1}^{I} \sum_{k=1}^{K} \delta_{ikp} \geq \sum_{r=1}^{R} DRCP_{rp} \quad \forall p \tag{20}$$

$$\sum_{j=1}^{J} \sum_{k=1}^{K} G_{jkm} \geq \sum_{f=1}^{F} DM_{fm} \quad \forall m \tag{21}$$

$$\sum_{j=1}^{J} \sum_{k=1}^{K} G_{jkm} \leq n_{mp} \left(\sum_{i=1}^{I} \sum_{j=1}^{J} \Phi_{ijp} \right) \quad \forall m, p \tag{22}$$

$$\sum_{j=1}^{J} \sum_{r=1}^{R} O_{jrm} \leq n_{mp} \left(\sum_{i=1}^{I} \sum_{j=1}^{J} \Phi_{ijp} \right) \quad \forall m, p \tag{23}$$

$$\mathrm{NRS}_{\min} \leq \sum_{i=1}^{I} \gamma_{ip} \leq \mathrm{NRS}_{\max} \quad \forall p \tag{24}$$

$$\mathrm{NDS}_{\min} \leq \sum_{j=1}^{J} \alpha_{jm} \leq \mathrm{NDS}_{\max} \quad \forall m \tag{25}$$

$$\mathrm{NPS}_{\min} \leq \sum_{k=1}^{K} \beta_{km} \leq \mathrm{NPS}_{\max} \quad \forall m \tag{26}$$

$$\mathrm{NRCS}_{\min} \leq \sum_{r=1}^{R} \lambda_{rm} \leq \mathrm{NRCS}_{\max} \quad \forall m \tag{27}$$

$$\sum_{f=1}^{F} T_{flm} = \sum_{c=1}^{C} V_{lcm} \quad \forall l, m \tag{28}$$

$$\sum_{k=1}^{K} G_{jkm} + \sum_{r=1}^{R} O_{jrm} \leq Y_{jm} \quad \forall j, m \tag{29}$$

$$\sum_{j=1}^{J} \Phi_{ijp} + \sum_{k=1}^{K} \delta_{ikp} \leq X_{ip} \quad \forall i, p \tag{30}$$

$$\sum_{f=1}^{F} Q_{kfm} + \sum_{r=1}^{R} S_{krm} \leq \theta_{km} \quad \forall k, m \tag{31}$$

$$\sum_{l=1}^{L} T_{flm} \leq \mu_{fm} \quad \forall f, m \tag{32}$$

$$\Phi_{ijp}, \delta_{ikp}, G_{jkm}, O_{jrm}, Q_{kfm}, S_{krm}, T_{flm}, V_{lcm}, \mu_{fm}, X_{ip}, Y_{jm}, \theta_{km}, \tau_{rm} \geq 0 \quad \forall i, j, k, f, r, p, m, l, c \qquad (33)$$

$$\alpha_{jm}, \beta_{km}, \gamma_{ip}, \lambda_{rm} = \{0, 1\} \quad \forall i, j, k, p, m \qquad (34)$$

4.1. Objective function

We want to demonstrate a model in reverse supply chain in a way to minimize the chain costs. We should introduce a model which minimizes the transportation cost of products and parts between centers and at the same time minimizes the fixed opening cost of sites and operation's cost on parts and supply maintenance costs and remanufacturing costs. By attention to the definition of indices, parameters, and decision variables; the objective function consists of: minimizing the costs of transportation of products and parts, the fixed opening cost of centers and operations costs on parts and the supply maintenance costs, remanufacturing costs in reverse supply chain (Equation 1).

4.2. Constraints

Equations 2 and 3 state that the amount of shipping products from any returning center (if it is opened) into the disassembly, and processing centers for each product should be equal or smaller than the capacity of that returning center. Equation 4 states that the amount of products which will be collected in the returning center should be equal or smaller than the capacity of that returning center. Equations 5 and 6 state that the amount of sent parts from any disassembly centers and recycling centers should be equal or smaller than the capacity of the same disassembly center for each part. Equation 7 is stating that the amount of a part which is in the disassembly center should be equal or smaller than the capacity of the same disassembly center. Equations 8 and 9 state that the amount of shipping parts from any processing centers (if it is opened) into the manufacturing centers and recycling centers should be equal or smaller than the capacity of the same processing centers for each part. Equation 10 states that the amount of a part which is in the processing center should be equal or smaller than the capacity of the same processing center. Equation 11 states that the amount of sent parts from any manufacturing center into the distribution centers should be equal or smaller than the capacity of the same manufacturing center for each part. Equation 12 states that the amount of part in each manufacturing center should be equal or smaller than the capacity of the same manufacturing center. Equation 13 states that the amount of sent parts from any distribution center to the client should be equal or smaller than the capacity of the same distribution center for clients. Equations 14 and 15 state the demand amount of manufacturing center for parts. Equation 16 states the part demand amount of distribution centers. Equation 17 indicates the client's part demand amount. Equations 18 and 19 state the part demand amount of recycling centers. Equations 20 and 21 state that the manufacturing and recycling center's demand, is for products and parts which are transported from the returning and disassembly centers into the processing center. Equations 22 and 23 are related to the balance of parts flow from the disassembly of products. Equations 24–27 are stating the min and max indices amount of returning, disassembling, processing, and recycling centers. Equation 28 states that the amount of sent parts from manufacturing centers to the distribution center is equal to the sent parts from distribution centers in to the client.

Equations 29 states that the amount of sent parts from each disassembly center into the processing and recycling centers should be equal or smaller than the parts amount in that disassembly center. Equation 30 states that the amount of sent products from each returning center into the disassembly, processing centers, should be equal or smaller than the product's amount in that returning center. Equation 31 states that the amount of sent parts from each of the processing centers into the manufacturing and recycling centers should be equal or smaller than the flow amount of parts in that processing center. Equation 32 states that the sent parts amount from any manufacturing center into the distribution centers should be equal or smaller than the parts flow amount in that manufacturing center. Equations 33 and 34 enforce the binary and non-negativity restrictions on the corresponding decision variables.

Table 1. Numerical results using Lingo 9 software

$\varphi(2,1,4)$	1.2	$T(2,3,3)$	16	$\theta(2,2)$	16
$\varphi(2,3,2)$	.9	$T(3,2,1)$	8	$\theta(2,3)$	63
$\varphi(3,3,1)$	.9	$T(3,2,2)$	8	$\theta(3,1)$	8
$\varphi(3,3,3)$	1.1	$T(4,1,1)$	13	$\theta(4,2)$	20
$\delta(2,1,2)$	34	$T(4,2,1)$	3	$\theta(4,3)$	7
$\delta(2,4,4)$	28	$T(4,2,2)$	5	$\gamma(2,1)$	1
$\delta(3,1,1)$	41	$T(5,1,1)$	8	$\gamma(2,2)$	1
$\delta(3,4,3)$	8	$T(5,1,2)$	16	$\gamma(2,3)$	1
$G(1,3,2)$	62	$T(5,3,3)$	17	$\gamma(2,4)$	1
$G(1,4,3)$	50	$V(1,1,2)$	6	$\gamma(3,1)$	1
$G(2,1,3)$	16	$V(1,1,3)$	6	$\gamma(3,2)$	1
$G(3,2,1)$	59	$V(1,2,2)$	17	$\gamma(3,3)$	1
$O(1,1,3)$	8	$V(1,3,1)$	1	$\gamma(3,4)$	1
$O(1,3,2)$	18	$V(1,4,1)$	20	$X(2,2)$	34.87
$O(1,3,3)$	4	$V(2,1,2)$	4	$X(2,4)$	29.2
$O(2,2,3)$	4	$V(2,2,3)$	16	$X(3,1)$	41.94
$O(2,3,1)$	18	$V(2,3,1)$	19	$X(3,3)$	9.09
$Q(1,1,1)$	9	$V(2,4,2)$	9	$\lambda(1,3)$	1
$Q(1,1,2)$	17	$V(2,4,3)$	14	$\lambda(2,1)$	1
$Q(1,2,1)$	18	$V(3,1,1)$	20	$\lambda(3,2)$	1
$Q(1,2,2)$	16	$V(3,1,3)$	13	$\tau(1,1)$	16
$Q(1,3,2)$	8	$V(3,2,1)$	4	$\tau(1,2)$	20
$Q(1,4,1)$	16	$V(3,3,2)$	3	$\tau(1,3)$	12
$Q(1,4,2)$	5	$V(3,3,3)$	20	$\tau(2,1)$	4
$Q(1,5,1)$	8	$\alpha(1,2)$	1	$\tau(2,2)$	11
$Q(2,1,3)$	19	$\alpha(1,3)$	1	$\tau(2,3)$	4
$Q(2,3,3)$	14	$\alpha(2,1)$	1	$\tau(3,1)$	18
$Q(2,4,3)$	9	$\alpha(2,2)$	1	$\tau(3,2)$	18
$Q(2,5,2)$	16	$\alpha(2,3)$	1	$\tau(3,3)$	4
$Q(2,5,3)$	17	$\alpha(3,1)$	1	$\mu(1,1)$	9
$Q(3,3,1)$	8	$Y(1,2)$	80	$\mu(1,2)$	17
$Q(4,2,3)$	7	$Y(1,3)$	62	$\mu(1,3)$	19
$S(1,1,1)$	16	$Y(2,1)$	18	$\mu(2,1)$	24
$S(1,2,1)$	4	$Y(2,3)$	20	$\mu(2,2)$	16
$S(1,2,2)$	11	$Y(3,1)$	59	$\mu(2,3)$	33
$S(2,1,3)$	4	$\beta(1,1)$	1	$\mu(3,1)$	8
$S(4,1,2)$	20	$\beta(1,2)$	1	$\mu(3,2)$	8
$T(1,1,3)$	6	$\beta(2,2)$	1	$\mu(3,3)$	14
$T(1,2,1)$	8	$\beta(2,3)$	1	$\mu(4,1)$	16
$T(1,2,3)$	13	$\beta(3,1)$	1	$\mu(4,2)$	5
$T(1,3,2)$	3	$\beta(4,2)$	1	$\mu(4,3)$	9
$T(2,1,2)$	7	$\beta(4,3)$	1	$\mu(5,1)$	8
$T(2,2,3)$	17	$\theta(1,1)$	71	$\mu(5,2)$	16
$T(2,3,1)$	24	$\theta(1,2)$	57	$\mu(5,3)$	17

It should be noted that one should decide whether to invest in reverse logistic system. Then, the proposed mathematical with the objective of cost minimization helps to point out the cost aspects of the reverse system and make an economic decision with more reliability.

5. Numerical experiment

We solved the presented mathematical model by using Lingo 9 software. In this multi-layers and multi-product model, we are attempting to minimize the costs of fixed opening facilities, transportation and shipping of products, and parts between centers and also the operations, supply, maintenance and remanufacturing costs as well as the product amount and sending parts into the centers and the amount of it would be calculated. To analyze the suggested model, we create a numerical example and solve the created example by Lingo software.

In this problem, we consider the index values between 3 and 5. We replace the inputs of problem in the model and by using the Lingo, we will solve the problem and finally the outputs are in hand with the corresponding objective function value in a negligible time.

The obtained objective function value is equal to 26121.20 unit of money. All the variables which were not zero (0) quantities are shown in Table 1.

After solving the model, we will find out that the decision variable $\alpha(2,2)$ gained one. This means that the disassembly center 2 should be opened for part 2. The decision variable $\lambda(1,3)$ obtained one, means that the recycling center 1 would be opened for part 3. Generally, when the decision variables α_{jm}, β_{km}, γ_{ip}, and λ_{rm} gained one, it indicates that the considered center to that decision variable will be opened for that part or product.

The decision variable $Q(4,2,3)$ is 7. This means that the amount of part 3 from processing center 4 into the manufacturing center 2 is 7. The decision variable $\tau(3,2)$ got 18, it means that the amount of part 2 in recycling center 3 is 15. $T(3,2,2)=8$ means that the amount of part 2 from manufacturing center 3 into distribution center 2 is 8.

It should be noted that the computational time is negligible since the model is linear. The proposed model is helpful in economic decision-making for investing in reverse supply chain. The activities within the backward flow are configured to make added value on the returned items. The results are effective in a macro decision-making for a reverse supply chain from economic view. Policy-makers can make use of the results to present a policy in the context of environment or economy with respect to lack of first materials and the essence of recycling. Some other factors are also perceived from the results namely the analysis on the effect of boycott in achieving some required materials and parts to produce a product. This issue forces the supply chain managers to conduct a comprehensive reverse chain to increase the productivity and at the same time by analyzing cost elements increase the profitability.

6. Conclusions

In this paper, a reverse supply chain was considered minimizing the total cost of transport, inspection, remanufacture, and maintenance. The presented model was an integer linear programming model for multi-layer, multi-product reverse supply chain. We solved the proposed model using Lingo 9 software. The advantages of the model were: obtaining exact solutions, no essence for linearization causing more computational efforts, capability of the program in large-sized problems, and simplicity of implementation in optimization software. Also, from the managerial aspect, the model was capable to help the policy-makers in the environmental and economical concepts streaming to green supply chain. The limitation of the model was its application in different industries which requires adapting the model with the specifications of a certain industry. As future research, including and integration pricing models is suggested.

Funding
This research is self-funded.

Author details
Hoda Mahmoudi[1]
E-mail: hfazl@iust.ac.ir
Hamed Fazlollahtabar[1]
E-mail: hamed.hero@gmail.com; hfazl@iust.ac.ir
[1] Faculty of Industrial Engineering, Iran University of Science and Technology, Tehran, Iran.

References

Barros, A. I., Dekker, R., & Scholten, V. A. (1998). A two-level network for recycling sand: A case study. *European Journal of Operational Research, 110*, 199–214. http://dx.doi.org/10.1016/S0377-2217(98)00093-9

Damghani, K., & Taghavifard, M. (2012). A three-stage fuzzy DEA approach to measure performance of a serial process including JIT practices, agility indices, and goals in supply chains. *International Journal of Services and Operations Management, 13*, 147–188. http://dx.doi.org/10.1504/IJSOM.2012.048828

Dekker, R., & Van Der Laan, E. A. (1999). Gestion des stocks pour la fabrication et la refabrication simultanées: synthèse de résultats récents [Inventory management for the simultaneous manufacturing and remanufacturing: Synthesis of recent results]. *Logistique & Management, 7*, 59–64.

Dev, N., & Shankar, R. (2012). Design of fractal information coordination system in a supply chain network. *International Journal of Services and Operations Management, 12*(1), 1–19. http://dx.doi.org/10.1504/IJSOM.2012.046672

Dobrzykowski, D., Tran, O., & Hong, P. (2011). Insights into integration for supply chain redesign in service and product-focused firms. *International Journal of Services and Operations Management, 8*, 260–282. http://dx.doi.org/10.1504/IJSOM.2011.038972

Du, F., & Evans, G. W. (2008). A bi-objective reverse logistics network analysis for post-sale service. *Computers & Operations Research, 35*, 2617–2634.

Fazlollahtabar, H., Mahdavi, I., & Mohajeri, A. (2013). Applying fuzzy mathematical programming approach to optimize a multiple supply network in uncertain condition with comparative analysis. *Applied Soft Computing, 13*, 550–562. http://dx.doi.org/10.1016/j.asoc.2012.08.016

Fleischmann, M., Van Nunen, J. A. E. E., & Gräve, B. (2003). Integrating closed-loop supply chains and spare-parts management at IBM. *Interfaces, 33*, 44–56. http://dx.doi.org/10.1287/inte.33.6.44.25189

Guide, V. D. R., Harrison, T. P., & Van Wassenhove, L. N. (2003). The challenge of closed loop supply chains. *Interfaces, 33*, 3–6. http://dx.doi.org/10.1287/inte.33.6.3.25182

Jayaraman, V., Guide, V. D. R., & Srivastava, R. A. (1999). A closed loop logistics model for remanufacturing. *Journal of the Operational Research Society, 50*, 497–508. http://dx.doi.org/10.1057/palgrave.jors.2600716

Jayaraman, V., Patterson, R. A., & Rolland, E. (2003). The design of reverse distribution networks: Models and solution procedures. *European Journal of Operational Research, 150*, 128–149. http://dx.doi.org/10.1016/S0377-2217(02)00497-6

Kim, K. B., Song, I. S., & Jeong, B. J. (2006). Supply planning model for remanufacturing system in reverse logistics environment. *Computers & Industrial Engineering, 51*, 279–287.

Kim, T., Hong, Y., & Goyal, S. K. (2009). Coordination through a quantity-incentive mechanism in a single-manufacturer-single-retailer supply chain. *International Journal of Services and Operations Management, 5*, 482–497. http://dx.doi.org/10.1504/IJSOM.2009.024581

Kroon, L., & Vrijens, G. (1995). Returnable containers: An example of reverse logistics. *International Journal of Physical Distribution & Logistics Management, 25*, 56–68.

Kumar, D., Singh, O., & Singh, J. (2011). Analysis of supplier related issues in supply chain practices. *International Journal of Services and Operations Management, 9*, 284–307. http://dx.doi.org/10.1504/IJSOM.2011.041101

Kumar, G., & Banerjee, R. N. (2012). An implementation strategy for collaboration in supply chain: An investigation and suggestions. *International Journal of Services and Operations Management, 11*, 407–427. http://dx.doi.org/10.1504/IJSOM.2012.046077

Lee, D., & Dong, M. (2008). A heuristic approach to logistics network design for end-of-lease computer products recovery. *Transportation Research Part E: Logistics and Transportation Review, 44*, 455–474. http://dx.doi.org/10.1016/j.tre.2006.11.003

Lee, J. E., Rhee, K. G., & Gen, M. (2007). Designing a reverse logistics network by priority-based genetic algorithm. In *Proceedings of International Conference on intelligent manufacturing logistics systems* (pp. 158–163). Kitakyushu, Japan.

Listes, O., & Dekker, R. (2005). A stochastic approach to a case study for product recovery network design. *European Journal of Operational Research, 160*, 268–287. http://dx.doi.org/10.1016/j.ejor.2001.12.001

Pati, R. K., Vrat, P., & Kumar, P. (2008). A goal programming model for paper recycling system. *The International Journal of Management Science: Omega, 36*, 405–417.

Salema, M. I. G., Barbosa-Povoa, A. P., & Novais, A. Q. (2007). An optimization model for the design of a capacitated multi-product reverse logistics network with uncertainty. *European Journal of Operational Research, 179*, 1063–1077. http://dx.doi.org/10.1016/j.ejor.2005.05.032

Slamanig, M., & Winkler, H. (2012). Management of product change projects: A supply chain perspective. *International Journal of Services and Operations Management, 11*, 481–500. http://dx.doi.org/10.1504/IJSOM.2012.046080

Sreekumar, S., & Mahapatra, S. (2011). Supplier selection in supply chain management: A fuzzy multi-criteria decision-making approach. *International Journal of Services and Operations Management, 8*, 108–126.

Thierry, M. C., Salomon, M., Van Nunen, J. A. E. E., & Van Wassenhove, L. N. (1993). *Strategic production and operations management: Issues in product recovery management* (Management Report Series 145). The Netherlands: Erasmus Universiteit/Rotterdam School of Management.

5

Performance analysis in the coal bagging plant of Garp Lignite Corporation

Mehmet Nuri Bahşiş[1] and Yaşar Kasap[2*]

*Corresponding author: Yaşar Kasap, Faculty of Engineering, Department of Mining Engineering, Dumlupınar University, Kütahya 43100, Turkey

E-mail: yasaroz71@hotmail.com

Reviewing editor: Wei Meng, Wuhan University of Technology, China

Abstract: Companies can contribute to a country's economy by means of efficient production in each of their units. The aim of this study was to examine the change in operational performance of the coal bagging plant of Garp Lignite Corporation in Turkey using the Malmquist total factor productivity index. The results showed that there was a decrease in the productivity of the bagging plant. This decrease could be due to the rapid growth of the market share of natural gas in heating and energy in recent years, the reduced demand for lignite, overuse use of energy as a result of the necessity to keep the plant active even when no bagging is performed, the current labor-intensive bagging process, and the difficulty in working continuously with qualified staff hired depending on the demand for coal. In conclusion, it is recommended that a fully automated coal bagging plant be built in the corporation.

Subjects: Automation Control; Control Engineering; Mining Engineering; Technology

Keywords: coal bagging; productivity; Malmquist total factor productivity index

1. Introduction

Mining industry, which has great importance due to its direct contributions to a country's economy and the inputs it gives to other sectors of economy, particularly to the manufacturing sector, also requires large investments, involves risks and uncertainties in both investment and production stages in geological, topographical, technological, and economic terms, and includes many hazards in terms of work safety. For this reason, in order for investments to be met as soon as possible and to make contributions to a country's economy, productivity, low-cost manufacturing, product quality, continuity of production, use of scientific methods, and environmentally friendly production are extremely important.

ABOUT THE AUTHOR

Yaşar Kasap has been working as an associate professor at Dumlupınar University in Mining Engineering Department since 2008. Kasap received her BSc in 1998, MSc in 2000, and PhD in 2008 in Mining Engineering from Eskişehir Osmangazi University, Eskişehir, Turkey. She worked as an assistant professor at Dumlupınar University from 2008 to 2013. Finally, she worked as an associate professor at Dumlupınar University in Mining Engineering Department from 2013 She is married with one child. Kasap is a member of the Chamber of Mining Engineers (CME). She is still working in the field of mining efficiency.

PUBLIC INTEREST STATEMENT

There has been a constant decrease in the production of the coal bagging plants from the year 2006 to the present day. Approximately, 26% of the energy demand of Turkey is met by the domestic sources, while the rest of this demand is met by various foreign sources. Consumption of the domestic lignite in residential heating is restricted in Turkey because it leads to air pollution due to its low calorific value and high sulfur content. Natural gas has replaced coal in recent in Turkey. Other causes of the observed decline in the coal industry are that this sector requires continuous investment, long-term planning, search and preparatory work, the climate change in recent years, efficient production, and rapid marketing. The aim of this study was to examine the change in operational performance of the coal bagging plant of Garp Lignite Corporation in Turkey.

Significant changes have occurred in both production (e.g. high production rate, work safety, low operating costs, high quality, and standardized products) and in the structure of the labor force. Developments in science and technology, and demand for the increase in staff qualifications and the quality of the products, have given priority to productivity that is one of the performance indicators. Productivity involves achieving production in higher quantity and quality with less cost and less equipment.

Companies can contribute to a country's economy by means of efficient production in each of their units. The existing operation processes should be automated professionally in order for corporations in the mining sector to operate in a modern and low-cost process with higher productivity and that their technology should meet the highest standards. The aim of this study was to examine the change in operational productivity of the bagging plant of Garp Lignite Corporation (GLC). Approximately, 85% of the lignite production in GLC is from opencasts and 15% is from underground mining plants. The annual production capacity is about 6.3 million tons, and 1 million tons of this figure is from underground production. The efficiency change of the bagging plant of the corporation between the years 2006 and 2012 was evaluated using the Malmquist total factor productivity index (MPI). Sensitivity analyses were made in order to make the year 2012, which was found to be inefficient, efficient and some recommendations were made in order to provide the corporation with guidelines for the next several years.

2. Methodology

Parametric efficiency measurements assume that the production functions of fully efficient units are known. Since the production function is never known in practice, Farrell (1957) suggested that the function could be estimated from sample data. The first proposal was evaluated by Charnes, Cooper, and Rhodes (1978).

The first step in non-parametric approach to measurement of efficiency, which is used for comparative efficiency analysis, is to determine enveloped surfaces (efficient frontier) that cover the linear combinations and efficient observations of the decision-making units (DMUs) which carry out the same production activities. Then the efficiency scores and radial distances (from the center) of inefficient units within the enveloped surface are calculated (Muñiz, 2002).

In order to calculate the technical efficiency (TE) for k^{th} DMU (the DMU in question), the following linear programming (LP) model is used.

$$\min \theta_k \tag{1}$$

$$\sum_{j=1}^{n} \lambda_j \cdot Y_{rj} - s_{rj}^{-} = Y_{rk} \qquad (r = 1, 2, ..., s), \tag{2}$$

$$\sum_{j=1}^{n} \lambda_j \cdot X_{ij} + s_{ij}^{+} = \theta_k \cdot X_{ik} \qquad (i = 1, 2, ..., m), \tag{3}$$

$$\lambda_j, s_{ij}^{+}, s_{rj}^{-} \geq 0 \qquad (j = 1, 2, ..., n). \tag{4}$$

In the model that is established for efficiency measurement to be performed under input minimization, the aim is to keep outputs constant but inputs at a minimum level (1).

Constraint (2) sets involve comparison of the outputs kept constant in data envelopment analysis (DEA) carried out under input minimization. With this constraint, r^{th} output of each j DMU will not be greater than the maximum linear combination of the units constituting the efficient frontier. The constraints where minimization is sought for the inputs in inefficient DMUs are shown in the Equation (3). Equation (5) represents the constraints to be negative (Kasap, 2011).

In order for a DMU to be considered efficient,

- optimal θ_k has to be equal to one and
- all slack variable scores have to be zero ($s^{+}_{ik}, s^{-}_{rk} = 0$).

The symbols used in the formulation of non-parametric LP are defined below:

N—the number of DMUs involved in comparison,
S—the number of outputs gained from the production,
M—the number of inputs used in the production,
$j = (1,2,....,n)$—set of all DMUs,
$k = (1,2,....,n)$—set of DMUs taken into consideration,
$r = (1,2,....,s)$—set of all outputs,
$i = (1,2,....,m)$—set of all inputs,
$y \in R^{+}_{s}$—vector of outputs $(y_1, y_2,...,y_s) = s \times n$,
$x \in R^{+}_{m}$—vector of inputs $(x_1, x_2,...,x_m) = m \times n$,
λ—the vector of density variables giving inputs–outputs weight averages = $k \times 1$,
λ_{jk}—the relative (compared to other units, j) weight value of "k" decision unit measured for efficiency in input-oriented process,
θ_k—the scaler variable (efficiency score) trying to decrease all inputs of k DMU, which is considered to obtain the best frontier,
Y_{rj}—the rth output amount produced by decision unit j,
Y_{rk}—the rth output amount produced by decision unit k,
X_{ij}—the ith input amount used by decision unit j,
X_{ik}—the ith input amount used by decision unit k.

As a form of static analysis, DEA performs analyses using data from DMUs in a single period. However, a DMU that is identified to be efficient before may lose its efficiency and reference quality. In this respect, in efficiency evaluation process, the MPI was developed to examine the changes that may occur over time.

The MPI, which was obtained by adding the functions of distance to the Farrell (1957) measure of TE, measures the variation in two units' total factor productivities as the proportion of the distances from a common technology. Distance function is used to define multi-input and multi-output production technologies without specifying objectives such as cost minimization or profit maximization. The input distance function considers a production technology by looking at a minimal proportional contraction of input vector, given an output vector, while the output distance function characterizes it a maximal proportional expansion of the output vector, given an input vector (Coelli, Rao, & Battese, 1998). This study used the output distance function because it was suitable for the analysis that was conducted to investigate the efficiency changes of the DMUs in the considered years.

Grifell-Tatjé and Lovell (1995) showed that using the assumption of variable returns to scale when calculating the distance functions required for MPI would not accurately measure the changes (gain or loss of productivity) in total factor productivity (TFP) index. For this reason, the index needs to be calculated under the assumption of constant returns to scale.[1]

Since the production in the coal bagging plant is performed depending on the demand by the market for coal, an input-oriented analysis was considered to be more appropriate.

The input distance function is

$$D_i(x,y) = \max\left\{\theta:(\frac{x}{\theta},y) \in L(y)\right\} \tag{5}$$

If *y* vector is an element of the possible production set of L(y) efficient frontier, the distance function $D_i(x, y)$ will have a value smaller than or equal to one. According to the input between *t* period and the subsequent (*t* + 1) period and within the framework of distance function, MPI is calculated as follows:

$$\text{MPI} = \sqrt{\frac{D_i^t(y^{t+1},x^{t+1})}{D_i^t(y^t,x^t)} \times \frac{D_i^{t+1}(y^{t+1},x^{t+1})}{D_i^{t+1}(y^t,x^t)}} \tag{6}$$

where $D_i(x, y)$ represents the distance of *t* period to *t* + 1 period technology. An MPI value greater than one indicates that there is an expansion of total factor productivity from *t* period to *t* + 1 period while an MPI value smaller than one shows a contraction in total factor productivity (Färe, Grosskopf, & Lovell, 1994).

MPI considers changes in productivity according to two separate components: TE change and technological change. TE change provides an assessment of the process in which DMUs approach the efficient frontier, whereas technological change (TC) provides the change of the efficient frontier over time (Grifell-Tatjé & Lovell, 1995; Mahadevan, 2002).

When Equation (6) is revised,

$$\text{MPI} = \underbrace{\frac{D_i^{t+1}\left(y^{t+1},x^{t+1}\right)}{D_i^t\left(y^t,x^t\right)}}_{\text{Technical Efficiency Change}} \times \underbrace{\sqrt{\frac{D_i^t\left(y^{t+1},x^{t+1}\right)}{D_i^{t+1}\left(y^{t+1},x^{t+1}\right)} \times \frac{D_i^t\left(y^t,x^t\right)}{D_i^{t+1}\left(y^t,x^t\right)}}}_{\text{Technological Change}} \tag{7}$$

where the ratio outside the square brackets measures the change in the output-oriented measure of Farrell TE between periods *t* and *t* + 1. In other words, the efficiency change is equivalent to the ratio of the TE in period *t* + 1 to the TE in period *t*. The remaining part of the index in Equation (2) is a measure of technical change. It is the geometric mean of the shift in technology between the two periods, evaluated at x^{t+1} and also at x^t (Coelli & Rao, 2005).

$$\text{MPI} = \text{TE} \times \text{TC}. \tag{8}$$

Non-parametric LP method is the most popular method that is used to estimate the distance functions that are required to form MPI. When there is a suitable panel data-set, the required distances can be calculated using non-parametric linear programs with this method. Four distance functions must be calculated to measure the changes TFP between the two periods for any i^{th} input and this requires the solution of four LP problems. The LPs required under the assumption of constant returns to scale are

$$\begin{aligned}\left[D_i^{t+1}\left(x_j^{t+1},y_j^{t+1}\right)\right]^{-1} &= \max\theta_k \\ &\sum_{j=1}^{n}\lambda_{jk}^{t+1}\cdot y_{rj}^{t+1} \geq y_{rk}^{t+1}\cdot\theta_k \\ &\sum_{j=1}^{n}\lambda_{jk}^{t+1}\cdot x_{ij}^{t+1} \leq x_{ij}^{t+1} \\ &\lambda_{jk}^{t+1} \geq 0\end{aligned} \qquad \begin{aligned}\left[D_i^{t}\left(x_j^{t},y_j^{t}\right)\right]^{-1} &= \max\theta_k \\ &\sum_{j=1}^{n}\lambda_{jk}^{t}\cdot y_{rj}^{t} \geq y_{rk}^{t}\cdot\theta_k \\ &\sum_{j=1}^{n}\lambda_{jk}^{t}\cdot x_{ij}^{t} \leq x_{ij}^{t} \\ &\lambda_{jk}^{t} \geq 0\end{aligned} \tag{9}$$

$$\left[D_i^{t+1}\left(x_j^t, y_j^t\right)\right]^{-1} = \max \theta_k \quad \sum_{j=1}^{n} \lambda_{jk}^{t+1} \cdot y_{rj}^{t+1} \geq y_{rk}^{t} \cdot \theta_k \quad \sum_{j=1}^{n} \lambda_{jk}^{t+1} \cdot x_{ij}^{t+1} \leq x_{ij}^{t} \quad \lambda_{jk}^{t+1} \geq 0$$

$$\left[D_i^{t}\left(x_j^{t+1}, y_j^{t+1}\right)\right]^{-1} = \max \theta_k \quad \sum_{j=1}^{n} \lambda_{jk}^{t} \cdot y_{rj}^{t} \geq y_{rk}^{t+1} \cdot \theta_k \quad \sum_{j=1}^{n} \lambda_{jk}^{t} \cdot x_{ij}^{t} \leq x_{ij}^{t+1} \quad \lambda_{jk}^{t} \geq 0 \tag{10}$$

Determining these defined distance values using Equations (9) and (10) for all time periods and years requires the solution of $n(3t-2)$, where n represents the number DMUs and t shows the number of periods, LP models (Färe et al., 1994). Since there were 7 years analyzed and 12 months (as periods of time) in this study, a total of 238 LP models were solved so that the analyses could be carried out.

3. Application

There are currently three bagging plants in GLC: Tunçbilek, BEKE and 1B. BEKE and 1B plants are not used because the capacity of Tunçbilek bagging plant is compatible with the amount of production. The total capacity of Tunçbilek Bagging Plant has now reached 6,000 tons/day.

The plant offers pure and dust-free products to customers by washing the run of mine coal coming from the GLC casts in the coal washing plants and bagging it. As can be seen in Figure 1, the coal coming to the bagging plant is first poured into the dumbs with a capacity of 100 tons, it is then sent to the screen through belt conveyors, and it is screened at +18–100 mm in size. The coal +100 mm in size are fed to the crushers as screen oversize and then back to the screen, while the coal −18 mm in size is sent to the dust silo. On the other hand, the coal +18–100 mm in size coal is fed to the coal-bunkers of bagging unit by belt conveyors.

The coal +18–100 mm in size that is sent to the coalbunkers in the plant is sent to the scales through belt scales under the bunker (eight on the right side and eight on the left side). The scales

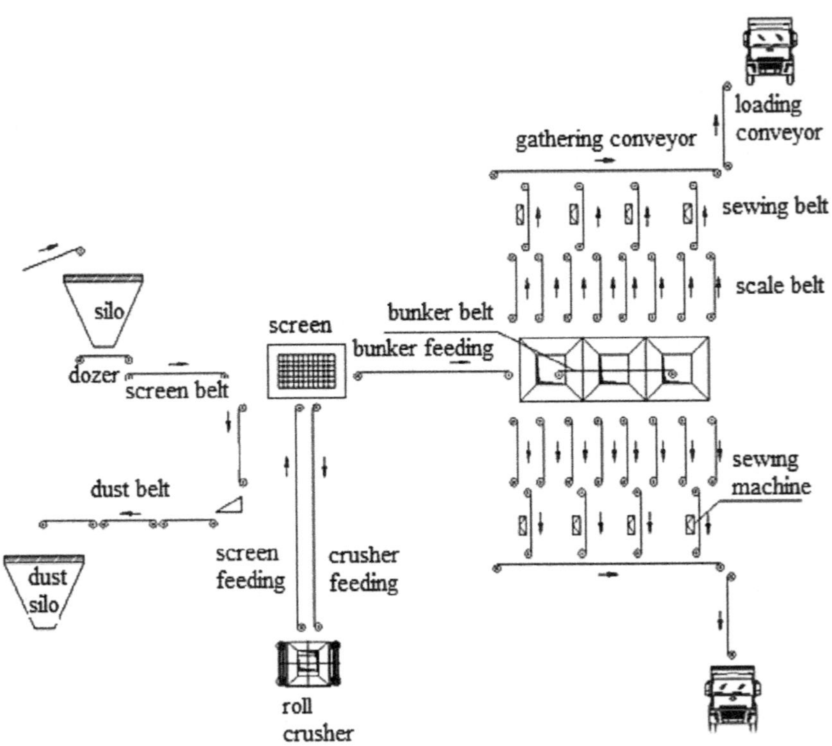

Figure 1. Flow diagram of unit 1 of Tunçbilek bagging plant (units 1, 2, and 3 are equivalent).

are electronic and weigh the amount of coal as 25 kg. When the coal in a scale reaches 25 kg, the system stops the scale-feeding belt and the coal is poured from the scale into the filling reduction. After a worker places an empty coal bag to the reduction, the coal in the scale is poured into the reduction by means of a switch and, from here, it is filled into the empty coal bag.

It takes approximately 12 s for a reduction to fill a coal bag. In order to increase production, on each coal-filling reduction are mounted two scales. When the worker responsible for the coal-filling reductions unloads the coal in one of the scales into an empty coal bag by means reduction, the other scale is loaded with coal. This increases coal-filling rate.

The worker responsible for the coal-filling reductions hooks 25 kg bags in 50 × 75 cm dimensions on the coal-filling reductions. The bags loaded with coal are sent to the sewing belt and a worker sews these bags with a sewing machine. After that, the bags are loaded to trucks through the loading line.

3.1. Efficiency change in the GLC coal bagging unit

The Malmquist TFP index was used to estimate the efficiency change in the GLC The aim of this study was to examine the change in operational productivity of the bagging plant of the GLC. When the conditions required for obtaining significant activity measurement results were considered, it was appropriate to choose seven years (2006–2012) as the DMUs in accordance with the number of inputs and outputs that were taken into consideration.

There would be differences in the yearly efficiency scores because the demand for coal increases in certain months of the year and, out of the raw ore lignite production in GLC, approximately 80% is from opencasts and 20% is from underground mining plants. Therefore, the periods of time that should be taken into consideration for the efficiency measurement method used were determined as months.

The amount of coal fed to the bagging units, the number of workers hired in the bagging units, and the amount of energy used in the bagging units were considered as the input, and the amount of bags of coal that were produced was considered as the output (Table 1).

While outputs can be kept constant in order to increase productivity in the non-parametric efficiency analysis, inputs can be kept constant in order to increase outputs as well. Although the Malmquist TFP index examines the efficiency changes of DMUs without input minimization or output maximization, because it is based on the non-parametric efficiency analysis, an increase in productivity was possible by reducing the amount of the coal fed to the plant, which was considered as an input. However, reducing the amount of the coal fed to the plant was not option so that the plant could operate at full capacity. Determining the inputs that are compatible with the structure of the models used is critical for obtaining accurate results. For this reason, the amount of coal that was fed to the plant were negated (1/the amount of the coal fed) and included in the analysis.

There are currently 24 coal-filling reduction and 24 sewing machines in the GLC coal bagging unit, which requires assigning 48 workers. Together with these 48 workers, there should be a total of 80 workers in a shift including eight loaders, four cleaners, three electricians, two welders, four

Table 1. The inputs and outputs of the bagging units

	Type	Unit	Description
Inputs	Coal that is fed	Tons	The quantity of coal that is fed to the bagging units
	Labor	Number/day	The average number of workers in the unit at a certain time
	Energy	kwh	The amount of energy used
Outputs	Bagged coal	Tons	The quantity of coal that is bagged

maintenance workers, one technician, one shift superintendent, four truck drivers, two loader operators, one sweeper operator, one shunter, and sewing machine worker. Therefore, the unit needs a total of 240 workers for the three shifts in a day. On the other hand, there could also be variations in the coal-filling reduction, which is operated according to the daily demand for bagged coal. This, consequently, causes a change in the number of workers in the plant at a specific time.

The amount of electric power used in the coal bagging units depends on production and, therefore, energy consumption increases depending on the number of units that are run when the amount of bagged coal increases.

According to the considered data, there was a difference between the quantity of the coal fed and the quantity of the coal bagged. The coal losses normally occur because improperly sewed, torn, or incorrectly loaded coal bags are sent back to the coal-washing plant.

The sets and parameters used in the efficiency analyses are defined below:

n—seven years of the coal bagging plant (2006–2012),

s—the number of outputs gained from the production (the quantity of bagged coal produced in the bagging units),

m—the number of inputs used in the production (the quantity of coal fed to the bagging units, the number of workers in the bagging units, the quantity of energy used in the bagging units),

$k \in \{1,2,....,7\}$—set of DMUs taken into consideration,

$j \in \{1,2,....,7\}$—set of all DMUs,

$r \in \{1\}$—set of all outputs,

$i \in \{1,2,3\}$—set of all inputs,

t—12 months in a year (January, February, ..., December).

The efficiency change of the bagging units between the years 2006 and 2012 was evaluated using the Malmquist TFP index. DEAP 2.1 written by Coelli (1996) was used in the analyses.

According to the productivity change within the seven-year period of the coal bagging unit in Table 2, there was a decrease in the mean Malmquist TFP (0.990 < 1). TC was effective on the scores of the inefficient ones among 2007, 2008, 2009, 2011, and 2012. In other words, there was a change in the efficiency frontier of the plant over the years. While there was a decrease in production due to the demand for coal, inefficiency occurred because there was no linear decrease in the quantity of the coal fed, the number of workers and the amount of energy consumed.

According to the 2012 Malmquist TFP index components, the sources of the inefficiency were both the overuse of the inputs taken into consideration and the inputs components not being in appropriate

Table 2. The yearly Malmquist TFP index components of the coal bagging unit

Years	Technical efficiency change (TE)	Technological change (TC)	Malmquist TFP change (MPI)
2006	1.001	1.092	1.093
2007	1.000	0.966	0.966
2008	1.000	0.969	0.969
2009	1.000	0.998	0.997
2010	1.000	1.000	1.000
2011	1.001	0.967	0.967
2012	0.989	0.956	0.945
Mean	0.999	0.992	0.990

scales and the efficient frontier were exposed to change in time (technological efficiency = 95.6% < 100%). While there was a decrease in the Malmquist TFP change from 2006 to 2010, the 100% score in the year 2010 indicates that there was no change in that year in comparison with the preceding year. On the other hand, the decrease in the Malmquist TFP change beginning from the year 2010 continued (96.7% in 2011 and 95.6% in 2012).

The reasons for the inefficiency in this year could be due to the following: wearing down of the bagging units in time, overconsumption of energy due to the continuous running of all equipment although the demand for coal was homogeneous in all months and the quantity of the coal that was fed was little, keeping the number of workers in this year constant, whereas the number of workers was determined depending on the production in the other years despite the homogeneity of demand; the greatest increase in the quantity of incorrectly bagged coal was in the year 2012, whereas there was a decrease in that quantity from 2006 to 2010.

There has been a constant decrease in the production of the coal bagging plants from the year 2006 to the present day. Approximately, 26% of the energy demand of Turkey is met by the domestic sources while the rest of this demand is met by various foreign sources. Consumption of the domestic lignite in residential heating is restricted in Turkey because it leads to air pollution due to its low calorific value and high sulfur content. The use of imported coal for heating emerged as an alternative in order to avoid the intense air pollution, especially in big cities toward the end of the 1970s. Natural gas has replaced coal in recent years with the increase of the natural gas network in Turkey. Other causes of the observed decline in the coal industry are that this sector requires continuous investment, long-term planning, search and preparatory work, efficient production, and rapid marketing.

In addition, the climate change in recent years has also affected the demand for coal. There has been a decrease in coal consumption due to relatively shorter and milder winter seasons.

Tunçbilek coal bagging plant was originally established as a single unit at the beginning. In 2008, two more identical units were added to Tunçbilek coal bagging plant. In 2007, all of the eight coal-filling reductions in the first unit in performed continuous production and 75% of the production was in the coal sales season. The energy consumed in the plant depends on production and, therefore, energy consumption increases in parallel to production. Because the three units of Tunçbilek coal bagging plant were simultaneously run in 2009, the amount of energy consumed was high in this year.

As in the case with the amount of energy, the number of workers varies depending on the production. For this reason, the loss of the qualified staff trained in bagging coal and new workers producing too much improperly bagged coal in the orientation process can be effective on inefficiency.

While there was a constant decrease in production, the lack of a decrease in the inputs at the same proportion and the increase in energy and labor costs day by day had a negative effect on the efficiency of GLC bagging plant.

Table 3 shows the efficiency changes that were analyzed based on the months within the seven-year period of the coal bagging unit. The study found that the obtained results were directly proportional to the demand developing according to the months. Since there is generally hardly any demand for coal in February, March, and April, the corresponding efficiency scores were around 70%. As the demand for coal generally increases gradually in May and June, the coal bagging plant starts to produce more bagged coal then. Production at full capacity generally starts in July and August. Because the religious holiday periods were in November in the years considered and there was no coal bagging work in these periods, there was hardly any efficiency change in November and December.

Table 3. Malmquist TFP monthly change of the coal bagging unit

Months	Technical efficiency change (TE)	Technological change (TC)	Malmquist TFP change (MPI)
January-February	0.933	0.832	0.777
February-March	0.860	0.815	0.701
March-April	0.683	1.121	0.766
April-May	1.100	1.097	1.206
May-June	1.374	1.136	1.561
June-July	1.028	0.951	0.977
July-August	0.892	1.094	0.976
August-September	1.184	0.977	1.157
September-October	1.091	1.039	1.134
October-November	0.946	1.078	1.020
November-December	1.060	0.844	0.895
Mean	0.999	0.993	0.990

There was a decrease in the efficiency levels in July, August, and December. However, there was an increase, albeit modest, in efficiency in May, June, September, October, and November. The highest increase was in June (56.1%).

The amount of overused energy, the number of extra workers, and losses in the bagging unit were found to be effective on inefficient months determined by the analysis results.

3.2. Sensitivity analysis

One of the advantages of non-parametric efficiency analysis is that it provides guidelines about what inefficient DMUs can do so as to become efficient. Using this feature, sensitivity analyses related to the efficiency analysis of the fully automated coal bagging plant were conducted for the

Table 4. The sensitivity analysis results of the current coal bagging unit

DMU	Efficiency value	Quantity of bagged coal produced (tons)	Quantity of coal fed (tons)		Number of workers (number/day)	Amount of energy (kwh)
	θ	%	Old	New	%	%
January	1.000	–	–	–	–	–
February	1.000	–	–	–	–	–
March	0.996	–	50,020	251,826	–	21
April	0.207	–	5,101	2,590,674	93	79
May	0.452	–	13,021	956,023	83	55
June	0.882	–	76,762	147,016	11	12
July	1.000	–	–	–		
August	0.798	–	73,982	109,123	20	20
September	0.973	–	110,736	114,194	3	3
October	1.000	–	–	–	–	–
November	1.000	–	–	–	–	–
December	1.000	–	–	–	–	–
Mean	0.859					

year 2012, which was found to be inefficient. Table 4 shows the non-parametric efficiency analysis results constant return to scale by taking the data for the 12 months of 2012 into consideration. As the efficient units ($\theta = 1.000$ and $s_{ik}^{+}, s_{rk}^{-} = 0$) use inputs in appropriate amounts and scales in order to reach the constant amount of production, Table 4 does not present any data about these units. Table 4 shows the % decrease values in the number of workers and amount of energy needed to make March, April, May, June, August, and September, which were determined to be inefficient, efficient. Because the % increase values of the amount of the coal fed were really high, the new feeding values were presented in the same table. The reason why the values in the input of the amounts of the coal fed increased was the fact that the negations of this input, which was considered in the analysis because reducing the inputs was a requirement to increase productivity, were obtained (1/the amount of the coal fed).

Since production is performed depending on the demand by the market, when the quantity of production is kept constant and the inputs considered are increased or decreased by the quantities given in Table 4, it would be possible for the coal bagging plant to be productive.

On the other hand, because the plant performed production depending on the demand by the market and the demand for coal has decreased over the recent years, the plant is required to be running even when there is no production, the plant cannot have enough qualified workers when necessary and developing technology has provided many advancements, it is recommended that GLC bagging plant should be upgraded to a fully automated bagging system.

4. Conclusion and recommendations

The aim of this study was to examine the change in productivity of the bagging plant of GLC between 2006 and 2012 using the Malmquist TFP index. The results indicated that there was a decrease in the productivity change according to the years examined (0.990 < 1). This decrease in productivity could be due to constant number of workers hired in the plant, constant amount of energy consumed in the plant, and losses in production despite the decrease in production as a result of the decrease in demand for coal. The analysis results based on the months within the seven-year period of the coal bagging unit revealed that there was a decrease in productivity in February, March, April, July, August, and December. However, there was an increase, albeit modest, in efficiency in May, June, September, October, and November. The highest increase was in June (56.1%). The study also found that these results were directly proportional to the demand increasing or decreasing depending on months.

The rapid growth of the share of natural gas in heating and energy in recent years has reduced demand for lignite. In addition, considering the formation and quality of coal, and economy of corporations, heating and industrial coal production have been affected adversely and, therefore, heating and industrial coal demand started to be met by coal import.

On the other hand, it seems likely for lignite to maintain its market share in the heating sector by increasing the number and quality of the plants to improve coal, preparing coal compatible with environmental protection criteria, and continuing to compete with imported coal by means of a pricing policy in market conditions. It also possible to increase the market share of lignite in the heating sector by developing smokeless boiler and stove technologies and revising legal regulations related to environmental protection in order to use low-calorie coal in the heating sector.

Considering the fact that each sector requires labor with specific qualifications, particularly regression in lignite mining has had adverse effects on qualified and experienced labor force needed by the sector. The GLC coal bagging plant has been producing bagged coal since the 2000s. Currently, the plant, which was designed according to the technology back then, requires too much labor and the plant consumes too much energy because it is required to be running even when there is no production. Also, there are inevitable production losses because the plant cannot have enough qualified workers when necessary as a result of hiring workers depending on the demand for coal.

An efficient staff policy and dealing with the problems of the staff efficiently would play a key role in increasing the productivity of the plant. However, it seems currently unlikely to develop an efficient staff policy as the coal bagging plant does not hire a constant number of workers, workers are hired depending on the need, and there is not sufficient qualified labor in general. The production capacity of the current GLC bagging plant is 6,000 ton/day. On the other hand, throughout the seven-year period (2006–2012) taken into consideration, the units have never operated at full capacity due to insufficient qualified labor force.

The study found that the maximum possible quantity of production would be 4,700 ton/day provided that the bagging units were run 24 h a day with all of their equipment. This value indicates that the plant operates approximately at 75% productivity. In fact, 28 conveyor belts, 3 vibrating screens, and 3 roll crushers are active even when there is no production. Their depreciation expenses should be taken into consideration as well as the energy they consume.

One of the advantages of non-parametric efficiency analysis, which the Malmquist TFP index is based on, is that it provides guidelines about what inefficient DMUs can do so as to become efficient. Using this feature, sensitivity analyses related to the efficiency analysis of the fully automated coal bagging plant were conducted for the year 2012, which was found to be inefficient. The analysis results showed that the months March, April, May, June, August, and September in the year 2012 were inefficient. This inefficiency was caused by the high number of workers hired and high amount of energy used in the plant despite the low amount of coal fed. Therefore, some recommendations were made so that the causes of inefficiency in 2012 can be eliminated and there is no inefficiency in the years following 2012.

Considering the fact that the existing operation processes should be automated professionally in order for corporations in the mining sector to operate in a modern and low-cost process with higher productivity and that their technology should meet the highest standards, it is recommended that a fully automated coal bagging plant be built in the corporation.

Acknowledgments

We would like to acknowledge the invaluable support of the Garp Lignite Corporation Directorate.

Funding

The authors received no direct funding for this research.

Author details

Mehmet Nuri Bahşiş[1]

E-mail: madenci21@mynet.com

Yaşar Kasap[2]

E-mail: yasaroz71@hotmail.com

[1] Mining Engineer, Garp Lignite Corporation, Kütahya, Turkey.

[2] Faculty of Engineering, Department of Mining Engineering, Dumlupınar University, Kütahya 43100, Turkey.

Note

1. In the assumption of constant return to scale, a radial increase in the input vector (an increase of all the input combinations by the same percentage) causes a same proportion of increase in the output vector. In other words, variations in production scale do not affect productivity (Charnes et al., 1978).

References

Charnes, A., Cooper, W., & Rhodes, E. (1978). Measuring the efficiency of decision making units. *European Journal of Operational Research, 2*, 429–444. http://dx.doi.org/10.1016/0377-2217(78)90138-8

Coelli, T. J. (1996). *A guide to DEAP version 2.1: A data envelopment analysis (computer) program* (Center for Efficiency and Productivity Analysis (CEPA) Working Paper 96/08). Armidale: University of New England.

Coelli, T. J., & Rao, D. S. P. (2005). Total factor productivity growth in agriculture: A Malmquist index analysis of 93 countries, 1980–2000 agricultural economics. *International Association of Agricultural Economists, 32*, 115–134, 01.

Coelli, T. J., Rao, D. S. P., & Battese, G. E. (1998). *An introduction to efficiency and productivity analysis, Kluwer, Boston.* http://books.google.com. http://dx.doi.org/10.1007/978-1-4615-5493-6

Färe, R., Grosskopf, S., & Lovell, C. A. K. (1994). *Production frontiers*. New York and Cambridge: Cambridge University Press.

Farrell, M. J. (1957). The measurement of productive efficiency. *Journal of the Royal Statistical Society. Series A (General), 120*, 253–290. http://dx.doi.org/10.2307/2343100

Grifell-Tatjé, E., & Lovell, C. A. K. (1995). A note on the Malmquist productivity index. *Economics Letters, 47*, 169–175. http://dx.doi.org/10.1016/0165-1765(94)00497-P

Kasap, Y. (2011). The Effect of work accidents on the efficiency of production in the coal sector. *South African Journal of Science, 107*, 77–85.

Mahadevan, R. (2002). A DEA approach to understanding the productivity growth of Malaysia's manufacturing industries. *Asia Pacific Journal of Management, 19*, 587–600. http://dx.doi.org/10.1023/A:1020577811369

Muñiz, M. A. (2002). Separating managerial inefficiency and external conditions in data envelopment analysis. *European Journal of Operational Research, 143*, 625–643. http://dx.doi.org/10.1016/S0377-2217(01)00344-7

6

Manufacturer-retailer inventory model for deteriorating items with price-sensitive credit-linked demand under two-level trade credit financing and profit sharing contract

Nita H. Shah[1]*

*Corresponding author: Nita H. Shah, Department of Mathematics, Gujarat University, Ahmedabad 380 009, Gujarat, India
E-mail: nitahshah@gmail.com

Reviewing editor: Zude Zhou, Wuhan University of Technology, China

Abstract: A supply chain of single manufacturer and single retailer is analyzed when units in warehouse are subject to deterioration at a constant rate. The demand is a decreasing function of selling price and increasing function of credit period offered by the retailer to the customers. Manufacturer follows a lot-for-lot production strategy. The manufacturer ensures delay payment to the retailer with agreement that the retailer will share a fraction of the profit realized during the credit period. The total joint profit of the supply chain is maximized with respect to replenishment time, selling price, and credit period. An algorithm is described to find the best strategy. Numerical examples are given to validate the proposed problem. Sensitivity analysis is carried out to examine important model parameters.

Subjects: Industrial Engineering & Manufacturing; Manufacturing Engineering; Production Engineering

Keywords: Single manufacturer and single retailer; deterioration; lot-for-lot production strategy; two-level trade credit financing; price-sensitive credit-linked demand; profit sharing contract

1. Introduction

The fact that the retailer must follow cash-on-delivery strategy is no more advocated in the business transactions. The manufacturer offers retailer a fixed credit period to boost the demand, and decrease his holding and deterioration cost. During this credit period, the retailers sell the items and

ABOUT THE AUTHOR

Nita H. Shah is a professor in the Department of Mathematics, Gujarat University, Ahmadabad, India. She holds a PhD in Inventory Control Management, Operations Research. Currently, she is engaged in research in inventory control and management, supply chain management, forecasting and information technology and information systems, neural networks, and sensors and image processing. She has more than 275 papers published in international and national journals. She is author of four books. She is serving as a member of the editorial board of *Investigation Operational*, *Journal of Social Science and Management*, *International Journal of Industrial Engineering and Computations* and *Mathematics Today*.

PUBLIC INTEREST STATEMENT

The proposed problem is interesting for the players working in a supply chain by sharing their interest, information of the product, investment, and also constraints. The decision is critical when manufacturer is using lot-for-lot production strategy to reduce the loss in terms of deterioration of the products. The manufacturer can push his demand by offering promotional tool of credit period to settle the account and also profit sharing. An algorithm is described to find the best strategy. The model can be used by business firm by deploying changes as per their firm policy.

earn interest on the generated revenue. The manufacturer charges interest on the unsold stock if the account is not settled during allowable credit period. This concept was first formulated by Goyal (1985). Refer to review articles by Chang, Teng, and Goyal (2008) and Shah, Soni, and Jaggi (2010) for available literatures on inventory modeling and trade credit. Giri and Maiti (2013) formulated model which considers that the retailer borrows loan from bank to settle the accounts and concluded that it is advantageous to do payment at the end of the allowable credit period. This offer entices to order more frequently which reduces manufacturer's inventory. On the other end, it increases holding cost, ordering cost, and deterioration cost if the retailer stocks more items which are deteriorating in nature. So the pay-off is needed between costs of two players.

Most of the citations in review articles are assuming that only supplier is offering a fixed credit period to the retailer. However, if the same intensive is extended to the customer by the retailer, the demand will escalate. Huang (2003) developed an EOQ model in which the supplier offers the retailer a credit period (upstream credit period), and retailer in turn provides a credit period (downstream credit period) to the customer. Huang (2006) formulated an inventory model under two-level trade credit with constrained floor space. Huang (2007) incorporated finite production capacity of the manufacture. Some more related articles are by Huang (2008), Jaggi, Goyal, and Goel (2008), Liao (2008), Shah, Soni, and Shah (2009), Teng, Chang, Chern, and Chan (2007), Teng and Goyal (2007), and their references.

Now-a-days, the researchers are analyzing integrated inventory models with trade credit financing as a promotional tool. Abad and Jaggi (2003) formulated a supplier–buyer inventory model under permissible delay in payments with price-sensitive demand. Sheen and Tsao (2007) discussed a vendor–buyer inventory model with trade credit and quantity discounts for freight cost. Sarmah, Acharya, and Goyal (2007) studied coordination and profit sharing between two players under permissible delay in payments. He and Huang (2013) developed inventory model for two players of a supply chain for non-instantaneous deterioration items where demand is a function of downstream credit-linked demand. Chung and Cárdenas-Barrón (2013) formulated simplified procedure when demand is stock-dependent and two-level trade credit is offered. Jiang, Skouri, Teng, and Ouyang (2014) gave a note on optimal replenishment policies for non-instantaneous deteriorating items with price and stock sensitive demand under permissible delay in payment. Aggarwal and Tyagi analyzed inventory and credit policies with date-terms credit-linked demand in a supply chain. Chung, Cárdenas-Barrón, and Ting (2014) developed an inventory model with non-instantaneous receipt and exponentially deteriorating items for an integrated three layer supply chain system under two levels of trade credit when demand is constant. Chen, Teng, and Skouri (2014) derived economic production quantity model for deteriorating items with up-stream full trade credit and down-stream partial trade credit when production is finite. Feng, Moon, and Ryu (2014) considered revenue-sharing contracts in a supply chain with reliability considerations. Jiang, Ouyang, Cardenas-Barron, and Goyal (2014) computed optimal credit period and purchase units for fixed life time deterioration of items under two-level trade credit scenario. Integrated vendor–buyer models are studied by Chen, Chen, Chiu, Choi, and Sethi (2010), Chen and Kang (2007), Chen and Kang (2010a, 2010b), Chiu, Choi, and Li (2009), Chiu, Choi, and Tang (2011), Ho, Ouyang, and Su (2008), Huang, Tsai, Wu, and Chung (2010), Jaber and Osman (2006), Kim, Hong, and Kim (2011), Ouyang, Ho, and Su (2008), Teng, Chang, and Chem (2012), Yen, Chung, and Chen (2012).

In this paper, we study the impact of two-level trade credit policy on a supply chain comprising of a single-manufacturer and single-retailer. The items in inventory system of both the players are subject to deterioration at a constant rate. The demand is considered to be increasing function of credit period which retailer offers to customer and decreasing function of selling price. It is assumed that the manufacturer offers a fixed credit period to the retailer with the contract that the retailer must share his profit during allowable credit period (Giri & Maiti, 2013). The manufacturer follows a lot-for-lot production policy. The joint profit of the supply chain is maximized with respect to replenishment time, selling price, and credit period to be offered to the customer by the retailer. Solution procedure is outlined to arrive at the best possible solution. Numerical examples are given to support the theoretical developments. The sensitivity of the optimal results with the model parameters is carried out to deduce managerial insights.

2. Notations and assumptions

The proposed problem is formulated with following notations and assumptions.

2.1. Notations

Inventory parameters for manufacturer

A_m	Setup cost per setup
h_m	Inventory holding cost per unit per annum
C_m	Purchase cost of an item per unit
P	Constant production rate
T_m	Time delay at which manufacturer starts production
M	Fixed credit period (in years) offered by the manufacturer to the retailer
I_{lm}	Loss of interest incurred due to offer of credit period
δ	Fraction of profit share by the manufacturer from the retailer during the credit period M; $0 < \delta < 1$
π_m	Profit per unit time of a manufacturer

Inventory parameters for retailer

A_r	Ordering cost per order
h_r	Inventory holding cost per unit per annum
C_r	Purchase cost of an item per unit
p	Selling price per unit (a decision variable)
I_b	Interest rate on the loan borrowed from the bank
I_e	Interest earned per unit per annum
τ	Fraction of customers allowed by the retailer to avail of credit period N; $0 \le \tau \le 1$
θ	Deterioration rate of items in warehouse; $0 \le \theta < 1$
$R(p, N)$	Retailer's demand rate (say); $R(p, N) = \alpha - \eta p + \beta N$; $\alpha > 0$ denotes scale demand, $\eta > 0$ denotes price elasticity markup and $\beta > 0$ denotes trade credit markup. (p and N are decision variables)
T	Replenishment time (a decision variable)
π_r	Profit per unit time of a retailer

Relation between inventory parameters

$$I_b > I_e, \quad N \le M$$

$$p .> C_r > C_m$$

$$T_m = \lambda T, \; 0 < \lambda < 1$$

Objective

Total joint profit per unit time of the supply chain

Maximize $\pi = \pi_m + \pi_r$
subject to $N, p, T \ge 0$

2.2. Assumptions

(1) A supply chain of single-manufacturer, single-retailer, and single-item is under consideration.

(2) The demand rate $R(p, N)$ for retailer is increasing function of credit period; N offered to customer and decreasing function of selling price; p. The function form of demand rate is $R(p, N) = \alpha - \eta p + \beta N$. $R(p, N)$ and R are interchangeable.

(3) The constant production rate $P\,(> R)$ to avoid shortages.

(4) Replenishment is instantaneous.

(5) The planning horizon is infinite to ensure long-term relationships.

(6) The manufacturer offers a delay period M to settle the accounts to the retailer with the contract that the retailer will share fraction of his profit incurred during the credit period M. The retailer takes a loan from a bank at the rate I_b to settle the accounts with manufacturer for the purchases made at the end of time M. The loan is repaid to the bank at the end of the cycle time.

(7) By offering credit period, the manufacturer incurs interest loss at the rate I_{lm} per unit per annum.

(8) The items in warehouse deteriorate at a constant rate, θ, $(0 \le \theta < 1)$. The deteriorated units can neither be repaired nor replaced during the cycle time.

3. Mathematical model

3.1. Model for retailer's inventory system

The retailer's inventory depletes due to credit-linked price-sensitive demand and deterioration of items in the warehouse. The rate of depletion of inventory at any instant of time $t \in [0, T]$ is given by the differential equation

$$\frac{dI_r(t)}{dt} = -R - \theta I_r(t), \quad 0 \le t \le T \text{ with } I_r(T) = 0.$$

The solution of differential equation is $I_r(t) = \frac{R}{\theta}(\exp(\theta(T-t)) - 1), \quad 0 \le t \le T$. The retailer's cycle starts with $Q = I_r(0) = \frac{R}{\theta}(\exp(\theta T) - 1)$ units in warehouse. The basic costs of the retailer are

- Purchase cost of procuring Q units per unit time is $PC_r = \frac{C_r R}{\theta T}(\exp(\theta T) - 1)$.
- Ordering cost per order is $OC_r = \frac{A_r}{T}$.
- Holding cost per unit time is $HC_r = \frac{h_r}{T}\int_0^T I_r(t)dt = \frac{h_r R}{\theta^2 T}(\exp(\theta T) - \theta T - 1)$.

Depending upon the lengths of credit periods M and N offered by the manufacturer and the retailer, respectively, and the cycle time T, following three cases are to be analyzed. Viz. (i) $N \le M \le T$, (ii) $N \le T \le M$, and (iii) $T \le N \le M$. Next, we detail out each case.

Case (i): $N \le M \le T$

Here, retailer generates revenue, $SR_r = \frac{p}{T}\int_0^T Rdt$. During [0, M], as per the contractual memorandum, the retailer gives δ% of profit to the manufacturer. So fraction of profit shared is, $FP_1 = \frac{\delta(p - C_r)RM}{T}$ and the remaining can be used to settle the account. To settle full account at M, the retailer takes a loan from the bank at the rate I_b per unit per annum and repays it at the end of cycle time T. So, the net interest charged by the bank per unit time is $ICB_r = \frac{I_b}{T}\left[C_r RT - pRM + \delta(p - Cr)MR\right](T - M)$ and interest earned by the retailer during the cycle time T per unit time is $IE_{r1} = \frac{pI_e}{T}\left[\int_0^M Rtdt + \int_0^{T-M} Rtdt\right]$. Due to offer of trade credit to the customer, the retailer incurs opportunity loss as $OL_{r1} = \frac{\tau p I_e}{T}\int_0^N Rtdt$. Hence, retailer's net profit, π_{r1} per unit time is

$$\pi_{r1}(N, p, T) = SR_r - PC_r - FP_1 - OC_r - HC_r - ICB_r - OL_{r1} + IE_{r1} \tag{1}$$

Case (ii): $N \le T \le M$

The profit to be shared is, $FP_2 = \delta(p - C_r)R$ and the interest earned by the retailer during the cycle time T per unit time is $IE_{r2} = \frac{PI_e}{T}\left[\int_0^T Rtdt + Q(M-T)\right]$. Due to offer of trade credit to the customer, the retailer incurs opportunity loss as $OL_{r2} = \frac{\tau pI_e}{T}\int_0^N Rtdt$. Here, the retailer has sufficient amount to pay for the purchases so retailer does not borrow any loan. Hence, retailer's net profit, π_{r2} per unit time is

$$\pi_{r2}(N,p,T) = SR_r - PC_r - FP_2 - OC_r - HC_r - OL_{r2} + IE_{r2} \tag{2}$$

Case (iii): $T \le N \le M$

The profit to be shared is, $FP_3 = \delta(p - C_r)R$ and the interest earned by the retailer during the cycle time T per unit time is $IE_{r3} = \frac{PI_e}{T}\left[\int_0^T Rtdt + Q(M-T)\right]$. Due to offer of trade credit to the customer, the retailer incurs opportunity loss as $OL_{r3} = \frac{\tau pI_e}{T}\left[\int_0^N Rtdt + Q(N-T)\right]$. Since, the cycle time ends before the credit period, the retailer does not have to pay interest charges and interest loss due to profit sharing with the manufacturer is zero. Hence, retailer's net profit, π_{r3} per unit time is

$$\pi_{r3}(N,p,T) = SR_r - PC_r - FP_3 - OC_r - HC_r - OL_{r3} + IE_{r3} \tag{3}$$

3.2. Model for manufacturer's inventory system

The rate of change of inventory in manufacturer's warehouse at any instant of time $t \in [0,T]$ is given by the differential equation $\frac{dI_m(t)}{dt} = P - \theta I_m(t), \quad t_m \le t \le T$ with $I_m(T) = 0$.

The solution of differential equation is $I_m(t) = \frac{P}{\theta}\left(1 - \exp(-\theta(t - t_m))\right), \quad 0 \le t \le T$. Therefore, holding cost per unit time is $HC_m = \frac{h_m}{T}\int_{T_m}^T I_m(t)dt$ and interest loss occurred per unit time due to allowable credit period M to the retailer is $OL_m = \frac{C_r PI_{lm}(T - T_m)M}{T}$. The setup cost per unit time is $OC_m = \frac{A_m}{T}$. Under contract, manufacturer receives δ% of profit earned by the retailer during the credit period. So, fraction of profit gained by the manufacturer per unit time is

$$FP_m = \begin{cases} FP_{m1} = \frac{\delta(p - C_r)RM}{T}, & M \le T \\ FP_{m2} = \delta(p - C_r)R & M < T \end{cases}$$

Hence, the manufacturer's net profit per unit time is

$$\pi_{m1}(T) = (C_r - C_m)R - OC_m - HC_m - OL_m + FP_{m1}, \quad M \le T \tag{4}$$

$$\pi_{m2}(T) = (C_r - C_m)R - OC_m - HC_m - OL_m + FP_{m2}, \quad M > T \tag{5}$$

3.3. Joint profit of the supply chain

The objective of this study is to maximize the profit of supply chain comprising of single-manufacturer and single-retailer. The joint profit of the integrated supply chain is:

$$\pi(N,p,T) = \begin{cases} \pi_1(N,p,T) = \pi_{r1}(N,p,T) + \pi_{m1}(T), & N \le M \le T \\ \pi_2(N,p,T) = \pi_{r2}(N,p,T) + \pi_{m2}(T), & N \le T \le M \\ \pi_3(N,p,T) = \pi_{r3}(N,p,T) + \pi_{m2}(T), & T \le N \le M \end{cases} \tag{6}$$

For given values of p and N, we have $\pi_1(M/p, N) = \pi_2(M/p, N)$ and $\pi_2(N/p, N) = \pi_3(N/p, N)$. Thus, π_i, $i = 1, 2, 3$ are well-defined functions of three decision variables. Next, we outline steps to find the best solution for the supply chain.

Algorithm

Step 1: Set numerical values for the inventory parameters.

Step 2: Set $\frac{\partial \pi_i}{\partial N} = 0$, $\frac{\partial \pi_i}{\partial p} = 0$ and $\frac{\partial \pi_i}{\partial T} = 0$, $i = 1, 2, 3$ and solve by Maple 14 simultaneously for N, p and T. Find out the appropriate scenario and for that obtained joint profit.

Step 3: Check sufficiency condition graphically.

4. Numerical examples

Example 1: Consider

$\alpha = 300$, $\beta = 5$, $\eta = 0.5$, $\delta = 0.40$, $\tau = 0.75$, $\lambda = 0.03$, $\theta = 0.15$, $A_r = 150$, $A_m = 300$, $h_r = 0.3$, $h_m = 0.5$, $C_r = 40$, $C_m = 10$, $p = 1000$, $I_e = 9\%$, $I_b = 15\%$, $I_{lm} = 18\%$, $M = 0.3$ in appropriate units. The maximum profit is $\pi_1 = \$44754.29$ for cycle time $T = 0.304$ years by giving credit period $N = 0.156$ years to the customer at retail price $p = \$292.13$ per unit. For obtained values:

$$\begin{vmatrix} \frac{\partial^2 \pi_1}{\partial T^2} & \frac{\partial^2 \pi_1}{\partial T \partial p} & \frac{\partial^2 \pi_1}{\partial T \partial N} \\ \frac{\partial^2 \pi_1}{\partial T \partial p} & \frac{\partial^2 \pi_1}{\partial p^2} & \frac{\partial^2 \pi_1}{\partial p \partial N} \\ \frac{\partial^2 \pi_1}{\partial T \partial N} & \frac{\partial^2 \pi_1}{\partial p \partial N} & \frac{\partial^2 \pi_1}{\partial N^2} \end{vmatrix} = -6.623616594 \times 10^8 < 0,$$ suggesting the concavity of the profit.

Figures 1–3 show concavity of the profit function with respect to two variables.

Example 2: Take $\beta = 3$ and other parametric values as given in Example 1. The scenario $N \le T \le M$ gives maximum profit as $\pi_2 = \$44825.81$ at $N = 0.071$, $p = 290.05$, $T = 0.229$. One can check the maximization of the profit for obtained values in a similar way as given in Example 1. Figures 4–6 show concavity of the profit function.

Example 3: Take $\beta = 10$ and $M = 0.4$ with other parametric values as given in Example 1 results profit to be $\pi_3 = \$44412.11$ at $N = 0.348$, $p = 293.66$, $T = 0.249$ from the scenario $T \le N \le M$. One can check the maximization of the profit for obtained values in a similar way as given in Example 1. Figures 7–9 show concavity of the profit function.

If the retailer does not offer credit period N to the customer, then the maximum profit is \$43758.87 with selling price \$247.83 and replenishment time 0.281 years using data as given in Example 1. The decrease in profit is due to decrease in demand. In Figure 10, we compare profit of the players and supply chain when sharing is considered and sharing is not considered.

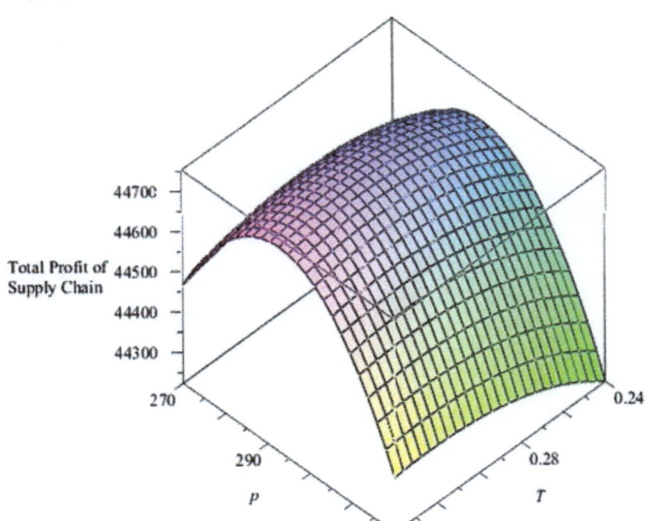

Figure 1. Concavity of total profit of the supply chain when $N = 0.156$ for $N \le M \le T$.

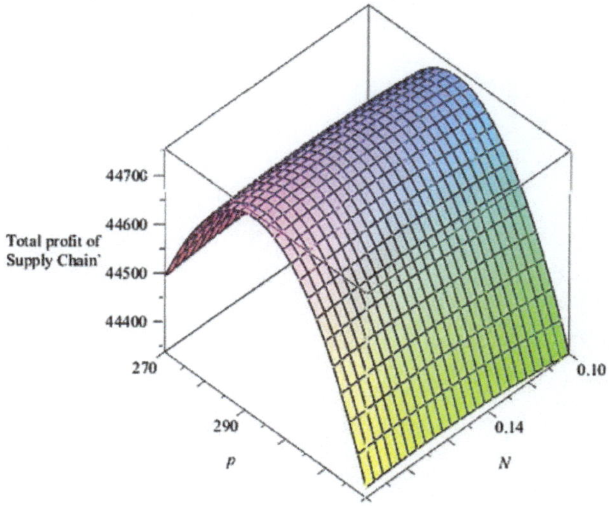

Figure 2. Concavity of total profit of the supply chain when $T = 0.304$ for $N \leq M \leq T$.

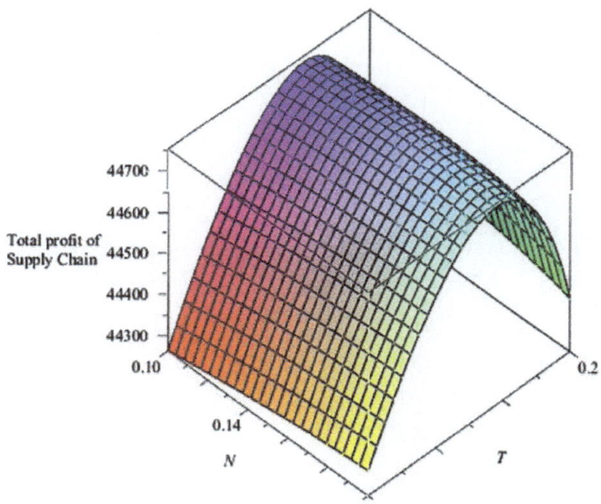

Figure 3. Concavity of total profit of the supply chain when $p = 292.13$ for $N \leq M \leq T$.

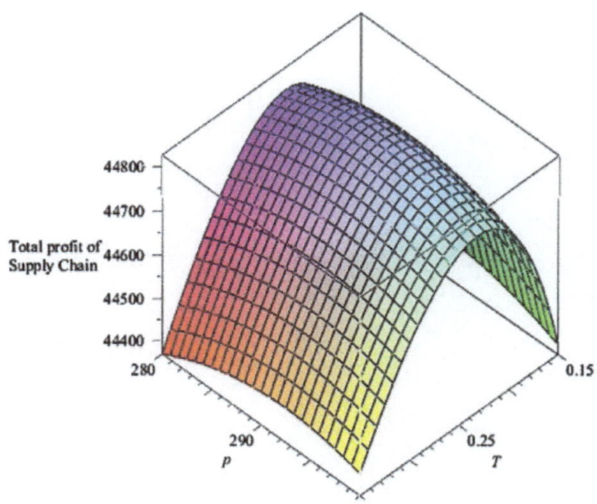

Figure 4. Concavity of total profit of the supply chain when $N = 0.071$ for $N \leq T \leq M$.

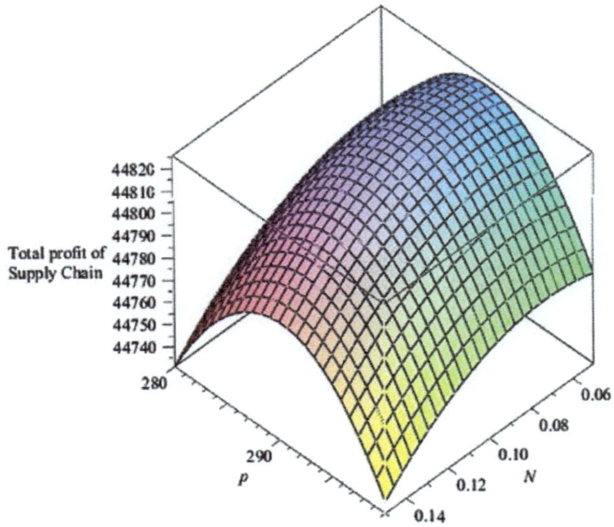

Figure 5. Concavity of total profit of the supply chain when $N = 0.229$ for $N \leq T \leq M$.

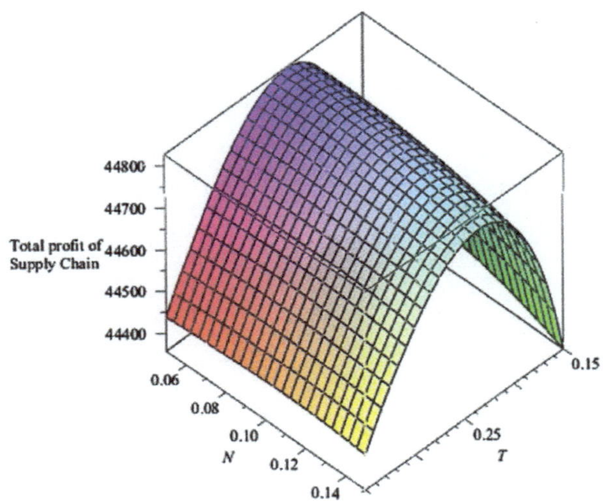

Figure 6. Concavity of total profit of the supply chain when $p = 290.05$ for $N \leq T \leq M$.

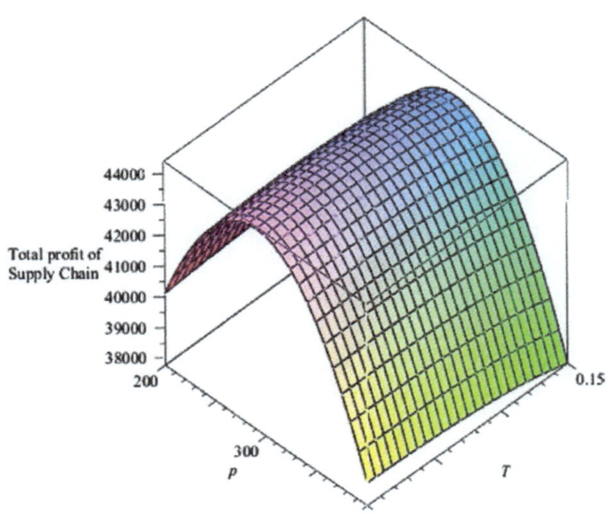

Figure 7. Concavity of total profit of the supply chain when $N = 0.348$ for $T \leq N \leq M$.

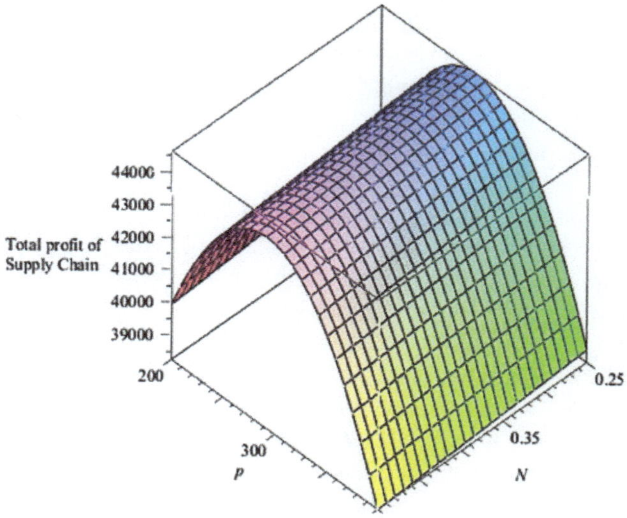

Figure 8. Concavity of total profit of the supply chain when $T = 0.249$ for $T \leq N \leq M$.

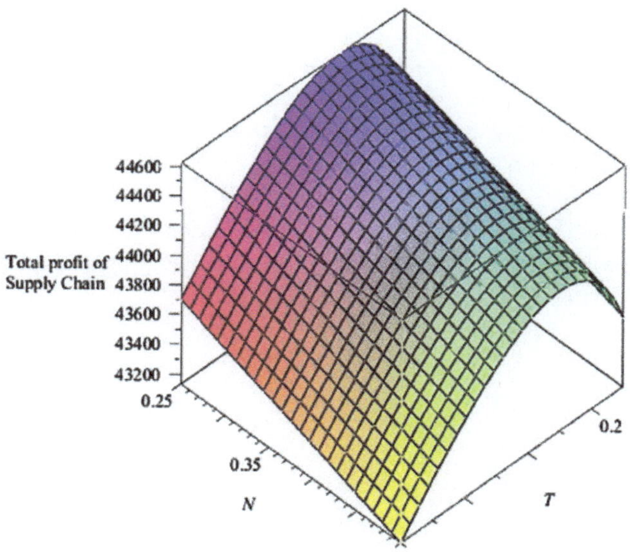

Figure 9. Concavity of total profit of the supply chain when $p = 293.66$ for $T \leq N \leq M$.

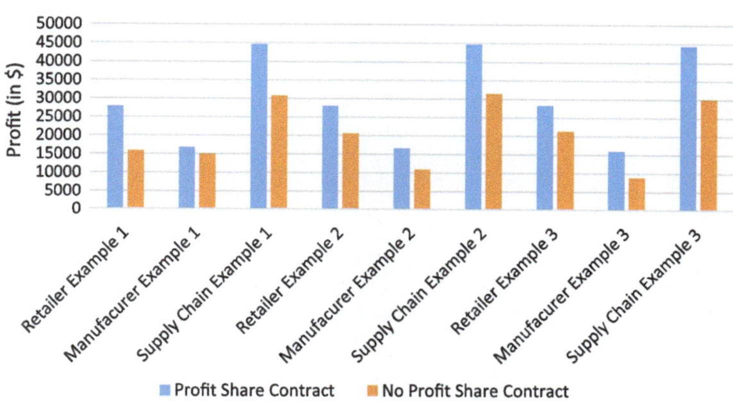

Figure 10. Profit of the players in different scenario.

It is observed from Figure 10 that the retailer as an individual player has more profit as compared to manufacturer. This suggests that the retailer will like to be dominant player of the business transactions. To increase manufacturer profit, the strategy of profit sharing will be advantageous to both the players which results the better profit of the supply chain.

5. Sensitivity analysis

In this section, we consider the data given in Example 1. Varying one parameter at a time by −20%, −10%, + 10%, + 20%, we study its effects on optimal solutions and total profit of the supply chain. The changes in the trade credit, *N* offered by the retailer to the customer, retail price, *p* set by the retailer, cycle time, *T* and joint profit of the whole system are studied.

From Figure 11 it is observed that increase in scale demand decreases credit period offered to the customer by the retailer significantly. Similar observations are for profit sharing fraction, rate of interest earned, and price elasticity markup. It also suggests that the retailer's purchase cost decreases the credit period. This supports that the customers will be attracted if they have to pay less as the purchase cost. The interest on the bank loan increases the credit period because this will facilitate the retailer to elongate the settlement of the payment.

Figure 12 depicts variations in the cycle time with respect to variations in inventory parameters. Increase in retailer's purchase cost decreases cycle time significantly. The scale demand, price elasticity markup, profit sharing fraction, and interest earned reduces the cycle time of the supply chain. The bank interest and setup cost incurred by the manufacturer increases the cycle time. This suggests that both the players should have proper sharing information to stabilize the trade-off between retailer's ordering quantity and manufacturer's setup cost. The retailer should try to reduce the loan amount for settlement of the accounts against the purchases.

The selling price is positively sensitive to scale demand and negatively sensitive to price markup and retailer's purchase cost as shown in Figure 13. Other inventory parameters have standard impact as discussed by cited articles in the references.

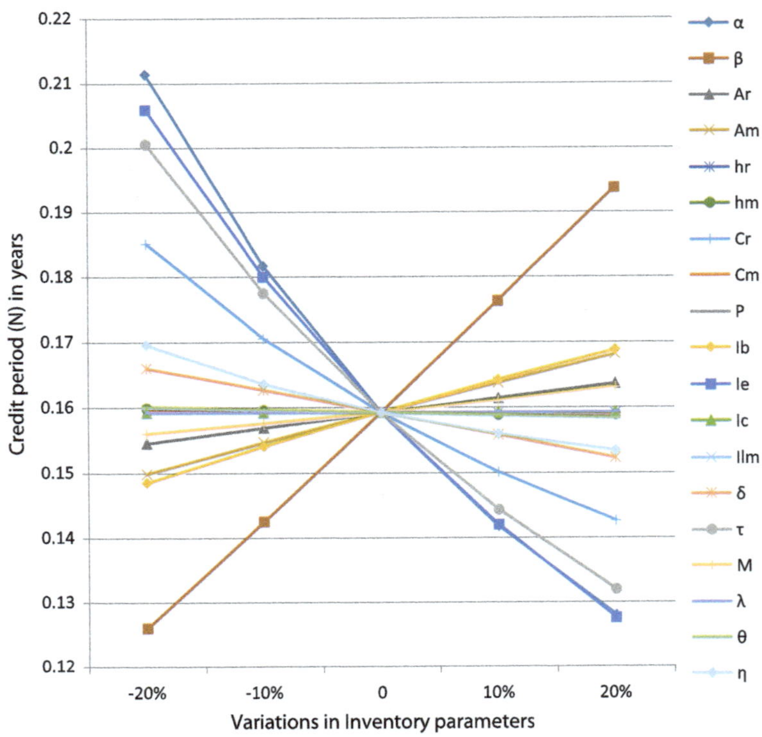

Figure 11. Changes in credit period *N* with variations in inventory parameters.

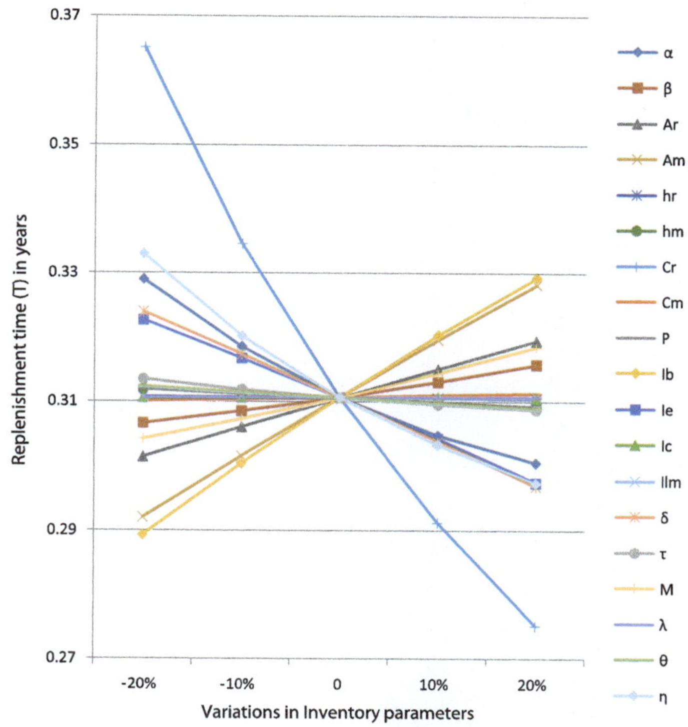

Figure 12. Changes in replenishment time with variations in inventory parameters.

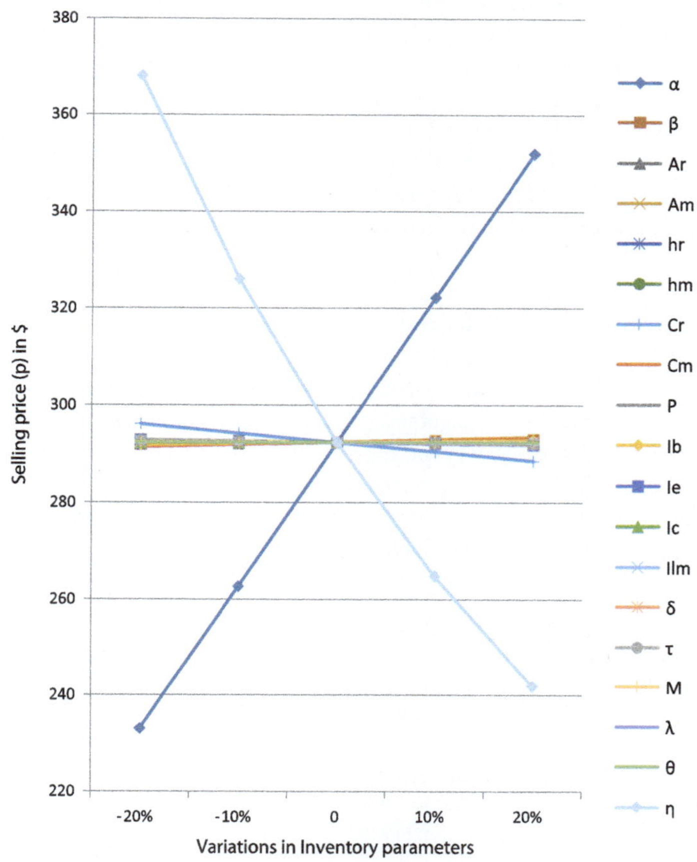

Figure 13. Changes in selling price with variations in inventory parameters.

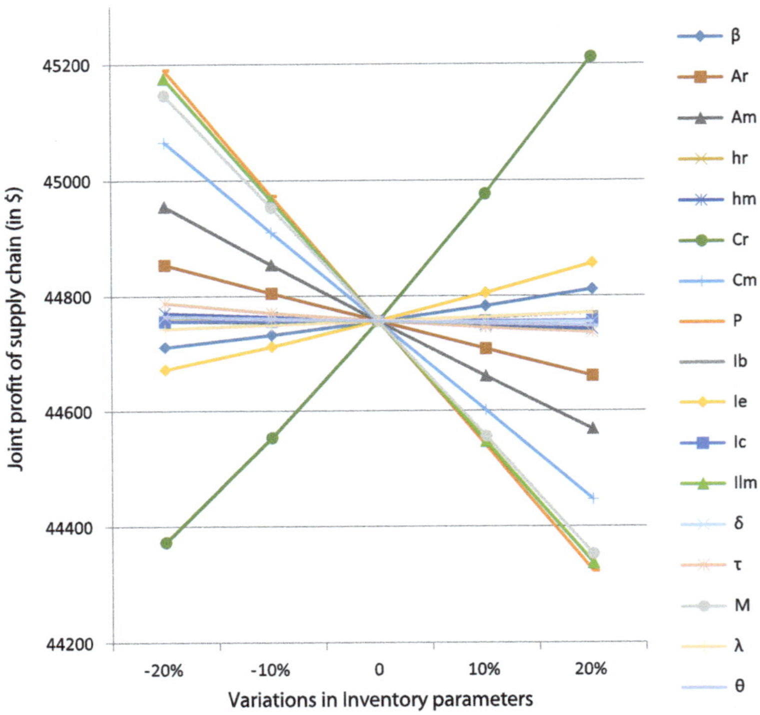

Figure 14. Changes in profit of supply chain with variations in inventory parameters.

Figure 14 is about profit variations with respect to inventory parameters. The retailer's purchase cost set by the manufacturer increases joint profit significantly. The credit period offered by the manufacturer to the retailer decreases the profit of the supply chain. The profit decreases with increase in the production rate. This suggests that the manufacturer and retailer should work on the production rate, credit period, and retailer's purchase cost to increase the profit of the supply chain. This strategy will also control the setup cost of manufacturer. The retailer should put large order to save in ordering cost and as a result profit of the supply chain can be increased significantly. It is observed that increase in customer eligible for delay payments results decrease in the profit of the supply chain.

6. Conclusions

This study deals with two-level trade credit financing for two players of the supply chain viz. manufacturer and the retailer. The market demand is dependent on retail price and credit period offered to the customer by the retailer. The study considers sharing of profit between the players during the credit period. The units in inventory system are subject to deterioration at a constant rate. Numerically, it is observed that a two-level trade credit strategy with profit sharing contract increases profit of both the players and also that of the supply chain. The proposed model can be extended to incorporate preservation technology investment to control the deterioration of items in the inventory system. The problem can be studied for stochastic demand which will explore real market practices.

Acknowledgments

Author likes to thank anonymous reviewers' suggestions for improvement of the readability of the manuscript. The department is under the scheme of DST-FIST.

Funding

The author received no direct funding for this research.

Author details

Nita H. Shah[1]

E-mail: nitahshah@gmail.com

[1] Department of Mathematics, Gujarat University, Ahmedabad 380 009, Gujarat, India.

References

Abad, P. L., & Jaggi, C. K. (2003). A joint approach for setting unit price and the length of the credit period for a seller when end demand is price sensitive. *International Journal of Production Economics, 83*, 115–122. http://dx.doi.org/10.1016/S0925-5273(02)00142-1

Chang, C. T., Teng, J. T., & Goyal, S. K. (2008). Inventory lot-size models under trade credits: A review. *Asia-Pacific Journal of Operational Research, 25*, 89–112. http://dx.doi.org/10.1142/S0217595908001651

Chen, H., Chen, Y. H., Chiu, C. H., Choi, T. M., & Sethi, S. (2010). Coordination mechanism for the supply chain with leadtime consideration and price-dependent demand. *European Journal of Operational Research, 203*, 70–80. http://dx.doi.org/10.1016/j.ejor.2009.07.002

Chen, L. H., & Kang, F. S. (2007). Integrated vendor–buyer cooperative inventory models with variant permissible delay in payments. *European Journal of Operational Research, 183*, 658–673. http://dx.doi.org/10.1016/j.ejor.2006.10.035

Chen, L. H., & Kang, F. S. (2010a). Integrated inventory models considering the two-level trade credit policy and a price-negotiation scheme. *European Journal of Operational Research, 205*, 47–58. http://dx.doi.org/10.1016/j.ejor.2009.11.028

Chen, L. H., & Kang, F. S. (2010b). Coordination between vendor and buyer considering trade credit and items of imperfect quality. *International Journal of Production Economics, 123*, 52–61. http://dx.doi.org/10.1016/j.ijpe.2009.06.043

Chen, S. C., Teng, J. T., & Skouri, K. (2014). Economic production quantity models for deteriorating items with up-stream full trade credit and down-stream partial trade credit. *International Journal of Production Economics, 155*, 302–309. http://dx.doi.org/10.1016/j.ijpe.2013.07.024

Chiu, C. H., Choi, T. M., & Li, D. (2009). Price wall or war: The pricing strategies for retailers. *IEEE Transactions on Systems, Man, and Cybernetics, Part A—Systems and Humans, 39*, 331–343. http://dx.doi.org/10.1109/TSMCA.2008.2010753

Chiu, C. H., Choi, T. M., & Tang, C. S. (2011). Price, rebate, and returns supply contracts for coordinating supply chains with price-dependent demands. *Production and Operations Management, 20*, 81–91. http://dx.doi.org/10.1111/poms.2010.20.issue-1

Chung, K. J., & Cárdenas-Barrón, L. E. (2013). The simplified solution procedure for deteriorating items under stock-dependent demand and two-level trade credit in the supply chain management. *Applied Mathematical Modelling, 37*, 4653–4660. http://dx.doi.org/10.1016/j.apm.2012.10.018

Chung, K. J., Cárdenas-Barrón, L. E., & Ting, P. S. (2014). An inventory model with non-instantaneous receipt and exponentially deteriorating items for an integrated three layer supply chain system under two levels of trade credit. *International Journal of Production Economics, 155*, 310–317. http://dx.doi.org/10.1016/j.ijpe.2013.12.033

Feng, X., Moon, I., & Ryu, K. (2014). Revenue-sharing contracts in an N-stage supply chain with reliability considerations. *International Journal of Production Economics, 147*, 20–29. http://dx.doi.org/10.1016/j.ijpe.2013.01.002

Giri, B. C., & Maiti, T. (2013). Supply chain model with price- and trade credit sensitive demand under two-level permissible delay in payments. *International Journal of Systems Science, 44*, 937–948. http://dx.doi.org/10.1080/00207721.2011.649367

Goyal, S. K. (1985). Economic order quantity under conditions of permissible delay in payments. *Journal of the Operational Research Society, 36*, 335–338. http://dx.doi.org/10.1057/jors.1985.56

He, Y., & Huang, H. (2013). Two-level credit financing for noninstantaneous deterioration items in a supply chain with downstream credit-linked demand. *Discrete Dynamics in Nature and Society*, 917958, 22. doi:10.1155/2013/917958

Ho, C. H., Ouyang, L. Y., & Su, C. H. (2008). Optimal pricing, shipment and payment policy for an integrated supplier-buyer inventory model with two-part trade credit. *European Journal of Operational Research, 187*, 496–510. http://dx.doi.org/10.1016/j.ejor.2007.04.015

Huang, Y. F. (2003). Optimal retailer's ordering policies in the EOQ model under trade credit financing. *Journal of the Operational Research Society, 54*, 1011–1015. http://dx.doi.org/10.1057/palgrave.jors.2601588

Huang, Y. F. (2006). An inventory model under two levels of trade credit and limited storage space derived without derivatives. *Applied Mathematical Modelling, 30*, 418–436. http://dx.doi.org/10.1016/j.apm.2005.05.009

Huang, Y. F. (2007). Optimal retailer's replenishment decisions in the EPQ model under two levels of trade credit policy. *European Journal of Operational Research, 176*, 1577–1591. http://dx.doi.org/10.1016/j.ejor.2005.10.035

Huang, Y. F. (2008). An EOQ model under retailer partial trade credit policy in supply chain. *International Journal of Production Economics, 112*, 655–664. http://dx.doi.org/10.1016/j.ijpe.2007.05.014

Huang, C. K., Tsai, D. M., Wu, J. C., & Chung, K. J. (2010). An integrated vendor-buyer inventory model with order-processing cost reduction and permissible delay in payments. *European Journal of Operational Research, 202*, 473–478. http://dx.doi.org/10.1016/j.ejor.2009.05.035

Jaber, M. Y., & Osman, I. H. (2006). Coordinating a two level supply chain with delay in payments and profit sharing. *Computers & Industrial Engineering, 50*, 385–400.

Jaggi, C. K., Goyal, S. K., & Goel, S. K. (2008). Retailer's optimal replenishment decisions with credit-linked demand under permissible delay in payments. *European Journal of Operational Research, 190*, 130–135. http://dx.doi.org/10.1016/j.ejor.2007.05.042

Jiang, W., Ouyang, L. Y., Cárdenas-Barrón, L. E., & Goyal, S. K. (2014). Optimal credit period and lot size for deteriorating items with expiration dates under two-level trade credit financing. *European Journal of Operational Research, 237*, 898–908.

Jiang, W., Skouri, K., Teng, J. T., & Ouyang, L. Y. (2014). A note on "optimal replenishment policies for non-instantaneous deteriorating items with price and stock sensitive demand under permissible delay in payment". *International Journal of Production Economics, 155*, 324–329.

Kim, J., Hong, Y., & Kim, T. (2011). Pricing and ordering policies for price-dependent demand in a supply chain of a single retailer and a single manufacturer. *International Journal of Systems Science, 42*, 81–89. http://dx.doi.org/10.1080/00207720903470122

Liao, J. J. (2008). An EOQ model with noninstantaneous receipt and exponentially deteriorating items under two-level trade credit. *International Journal of Production Economics, 113*, 852–861. http://dx.doi.org/10.1016/j.ijpe.2007.09.006

Ouyang, L. Y., Ho, C. H., & Su, C. H. (2008). Optimal strategy for an integrated system with variable production rate when the freight rate and trade credit are both linked to the order quantity. *International Journal of Production Economics, 115*, 151–162. http://dx.doi.org/10.1016/j.ijpe.2008.04.012

Sarmah, S. P., Acharya, D., & Goyal, S. K. (2007). Coordination and profit sharing between a manufacturer and a buyer with target profit under credit option. *European Journal of Operational Research, 182*, 1469–1478. http://dx.doi.org/10.1016/j.ejor.2006.09.047

Shah, N. H., Shukla, K. T., & Shah, B. J. (2009). Deteriorating inventory model for two-level credit linked demand under permissible delay in payment. *Australian Society of Operational Research Bulletin, 28*, 27–36.

Shah, N. H., Soni, H., & Jaggi, C. K. (2010). Inventory models and trade credit: Review. *Control & Cybernetics, 39*, 867–884.

Sheen, G. J., & Tsao, Y. C. (2007). Channel coordination, trade credit and quantity discounts for freight cost. *Transportation Research Part E: Logistics and Transportation Review, 43*, 112–128. http://dx.doi.org/10.1016/j.tre.2005.08.001

Teng, J. T., Chang, C. T., & Chem, M. S. (2012). Vendor-buyer inventory models with trade credit financing under both non-cooperative and integrated environments. *International Journal of Systems Science, 43*, 2050–2061. doi:10.1080/00207721.2011.564322

Teng, J. T., Chang, C. T., Chern, M. S., & Chan, Y. L. (2007). Retailer's optimal ordering policies with trade credit financing. *International Journal of Systems Science, 38*, 269–278. http://dx.doi.org/10.1080/00207720601158060

Teng, J. T., & Goyal, S. K. (2007). Optimal ordering policies for a retailer in a supply chain with up-stream and down-stream trade credits. *Journal of the Operational Research Society, 58*, 1252–1255. http://dx.doi.org/10.1057/palgrave.jors.2602404

Yen, G. F., Chung, K. J., & Chen, T. C. (2012). The optimal retailer's ordering policies with trade credit financing and limited storage capacity in the supply chain system. *International Journal of Systems Science, 43*, 2144–2159. doi:10.1080/00207721.2011.565133

7

Implementation of hybrid parallel kanban-CONWIP system: A case study

Joshua Prakash[1*] and Jeng Feng Chin[2]

*Corresponding author: Joshua Prakash, Faculty of Engineering and Green Technology, Universiti Tunku Abdul Rahman, Jalan Universiti, Bandar Baru Barat, Kampar, Perak 31900, Malaysia
E-mail: joshuaj@utar.edu.my

Reviewing editor: Zude Zhou, Wuhan University of Technology, China

Abstract: The most common form of production control strategy in lean management is the pull system. One emerging form of pull system uses kanban and CONWIP systems to handle products with different demand patterns. Case studies have protractedly depicted the actual implementation of pull systems; however, the use of hybrid systems is rare. This paper examines the procedures involved in implementing a hybrid system in a low variety/low volume shop floor. This paper presents discussions on shop floor constraints in the proposed system and how the simplicity of a pull system is able to reduce work-in-process inventory by 23%. Guidelines for the replication of the system for similar production environments are also provided. The case study proves that pull systems can be successfully implemented in production environments that do not conform to the typical prerequisites of the kanban system.

Subjects: Industrial Engineering & Manufacturing, Manufacturing Engineering, Production Engineering, Production Systems & Automation

Keywords: hybrid kanban, CONWIP, pull system implementation, pull system case study

1. Introduction

Many case studies have proven the feasibility of pull systems (Cao & Chen, 2005; Horbal, Kagan, & Koch, 2008; Mukhopadhyay & Shanker, 2005). One common element among these cases is their well-controlled process cycle times, which lead to low fluctuations. High variability in cycle times will complicate the determination of the number of kanban cards, lot size, and economic safety stock.

Employing a pull system with cycle time variation in an industry with a heavy reliance on manual tooling will be difficult. This particular environment is further characterized by stringent quality specifications,

ABOUT THE AUTHORS

Joshua Prakash is an assistant professor in the Faculty of Engineering and Green Technology, Universiti Tunku Abdul Rahman, Malaysia. He holds both BEng and PhD in manufacturing engineering obtained from Universiti Sains, Malaysia. The recent research projects include production control, manufacturing simulation, and lean manufacturing.

Jeng Feng Chin obtained his Doctor of Philosophy (PhD) from University of Birmingham, UK and currently is a senior lecturer in the School of Mechanical Engineering, Universiti Sains Malaysia. The research areas of interest include production planning and control, lean manufacturing and intelligent manufacturing system.

craft work that requires visual inspection, and non-generalized operating procedures. In such a case, backorders are intolerable and a minimum finished goods inventory level is imposed by a customer.

This paper attempts to provide insights and guidelines on the implementation of a hybrid system in a low variety/low volume environment. This paper is organized as follows. Section 2 reviews literature on pull systems. Section 3 provides the methodology of study. Section 4 discusses the company background. Sections 5 and 6 discuss the development of current-state value stream map and the steps executed, respectively. Section 7 discusses the problems encountered and solutions used. Section 8 concludes.

2. Literature review

A push system is a form of production control wherein a product is produced by a workstation that strictly follows a predetermined schedule regardless of immediate downstream needs. A push system is usually implemented to use the processing capacity of a factory (Burke & Wilks, 2006). This system allows individual workstations to exploit different operating capacities. Hence, the workload assignments in each workstation are based solely on capacity; this setup ignores the possibility of starvation and blocking. Push system also neglects the additional costs accrued by transportation, storage, and staffing in the long term (Womack & Jones, 1996).

By contrast, a pull system is a form of production control wherein a product is produced by a workstation according to its immediate downstream needs. In a pull system, waste elimination is the essence of production planning (Jacobsen, 2011) because any product delivered by a workstation regardless of immediate downstream needs is considered waste. However, this strategy is unfavorable in a production environment with significant setups (Spearman, Woodruff, & Hopp, 1990). In the following literature, several notable pull systems are discussed.

Toyota Motor Corporation pioneered the use of kanban in its production environment as one of the methods to control the use of limited resources. In kanban, cards are used to initiate production between workstations. When certain parts are required from a particular workstation, card(s) is/are passed to its immediate upstream workstation to fulfill that need (Liker, 2004). In several cases, the modification of the original kanban system serves production environments with specific needs. The adaptive kanban system of Tardif and Maaseidvaag (2001) is applied to production environments with varying part batch sizes. The push–pull system of Huang and Kusiak (1998) is applied to production environments wherein each production or assembly workstation operates on either a push or pull system.

Other forms of pull systems have also emerged. However, the capabilities of the modified kanban system are still insufficient to cope with the stringent requirements of a production environment. The CONWIP system by Spearman et al. (1990) was developed to include scrap loss or job orders with short production runs. In practice, when certain parts are required at the end of a line, card(s) is/are passed to the beginning of the line to fulfill that need. CONWIP system limits WIP inventories within a line, rather than between workstations, as in the case of the kanban system.

Integrating both push and pull system is also a viable option. A vertically integrated hybrid system has two levels, namely, an upper level push system and a lower level pull system (Cochran & Kim, 1998). A horizontally integrated hybrid system (HIHS) consist of one level only; a portion of production workstations operates on a pull system, whereas remaining workstations operate on a push system (Cochran & Kaylani, 2008). A parallel integrated hybrid system (PIHS) has one level; both push and pull systems operate in parallel with independent product types (Cheikhrouhou, Hachen, & Glardon, 2009). Several push and pull systems may exist in a push–pull system, whereas only one push system and one pull system exist in HIHS. Integrated systems require adopting a holistic view to ensure the individual systems can work well as a unit (Pham, Pham, & Thomas, 2008). Failure to address the constituent elements (e.g. point of integration between push and pull systems) can cause the entire system to function as a push system.

Literature on pull system modeling and simulation is rich and can be used to build and enhance theoretical understanding. However, literature on the implementation and comprehensive coverage of case studies are limited. Mukhopadhyay and Shanker (2005) describe the implementation of the original kanban system, which was introduced to a continuous production line of a tire manufacturing plant. The case study describes the intricacies involved in the detailed steps of using kanban, 5S, and SMED on the shop floor. Other cases that implemented the kanban system have also been described in literature (Araya, 2012; Gross & McInnis, 2003; Li, 2009).

Slomp, Bokhorst, and Germs (2009) describe the implementation of the CONWIP system. They applied the CONWIP system, together with the first-in-first-out (FIFO) loading rule, to a switchgear-manufacturing company. Their results show that this approach is feasible. The paper illustrated the design and phased-implementation of a system in a "built-to-ship" production environment. Ryan, Baynat, and Choodineh (2000), Huang, Wang, and Ip (1998) and Cao and Chen (2005) discuss other CONWIP implementation cases. Although many practical cases prove that pull systems are effective, no study has investigated the implementation of a modified kanban and hybrid system.

Another issue that has arisen is the implementation sequence of lean principles. Åhlström (1998) deduces the implementation principles on the basis of a case study over the span of two and a half years. In this study, lean principles are implemented in apparent sequences, but management efforts and resources need to be devoted for a set of principles in parallel. Li (2010) investigates the coordination among layout reconfiguration, quality improvement, and setup time reduction to adopt the CONWIP system in job shops. Simulations are used to coordinate these factors in a model obtained from literature and known manufacturing processes. The simulation results are used to derive guidelines for reforming traditional job shop practices and progressing through layout reconfiguration to realize the benefits of a CONWIP system. In certain cases, the implementation sequence of lean principles is required for a specific product type rather than a substantial fraction of the shop floor activity. The case study presented examines the implementation steps for a hybrid system. The hybrid system employed is an extension of PIHS, wherein two forms of pull systems, namely, kanban and CONWIP systems, operate in parallel for two independent product types within the same product family.

3. Methodology

An implementation methodology explores the practicality of a solution developed from simulation studies. In the absence of such a methodology, inaccuracies overlooked during simulation studies may result in a magnified flaw in future pull system research. Thus, this paper also aims to contribute a sound methodology that will benefit future simulation studies.

The method used in this study consists of the "plan-do-check-act" successive cycle. Such a method is used as it is a widely adopted process improvement plan and bears concurrency with principles of lean manufacturing (Rother, 2009). In "plan," value stream mapping (VSM) is a tool that assists in understanding the flow of material and information as a product travels through the shop floor (Rother & Shook, 1999). VSM is selected as part of the methodology because it visualizes and communicates the flow of *p* effectively. In "plan," one crucial task is identifying the root cause of the problems observed from the current state VSM. This identification also aids in determining the subsequent improvement steps as dictated in the "do" stage. In "do," the improvement steps as identified in the "plan" phase, is materialized. Within this phase itself, the improvements steps are carried out in a structured and sequential manner in order for the team members involved are able to maintain the system over the long run.

In "check," the performance of the new process is measured. Problems that arise are investigated and a set of corrective actions are developed and implemented. This requires a feedback to the "do" phase, where once implemented, the system is checked and refined until no short-term problems are encountered. This gives rise to the "act" phase, where the cycle repeats at a finer scope at each successive run until the system exhibits acceptable performance. Figure 1 depicts the flow of the methodology. Note that in Figure 1, the steps are presented in accordance with the improvement

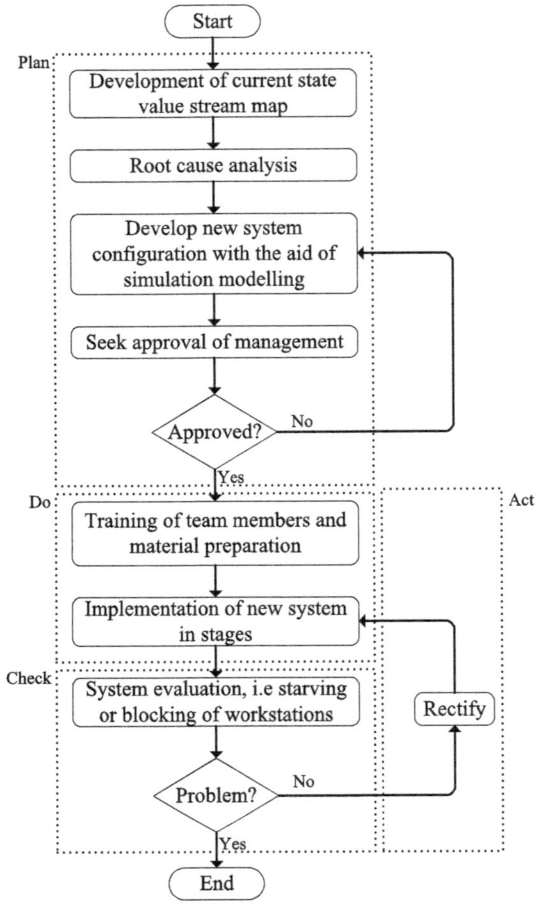

Figure 1. Flow of the methodology.

steps in the subsequent sections. This intends to give readers an entirety of the improvement process and feedback where necessary.

4. Company and project background

Company X is a supplier of composite subassemblies to a well-known commercial aircraft. Production began in early 2009 in a 242,000 square feet facility. The production operates 52 weeks a year, five days a week, and one shift a day, with an average of 8 h per shift (overtime for backlogs). To date, company X has 250 employees and produces 13 sets of subassembled products. The pilot project is led by a mechanical engineer who is also the lean engineer of the company. Regular team members consist of three research students from a local public university and five stage leaders from each production stage. Weekly meetings among the mechanical engineer, research students, and stage leaders are held to discuss the progress and milestones of the project.

To introduce the pull system to the shop floor, product *p* is selected as a pilot project. *p* is one of the subassembled composite panels located at the underside of the aircraft wing. Reasons for the selection of *p* as pilot include its relatively small size (28 in × 28 in/panel) compared with other product types, its relatively small pack size, a stable supply of raw composite panels. A pack size is one set containing two non-identical panels; one for the right-hand wing and one for the left-hand wing. *p* contains four product varieties, namely, p_1, p_2, p_3, and p_4, each with different demands.

5. Development of the current-state value stream map

Demand forecast is received one year in advance and revised two months before the scheduled shipment. The revised schedule will be sent to three production stages, namely, subassembly, paint shop, and final assembly. This schedule finalizes a weekly raw material delivery schedule to inbound sections. Finished goods are packed and stored in the outbound section. Raw materials are delivered from inbound sections to subassemblies by using a reorder point system. When a minimum of four sets are present in the sub-assembly, the inbound section begins replenishment by delivering additional six sets. Other items follow a similar process.

The processes in each panel in a set are identical. The difference lies within their work content. In a subassembly, the panels undergo five processes, namely, in-jig, out-jig, extended curing, out-jig wet, mild curing, and touch-up. These stages involve hole drilling, fitting insertion, sealant fixing, air curing, and inspection. Subsequently, the panels are sent to the paint shop, where the panels undergo five processes, such as part preparation 1, primer coating, part preparation 2, top coating, and touch-up. These stages involve hole filling, oven curing, masking, primer coating, oven curing, top coating, additional curing, demasking, touch-up, and inspection. Finally, the panels are sent to the final assembly, where the remaining components are assembled prior to final touch-up and inspection. The workstations in the sub and final assemblies are arranged in a dedicated work-cell layout. Shared facilities in the paint shop constitute a generalized machine layout. The difference among p_1, p_2, p_3, and p_4 lies in the paint shop, wherein each product variety receives different top coating and touch-up colors. Cycle times, WIP inventory between processes, number of operators involved, and flow of information are presented in the current-state VSM of p in Figure 2.

From Figure 2, the occurrence of high WIP inventory between processes is traced to the scheduling system used. The assignment of daily production targets to successive processes initiates production to proceed regardless of downstream needs, similar to general push systems. This effect is magnified in the immediate upstream buffer of the bottleneck workstation. In-jig continues to process at its usual pace with the intention of avoiding possible starvation at out-jig because of the substantially long curing time (24 h) even when demand is low. The WIP inventories eventually accumulate between in-jig and out-jig because of the poor estimation of downstream needs. The batch production in the primer coating and top coating also waits for the right quantity before processing resumes.

The start and end of each production stage is recorded on a job card by the operator. Each set is assigned a job card. In each job card, the start and end phases recorded upon the completion of processing are not used for future planning. A more experienced operator will be able to complete a task at a shorter time than his/her peers. Disregarding this information in future planning can lead to substantial waiting time for the WIP inventory before being moved to subsequent processes. Break times are set as a pacemaker, whereby a given quantity of operations are to be completed before each break period because of the uncertainty in the start and end phases. The presence of these wastes leads to the low ratio of value-added time to non-value-added time (≈.15).

The performance target established by team members is a reduction in flow time by a minimum of four days while maintaining 100% service level. The boundaries for flow time reduction ranges from fresh sets of panels delivered from inbound sections to subassembly up to completed sets of panels in the final assembly for delivery to the outbound section. Continuous flow is not favored between production stages because of relatively large distances, which diminish its effectiveness. This limitation calls for a buffer (supermarket) system to be placed between production stages. Additionally, working on one set at a time is unfavorable because material is shared between sets. In the sub and final assemblies, adhesives and lacquer are shared between two sets and paint is shared among four sets in the paint shop. Given that material cost savings for one set at a time is not justified, the current method of producing two sets simultaneously is maintained. Consequently, no addition or removal of operators is involved in all processes.

6. Phased implementation

The four phases for the implementation of hybrid parallel kanban-CONWIP system were chosen such that each phase would show certain performance advantages as well as help workers to

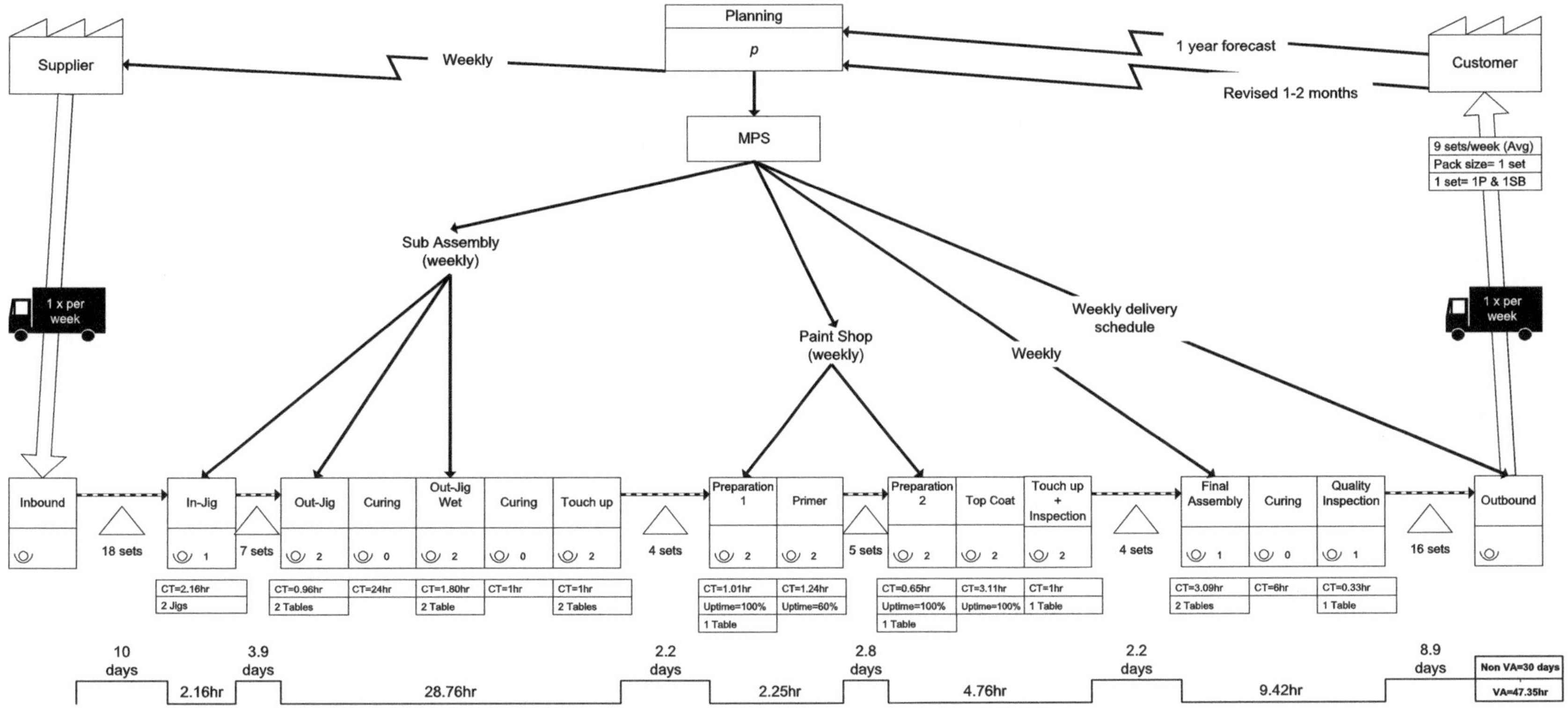

All stations working time = working time (8-5.30) - break time (10-10.30) – lunch (average of 1hr per day)=8.1 hrs

Figure 2. Current state VSM of *p*.

gradually learn the features of the system. We named the four phases: 1—system definition phase, 2—team member training and material preparation phase, 3—implementation of kanban phase, and 4—implementation of CONWIP phase.

6.1. The system definition phase

Demand of p_1 constitutes 90% of the total demand, whereas p_2, p_3, and p_4 comprise 10% of total demand. Hence, p_1 (high runner or *HR*) follows a kanban system. The takt time of p_2, p_3, and p_4 is 40 h/set. The inventory of p_2, p_3, and p_4 are not maintained because the demand does not justify stocking. With approximately equal quantities of p_2, p_3, and p_4 produced in any given week (i.e. p_2, p_3, or p_4 each has one order), p_2, p_3, and p_4 (low runner or *LR*) follow the CONWIP system.

Figure 3 depicts the flow of *HR* and *LR* in the parallel kanban-CONWIP system according to the examples obtained from Krieg (2005). When one set of *HR* in B_4 is withdrawn, the kanban card attached to *HR* is detached and placed at the final assembly kanban board. When a kanban card is added at this kanban board, one set of *HR* is pulled from B_3 and begins the final assembly process for quality inspection. When the processing is completed, *HR* is pushed to B_4 and the kanban card at the final assembly kanban board is attached to *HR*. When one set of *HR* in B_3 is withdrawn, the kanban card attached to *HR* is detached and placed at the paint shop kanban board. When a kanban card is added at this kanban board, one set of *HR* is pulled from B_2 and begins process preparation 1 to primer. When processing is completed, processing continues with preparation 2 to inspection. When processing is completed, *HR* is pushed to B_3, and the kanban card at the paint shop kanban board is attached to *HR*. When one set of *HR* in B_2 is withdrawn, the kanban card attached to *HR* is detached and placed at the subassembly kanban board. When a kanban card is added at this kanban board, one set of *HR* is pulled from B_1 and begins process in-jig. When the processing is completed, processing continues with out-jig for touch-up. When processing is completed, *HR* is pushed to B_2, and the kanban card at the subassembly kanban board is attached to *HR*.

A CONWIP card is placed at the paint shop kanban board by the planning department when an order for *LR* is placed. When a CONWIP card is added at this kanban board, one set of *HR* is pulled from B_3 and begins process preparation 1 to primer. When processing is completed, processing continues with preparation 2 to inspection. When the processing is completed, *HR* is pushed to B_3 and a CONWIP card at paint shop kanban board is attached to *HR*. By following this process, the final assembly to quality inspection is performed. When processing is completed, *HR* is pushed to B_4. Upon shipment, the CONWIP card is returned to the planning department. The remaining upstream operations resemble the *HR* sequence.

6.2. Training of team members and material preparation phase

A computer simulation model of the future system is developed to provide insights on the functionality of the proposed system and to discuss several performance measure of interest. A visual simulation enables the viewing of material flow that is difficult to imagine. WITNESS Simulation 2007 possesses the ability to display material flow between workstations. Input order quantity and supermarket sizes vary and their effects on system performance measures, such as flow time and total WIP inventory, are analyzed. The results reveal that flow time/set remains almost constant and WIP is maintained (as long as the number of kanban and CONWIP cards is fixed) with input order quantity varying within the forecasted demand range.

To gain acceptance from the operations manager and product development engineer of *p*, an animation model is extracted from the simulation model. The animation simplifies the visual aspect of the simulation model from a layman's perspective. Additionally, a training course on the benefits of the pull system and the required contributions of operators to the system is conducted. A manual simulation activity for operators enables a hands-on experience on the flow of cards and their effects on running individual processes. In manual simulation, a layout of the shop floor and respective locations of the kanban boards are projected onto a large whiteboard. Markers that represent each set are positioned according to the actual quantity and position for a given day. Cards are

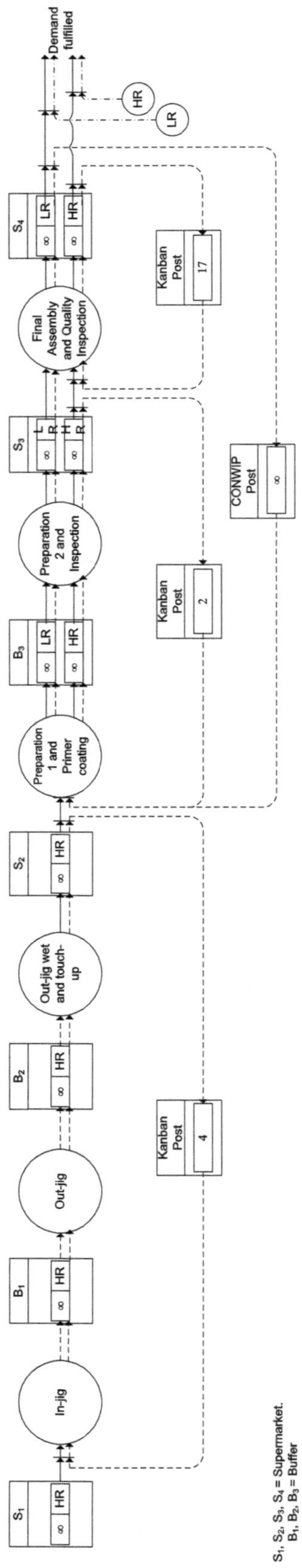

Figure 3. Parallel kanban-CONWIP system of *p*.

represented by a different colored marker. The operators simulate the movement of the markers in a real time situation. Cases such as cards being transferred late and authorized processing without a card are also simulated to demonstrate their implications. Corrective actions and contingency plans are developed and imparted to the operators.

The attachment of cards to the WIP inventory is vital. The three options for card attachments are on the panels, insert as part of the job card, and on material transfer. Attachments of cards on the panels are unfavorable because it imposes on the processing operations. Attachments of cards to the job card are also not favorable because the cards are less visible and can easily be displaced from the job card. Attachments of cards to the material transfer (trolley) are considered the most favorable because when panels are transferred between operations, the material transfer set is used to ensure that the cards are constantly visible throughout the processing cycle.

The cards are 8 cm × 20 cm laminated card boards. Each card represents one set. The card is designed to contain crucial information, such as card number, product type, and card circulation region. Kanban cards are presented in blue, whereas CONWIP cards are in pink. Kanban cards between workstations are recognized with the colored strip used. Prior to issuance of CONWIP card to the paint shop, the planning department writes on a strip of masking tape stating the product type required and fixes the tape on the CONWIP card. A kanban board is placed near the point of exit at each production stage to ensure that upon supermarket replenishment, cards are placed in the allocated compartment. Card compartments that contain kanban and CONWIP cards are placed on the material transfer. Card compartment materials are also carefully selected to withstand high temperatures (maximum of 60°C) in the oven. Standard operating procedures (SOPs) depicting the transfer of cards and its relation to each operation are placed at each production stage. This procedure enables operators to understand the sequences of transfer, particularly if a new operator is involved in the card transfer. Figure 4(a) shows an example of the kanban card design. Figure 4(b) shows the kanban card with its card compartment in place. Figure 4(c) shows the placement of a kanban board in the paint shop. Figure 5 shows the subassembly SOP.

6.3. Implementation of the kanban phase

The planning department is required to send the *HR* weekly production plan to the outbound section. The daily demand of other production stages is triggered by individual immediate downstream production stages. The implementation is purposely planned on a low demand season to ensure the allocation of sufficient time for production readjustment during unforeseen events. Each production stage SOP is displayed close to the kanban board to assist operators in conditions for transferring kanban cards.

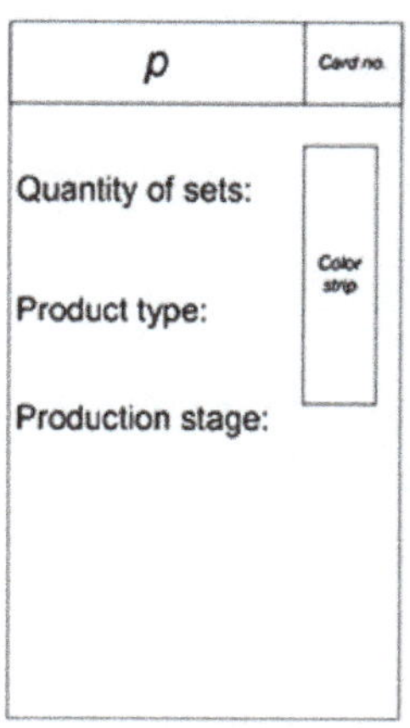

(a)

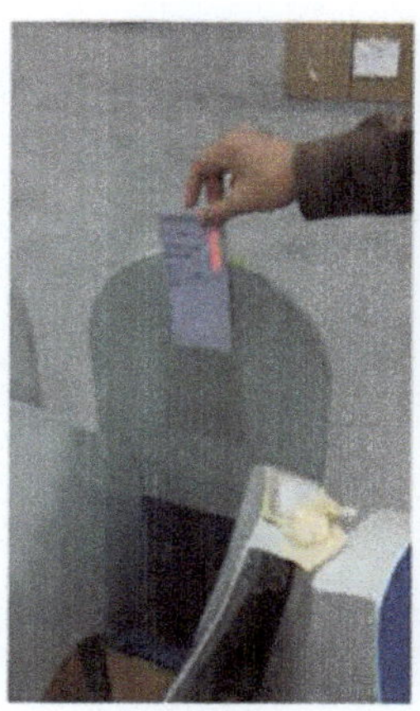

(b)

(c)

Figure 4. (a) Kanban card design; (b) Kanban card with card compartment in place; and (c) Paint shop kanban board.

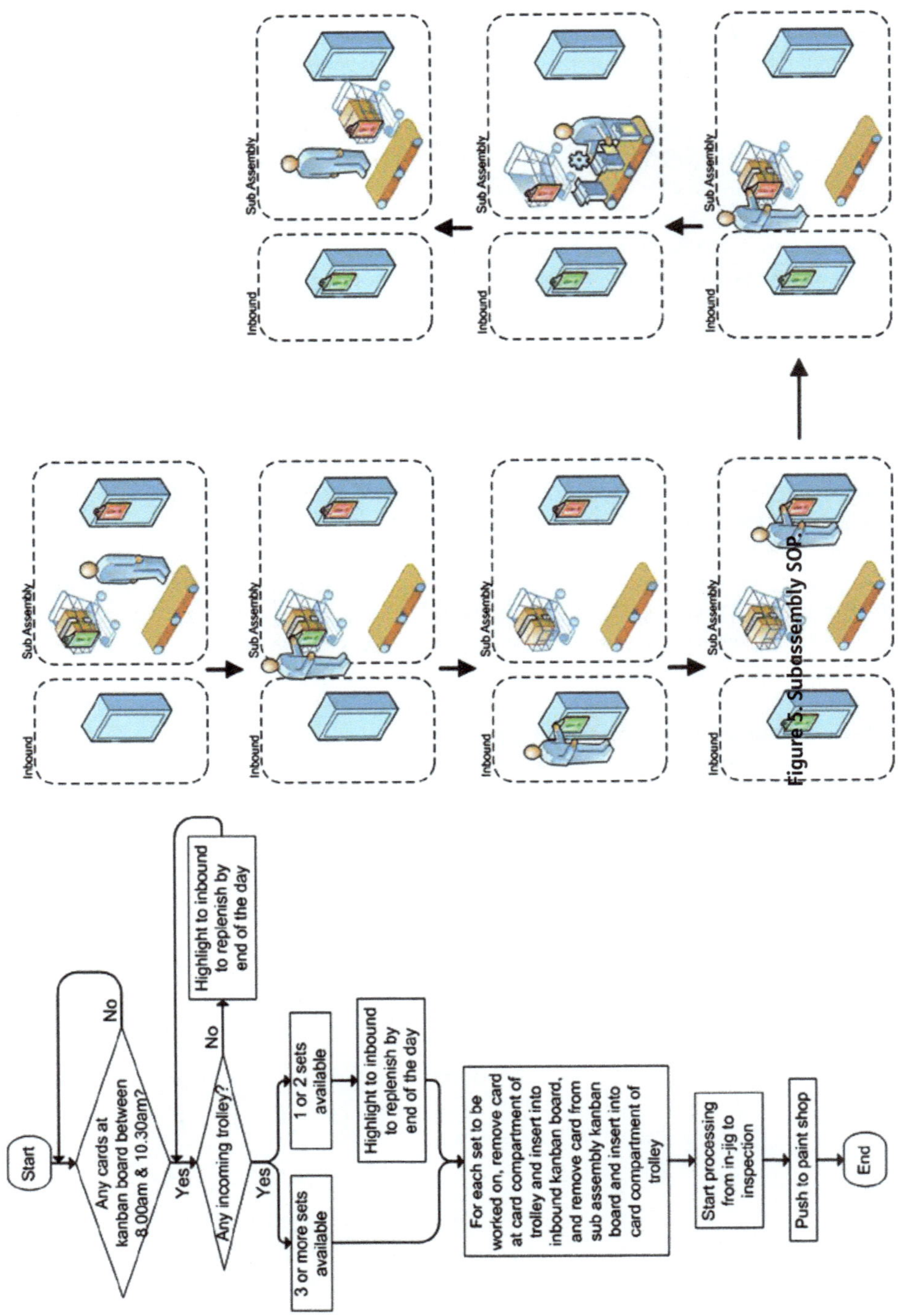

Figure 5. Subassembly SOP.

Each material transfer is attached with a kanban card and placed in the card compartment. The attachment of kanban cards is also performed in stages beginning from the outbound section to the inbound section. The purpose of implementing the transfer in stages is to account for manageable tasks, instead of crowding tasks in one step. A step-by-step set of tasks, goals, and milestones are created with these increment efforts.

Once the system has operated for two weeks, free cards that are constantly not in use will be removed. The presence of these cards indicates processing beyond capacity. The quantities of cards are gradually reduced to the final supermarket sizes. First, the quantity of cards at one particular supermarket is fixed at one value. Within a week, if the downstream process does not show any significant starvation with a lower WIP turnover, one card will be removed from the system in the next shift. During this transition, sets without cards are processed first in any supermarket to allow such sets to move downstream quickly and to minimize the potential interference to the kanban system. For constant starvation, root-cause investigation and later appropriate immediate improvement are proposed. If starvation remains, a card is added. This addition occurs in the paint shop to match the WIP to the batch quantity required. Table 1 summarizes the resizing of kanban cards at each production stage throughout the implementation period. Such an approach is reactive and ensures the stability to the production at the shop floor.

6.4. *Implementation of the CONWIP phase*

The planning department is required to send the CONWIP cards to the paint shop according to the weekly plan. Upon *LR* shipment, the outbound section returns the CONWIP cards to the planning department. One issue to be addressed is the loading rules in the paint shop and final assembly. *HR* and *LR* are present in both stages. FIFO is adopted in the final assembly because the required kitting is common for both *HR* and *LR*. In the paint shop, *HL* rule (priority to *HR* over *LR*) is adopted because of the difference in the paint used. *HR* uses a more common paint and is processed first; thus, *HR* can be completed in batches with other available product families.

7. System evaluation and improvement

Kanban cards are remained at the kanban board as long as starvation is experienced. Although the effect of starvation is magnified in such a case, starvation perpetuates the importance of maintaining a minimum quantity of sets at each workstation. This process also ensures that the kanban system implemented does not deplete over time.

Feedbacks from stage leaders and operators are subsequently collected. A number of discrepancies between the proposed and the implemented systems are observed and require rectification. The paint shop frequently experiences starvation. This delay in receiving parts leads to delays in processing and impedes the transfer of cards to upstream production stages, thereby perpetuating starvation to upstream workstations. One solution proposed involves sending cards from the paint shop to the subassembly on the basis of the anticipated work to be completed instead of the actual work to be completed. In practice, when the paint shop receives two sets from subassembly, it will instantly deliver the two cards to the subassembly while waiting for additional two sets before processing can begin. Within two weeks of implementing this system, the paint shop shows a gradual reduction in starvation.

Table 1. Reduction/increment of kanban cards

Week	Subassembly	Paint shop	Final assembly	Total
1	10	6	6	22
2	9	7	5	21
3	8	8	4	20
4	7	9	4	19
5	6	8	4	16

Another problem encountered is that operators are often not prompt in sending cards to upstream production stages. Operators usually conduct transfers at their convenience, such as during breaks. To solve this problem, a second session of manual simulation is performed with emphasis on the effect of card-circulating rule violations. Through this simulation, the operators relented and recommitted themselves to the pull system.

Finally, the material used for the card compartments was not durable when exposed to constant temperature changes. Consequently, the card compartments were easily detached from the material handling after a period of time. A cloth-based card compartment was adopted. In addition, spare card compartments were prepared for instant replacement of damaged cards. Table 2 summarizes the improvement achieved between the current and future state systems, while Figure 6 shows the future state VSM, referring to the newly improved system.

8. Conclusion

This paper presents the implementation sequences of a hybrid parallel kanban-CONWIP system in an environment where manual labor is intensive and back order is intolerable. The first step involved is the plotting of the current-state VSM, with the need to identify the underlying causes of problems arising on the shop floor. Once identified, several improvement plans are developed. The need to consult relevant people related to the improvement plan is a necessity because the improvement plan filters out plans that are unfeasible. Once consensus is reached on a plan, the preparation leading up to the implementation is executed. Preparation consumes a substantial amount of time to ensure that minimal hiccups are encountered upon the implementation of the final system. Among the steps involved in the preparation include the preparation of materials and the training of operators. Thereafter, the system is implemented in stages. Feedback from operators is required to detect and solve unforeseen problems. Product *p* has shown a decrease in flow time from 30 to 23 days, an improvement of 7 days (23%). One observation from the implementation steps is that several circumstances are overlooked during the planning phase. Decisions such as the actions required if the threshold is exceeded are difficult to forecast.

With such positive results, the management has encouraged the application of the pull system of *p* to the inbound section. The reduction of the number of fresh goods inventory of *p* is desired. One method proposed is the progressive reduction of fresh goods inventory. This method is preferred compared to the integration of pull systems with master production schedules because of the complexity of the mathematical formulation involved. In this progressive reduction, five sets of *p* are reserved for emergency use. The remaining sets are consumed according to demand. In the event that non-reserved sets are insufficient, the reserved sets are consumed. Older reserved sets are occasionally replaced with incoming sets to avoid the quality deterioration of fresh goods of *p*. The inbound section performance proceeds are monitored for one month. If no changes in reserved quantity are detected, the quantity is reduced by one and the cycle is repeated. The cycle stops when the threshold is reached, and the reserved sets are consumed. The simplicity of a pull system design has similar benefits to that of a complex pull system design, provided that a reliable team supports the system and planning is thorough.

Table 2. Lead time improvement of *p*

Period	Location					
	Between inbound and in-jig	Between in-jig and out-jig	Between touch up and preparation 1	Between primer and preparation 2	Between touch up and final assembly	Between quality inspection and outbound
Before implementation	10 days	3.9 days	2.2 days	2.8 days	2.2 days	8.9 days
After implementation	5.6 days	1.1 days	2.2 days	2.2 days	2.2 days	9.4 days

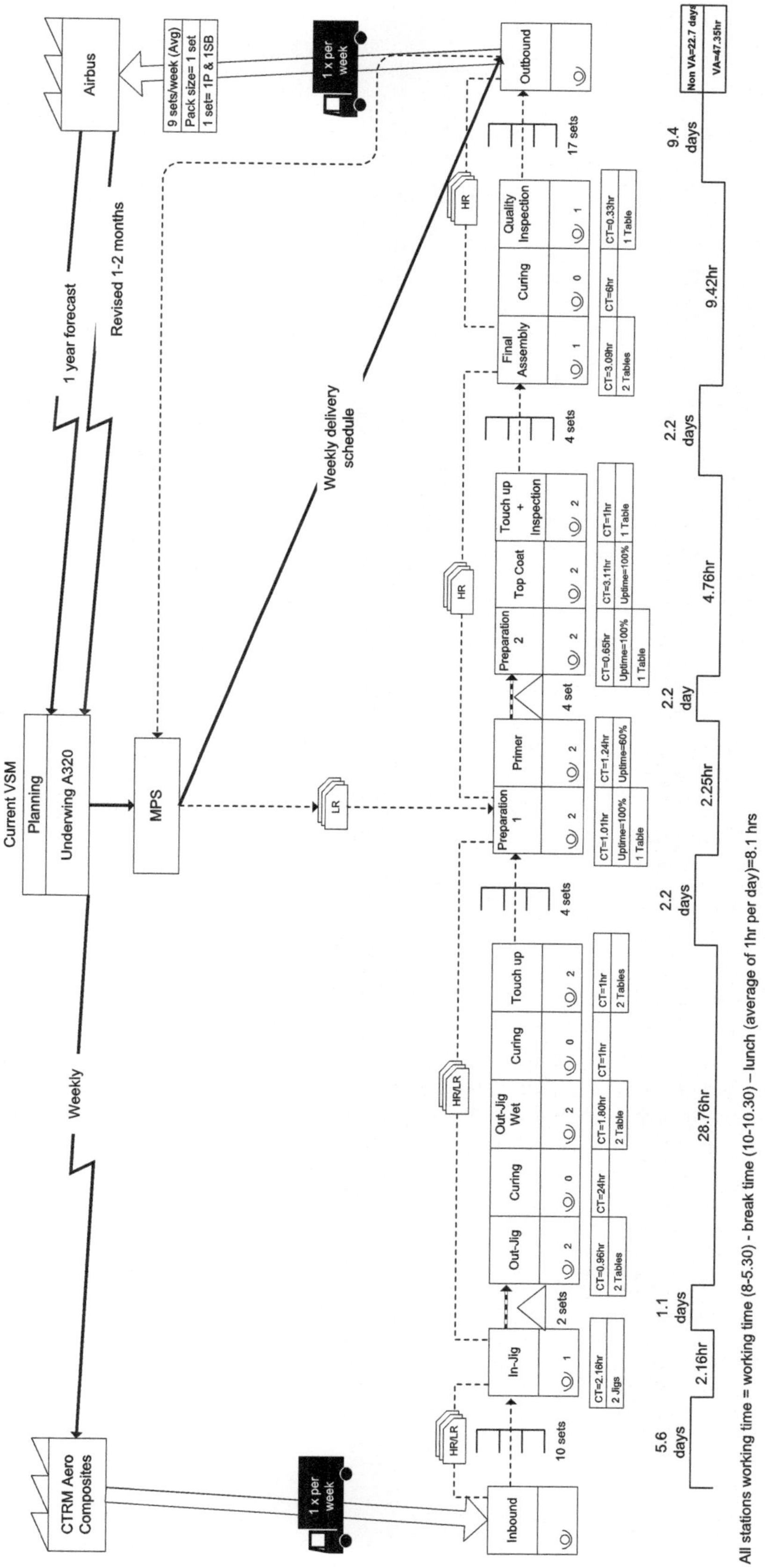

Figure 6. Future state VSM of *p*.

Funding
The authors wish to acknowledge Universiti Sains Malaysia (USM) for the full financial support through the short-term [grant number 60311042] and APEX [grant number 910345].

Author details
Joshua Prakash[1]
E-mail: joshuaj@utar.edu.my
Jeng Feng Chin[2]
E-mail: chinjengfeng@usm.my
[1] Faculty of Engineering and Green Technology, Universiti Tunku Abdul Rahman, Jalan Universiti, Bandar Baru Barat, Kampar, Perak 31900, Malaysia.
[2] School of Mechanical Engineering, Universiti Sains Malaysia (USM), Engineering Campus, 14300 Nibong Tebal, Seberang Perai Selatan, Penang, Malaysia.

References

Åhlström, P. (1998). Sequences in the implementation of lean production. *European Management Journal, 16*, 327–334. http://dx.doi.org/10.1016/S0263-2373(98)00009-7

Araya, J. M. (2012). *Value stream mapping adapted to high-mix, low-volume manufacturing environments* (PhD thesis), KTH Industrial Engineering and Management, Stockholm, Sweden.

Burke, L., & Wilks, C. (2006). *CIMA managerial level. Management accounting decision management.* Burlington, MA: Elsevier.

Cao, D., & Chen, M. (2005). A mixed integer programming model for a two line CONWIP-based production and assembly system. *International Journal of Production Economics, 95*, 317–326. http://dx.doi.org/10.1016/j.ijpe.2004.01.002

Cheikhrouhou, N., Hachen, C., & Glardon, R. (2009). A Markovian model for the hybrid manufacturing planning and control method "double speed single production line". *Computers & Industrial Engineering, 57*, 1022–1032. http://dx.doi.org/10.1016/j.cie.2009.04.013

Cochran, J. K., & Kaylani, H. A. (2008). Optimal design of a hybrid push/pull serial manufacturing system with multiple part types. *International Journal of Production Research, 46*, 949–965. http://dx.doi.org/10.1080/00207540600905547

Cochran, J. K., & Kim, S. S. (1998). Optimum junction point location and inventory levels in serial hybrid push/pull production systems. *International Journal of Production Research, 36*, 1141–1155. http://dx.doi.org/10.1080/002075498193561

Gross, J. M., & McInnis, K. R. (2003). *Kanban made simple: Demystifying and applying Toyota's legendary manufacturing process.* New York, NY: Amacon.

Horbal, R., Kagan, R., & Koch, T. (2008). Implementing lean manufacturing in high-mix production environment. In *Lean business systems and beyond* (pp. 257–267). Wroclaw: Springer.

Huang, C. C., & Kusiak, A. (1998). Manufacturing control with a push-pull approach. *International Journal of Production Research, 36*, 251–276. http://dx.doi.org/10.1080/002075498194038

Huang, M., Wang, D., & Ip, W. H. (1998). Simulation study of CONWIP for a cold rolling plant. *International Journal of Production Economics, 54*, 257–266. http://dx.doi.org/10.1016/S0925-5273(97)00152-7

Jacobsen, J. (2011). *Sustainable business and industry: Designing and operating for social and environmental responsibility.* Milwaukee, WI: ASQ Quality Press.

Krieg, G. N. (2005). *Kanban-controlled manufacturing systems.* Lecture notes in economics and mathematical systems. New York, NY: Springer-Verlag, Berlin Heidelberg.

Li, B. (2009). *Making a high-mix make-to-order production system lean* (PhD thesis), Massachusetts Institute of Technology, Cambridge, MA.

Li, J. W. (2010). Simulation study of coordinating layout change and quality improvement for adapting job shop manufacturing to CONWIP control. *International Journal of Production Research, 48*, 879–900. http://dx.doi.org/10.1080/00207540802478814

Liker, J. K. (2004). *The Toyota way: 14 management principles from the world's greatest manufacturer.* New York, NY: McGraw-Hill.

Mukhopadhyay, S. K., & Shanker, S. (2005). Kanban implementation at a tyre manufacturing plant: A case study. *Production Planning & Control, 16*, 488–499. http://dx.doi.org/10.1080/09537280500121778

Pham, D. T., Pham, P. T. N., & Thomas, A. (2008). Integrated production machines and systems—Beyond lean manufacturing. *Journal of Manufacturing Technology Management, 19*, 695–711. http://dx.doi.org/10.1108/17410380810888094

Rother, M. (2009). *Toyota Kata: Managing people for improvement, adaptiveness and superior results.* Cambridge, MA: McGraw-Hill.

Rother, M., & Shook, J. (1999). *Learning to see: Value stream mapping to add value and eliminate muda.* Cambridge, MA: Lean Enterprise Institute.

Ryan, M., Baynat, B., & Choodineh, F. (2000). Determining inventory levels in a CONWIP controlled job shop. *Institute of Industrial Engineers Transactions, 32*, 105–114.

Slomp, J., Bokhorst, J. A. C., & Germs, R. (2009). A lean production control system for high-variety/low-volume environments: A case study implementation. *Production Planning & Control, 20*, 586–595. http://dx.doi.org/10.1080/09537280903086164

Spearman, M. L., Woodruff, D. L., & Hopp, W. J. (1990). CONWIP: A pull alternative to kanban. *International Journal of Production Research, 28*, 879–894. http://dx.doi.org/10.1080/00207549008942761

Tardif, V., & Maaseidvaag, L. (2001). An adaptive approach to controlling kanban systems. *European Journal of Operational Research, 132*, 411–424. http://dx.doi.org/10.1016/S0377-2217(00)00119-3

Womack, J. P., & Jones, D. T. (1996). *Lean thinking: Banish waste and create wealth in your corporation.* New York, NY: Simon & Schuster.

8

Evaluation of factors important to enhance productivity

Sanjay Kumar[1]*, Manish Duhan[1] and Abid Haleem[2]

*Corresponding author: Sanjay Kumar, Department of Mechanical Engineering, International Institute of Technology and Management, Sonepat, Haryana, India
E-mail: skbhardwaj19711971@gmail. com

Reviewing editor: Zude Zhou, Wuhan University of Technology, China

Abstract: Productivity as a measure for output is important to industry and academia. In this research, factors to enhance productivity have been identified from the literature by reviewing various international and national sources to explore this evergreen field of research that is "productivity," which has always been an increasingly interesting area of research for researchers over decades or perhaps over centuries. In total, 15 number of factors have been identified to enhance productivity. Analytic hierarchy process approach has been appropriately chosen to rank these factors because of its simplicity and effectiveness. The tool has been used by taking perception of experts from the Indian manufacturing industry. Positive attitude and involvement of management, proactive employees, and good working conditions have been ranked as top three factors as per the experts' opinion. The ranking of factors to enhance productivity, categorization of the factors into four perspectives, and hierarchy of perspective and action plan as a final outcome of the paper may help academia and operations managers toward effective management of "operations and production activities of firms/supply chains."

Subjects: Economics; Finance; Business & Industry; Engineering & Technology; Environmental Studies & Management

Keywords: analytic hierarchy process (AHP); factors to enhance productivity; hierarchy of perspectives; manufacturing sector of India; productivity action plan; productivity perspectives; ranking of factors

ABOUT THE AUTHORS

Sanjay has completed his research under the supervision of Abid Haleem in the area of Supply Chain. Presently, he is the principal of IITM Murthal, Haryana, India. His pursuit for research is visible by his joint publication of more than 55 papers with Abid Haleem in the last 10 years, using methodologies like systems modeling and empirical research in different areas of industrial engineering.

Manish is a bright student who has contributed in getting the literature, survey, and other support activities.

Haleem, is with Jamia Millia Islamia, a central University of Indian Government, has headed Mechanical and Management departments. He has supervised more than 18 researchers and more than 170 publications to his credit in core and upcoming areas of industrial engineering, technology management, and operations research.

This paper is a result of more than a year's hard work of the three authors, who have also taken extensive support from learned experts.

PUBLIC INTEREST STATEMENT

Improving productivity is important to most of the stakeholders of the organization. This paper tries to address issues where organizations can learn to improve productivity. They can learn through this paper what are the important factors helping in improving productivity.

Analytic hierarchy process approach has been appropriately chosen to rank these factors. Here, perception of experts from the Indian manufacturing industry has been taken. Positive attitude and involvement of management, proactive employees, and good working conditions have been ranked as top three factors as per the experts' opinion. This paper may help academia and operations managers toward effective management of operations and production activities of organizations/supply chains.

1. Introduction

The wealth of nations and our living standards is being determined by productivity development (Brynjolfsson & Hitt, 1998). It has been found that the development of productivity has a positive impact on poverty reduction and it has been done by two ways. First, salaries and earnings have been increased by increasing the productivity which results in poverty reduction and second, the rate of increase in prices has been controlled by productivity profits (Bernard & Jensen, 2004). Further, increase in profits and sales has been observed with increase in productivity (Zoltners & Lorimer, 2000). An organization gets high profit margin when the input cost is lower and the lower input cost obtained when the productivity is high. Customer has the central role to play and customer requirements need to be determined by a team made by the high productive firms (Blackburn, Scudder, & Van Wassenhove, 1996). To achieve financial and other objectives, for example, brand image; vendor integration; goodwill; and customer satisfaction, the management of operations is very important (Kumar, Luthra, Haleem, & Garg, 2014).

2. Defining productivity

Productivity has been defined in numerous ways; we have attempted in this section to compile a few of the available information from different websites as presented in Table 1 (with important key words highlighted in "bold font").

It may be summarized from important key words, definitions, and concepts provided in Table 1 that: productivity is a ratio/measure/efficiency to evaluate outputs in terms of "extent of value addi tion"/"performance"/"economic gain" with respect to inputs in terms of resources pertaining to individual/sub-system/system/factory/firm/supply chain/sector/country/continent.

Table 1. Productivity concept and definitions from various websites

Concept	Computation/Implications	References
Average measure of **efficiency** of **production**	Ratio of output and inputs used in the production process	Retrieved February 12, 2015, from http://en.wikipedia.org/wiki/Productivity
Measure of **efficiency** of **person**, **machine**, **factory**, and/or **system** in converting inputs into useful outputs; Productivity may be a key determinant of cost efficiency	Productivity may be evaluated by dividing average output per period by the total costs incurred	Retrieved February 12, 2015, from http://www.businessdictionary.com/definition/productivity.html
Economic measure of output per unit of input; Inputs include labor and capital, there as output may be measured in revenues and other **GDP components**	Productivity measures may be examined collectively (across whole economy) or evaluated industry by industry to examine trends in wage levels, labor growth, and improvements in technologies	Retrieved February 12, 2015, from http://www.investopedia.com/terms/p/productivity.asp
Productivity is the effective use of **innovation and resources** to increase the **value-added content** of products/services	Productivity is the true source of competitive advantage that creates long-term economic viability and a better standard of living	Retrieved February 12, 2015, from http://www.hkpc.org/index.php?option=com_content&view=article&id=106&Itemid=270&lang=en
Productivity is often utilized at workplace; It may be described as **performance** of **individual worker/department/industry**	It can be evaluated in percentage; Sometimes, productivity is used in relation to farming	Retrieved February 12, 2015, from http://www.vocabulary.com/dictionary/productivity
Productivity is **ratio** of outputs and inputs	It may be referred to output volume from given inputs volume; If organization becomes more productive, then it may be implied as that organization is known more efficient	Retrieved February 12, 2015, from http://www.dineshbakshi.com/as-a-level-business-studies/operations-management/104-revision-notes/1409-meaning-of-productivity
Rate at which products/services are **offered / produced**	It can be referred as goal accomplishment measure	Retrieved February 12, 2015, from http://www.merriam-webster.com/dictionary/productivity

3. Identification of factors to enhance productivity

Fifteen factors have been identified after review of appropriate literature. Review of literature was carried out by searching keywords such as: factors to enhance productivity, productivity enablers, critical success factors of productivity, factors responsible for productivity increase, etc. Search was accomplished utilizing various primary and secondary sources. Information and communication technology played a very important role toward exploring the factors for enhancing productivity. The factors and frequency of occurrence in the literature as outcomes of this information exploring activity have been tabulated in Table 2.

4. Analytic hierarchy process (AHP) methodology application

Professor A.L. Saaty (the American operations researcher) introduced the AHP in 1970. Using this method, firstly, the complex problems decompose into some factors and then make a simple

Table 2. Identification of factors to enhance productivity from literature

S. No.	Factors to enhance Productivity	References
	Proactive employees	Campbell (2000), Cooper and Cartwright (1994), Erdogan and Bauer (2005), Burton, Conti, Chen, Schultz, and Edington (1999)
	Accountability	Tausky and Chelte (1983), Mamiseishvili and Rosser (2011), Cohen et al. (2012)
	Follow-up	Eriksson and Lindroos (2014), De Putter et al. (2012)
4.	Manage the work force but avoid micromanagement	Martin (2005), Head (2007), Wright (2000), Broedling et al. (1980), White (2010)
5.	Encourage, motivate, reward, and recognize	Bartol and Srivastava (2002), Fairbank and Williams (2001), Manzoor (2012), Chang and Mills (2013), Karia and Hasmi Abu Hassan Asaari (2006), Dike (2005)
6.	Reach out to employees by seeking them out	Eccles, Perkins, and Serafeim (2012), Gill, Koettl, and Packard (2013),
7.	Demand realistic targets	Kirk (1990), Caballero, Gómez, and Ruiz (2009), Donnellan and Hennessy (2007)
8.	Team work	Wheelan, Burchill, and Tilin (2003), Gallie, Zhou, Felstead, and Green (2012), O'leary, Mortensen, and Woolley (2011), Bartel, Beaulieu, Phibbs, and Stone (2014), Tohidi (2011)
9.	Break the monotony and rotate	Kumar, Kumar, and Haleem (2014), Kumar, Luthra, Haleem, Garg and Singh (2014), Diego-Mas, Asensio-Cuesta, Sanchez-Romero, and Artacho-Ramirez (2009), Synwoldt and Gellerstedt (2003), Nichols, Rosenberg, Majoni, and Mukanjari (2014)
10.	Courses and improvement options	De Grip and Sauermann (2013)
11.	Tools and equipment to raise productivity	Kumar, Kumar, Haleem, and Gahlot (2013), Cusumano (1988), Malmir, Khalili, and Damirchi (2012), Colak, Cetin, Yilmaz, Yildiz, and Korkmaz (2013)
12.	Availability of water, power, and other input supplies	Allcott, Collard-Wexler, and O'Connell (2014), Tersoo (2014)
13.	Positive attitude and involvement of management	Carmeli (2003), Karia and Hasmi Abu Hassan Asaari (2006), Alfes, Shantz, Truss, and Soane (2013), Holden (2011), Vail et al. (2012), Kumar, Luthra, Garg, and Haleem (2014)
14.	Encouragement to effective communication	Kupperschmidt (2000), Leblebici (2012), Ruck and Welch (2012), Daim et al. (2012), Aral, Brynjolfsson, and Van Alstyne (2012), Kumar, Singh, and Haleem (2014)
15.	Good working conditions	Luthra, Kumar, Kharb, Ansari, and Shimmi (2014), Locke and Romis (2012), Ali, Ali, and Adan (2013), Jaskiewicz and Tulenko (2012), Sandeep, Sanjay, Pardeep, and Abid (2013)

comparison and calculation between factors; accordingly, they can obtain the target weight of various programs. Best programs can be chosen by these target weights and this method also maintains consistency (Liqian & Jianming, 2012). For giving rankings, it is one of the most commonly used analytical methods. AHP has been extensively used in integrated manufacturing, in the layout design, evaluation of technology investment decisions, and related work. Following steps are included in AHP (Kumar, Luthra, & Haleem, 2015; Kumar, Luthra, Haleem, Mangla, & Garg, 2015; Kumar, Parashar, & Haleem, 2009; Saaty, 2000, 2008; Swaminathan & Anand, 2014):

Step 1: Pairwise comparisons made between decision alternatives and nine-point scales are used as criteria.

Step 2: A set of pairwise matrix are constructed and each element in the upper level compares with the elements of lower level.

Step 3: To ensure that the priority of elements is consistent, the maximum eigen values and the maximum eigen vectors are calculated.

4.1. Application of AHP for ranking of factors to enhance productivity

AHP methodology has been appropriately utilized to evaluate ranking of factors enhancing productivity in manufacturing firms and service organizations. Five experts from industry and academia were asked to reach to a consensus about rating of factors (as tabulated in Table 3). Two out of the five experts were senior professors researching the area of operations management for more than a decade. Other three experts were from manufacturing industry (one expert was holding top management position having a total experience of more than 30 years; and other two experts from industry were holding managerial positions with experience in a range of 8–15 years).

It may be further noticed that consistency ratio has been found acceptable (below 0.1). Resulting priority matrix and rank of various factors enhancing productivity have been presented in Table 4.

Table 3. Rating of factors assigned by experts

	1	2	3	4	5	6	7	8	9	10	11	12	13	14	15
1	1	3.00	2.00	4.00	2.00	3.00	4.00	1.00	5.00	2.00	1.00	1.00	0.50	2.00	1.00
2	0.33	1	1.00	2.00	1.00	2.00	1.00	1.00	4.00	2.00	0.50	0.50	0.33	0.50	0.33
3	0.50	1.00	1	0.50	1.00	3.00	1.00	1.00	3.00	0.50	0.33	0.33	0.20	1.00	1.00
4	0.25	0.50	2.00	1	0.33	1.00	0.50	0.50	2.00	1.00	0.33	0.50	0.20	1.00	0.25
5	0.50	1.00	1.00	3.00	1	3.00	1.00	1.00	1.00	1.00	0.50	0.50	0.33	0.50	0.50
6	0.33	0.50	0.33	1.00	0.33	1	0.33	0.33	0.50	0.33	0.25	0.25	0.17	0.33	0.20
7	0.25	1.00	1.00	2.00	1.00	3.00	1	1.00	2.00	0.33	0.33	0.33	0.25	0.33	0.50
8	1.00	1.00	1.00	2.00	1.00	3.00	1.00	1	2.00	1.00	1.00	1.00	0.50	1.00	1.00
9	0.20	0.25	0.33	0.50	1.00	2.00	0.50	0.50	1	0.33	0.25	0.25	0.20	0.25	0.20
10	0.50	0.50	2.00	1.00	1.00	3.00	3.00	1.00	3.00	1	0.50	0.50	0.33	0.50	0.50
11	1.00	2.00	3.00	3.00	2.00	4.00	3.00	1.00	4.00	2.00	1	1.00	0.50	1.00	1.00
12	1.00	2.00	3.00	2.00	2.00	4.00	3.00	1.00	4.00	2.00	1.00	1	0.50	1.00	1.00
13	2.00	3.00	5.00	5.00	3.00	6.00	4.00	2.00	5.00	3.00	2.00	2.00	1	4.00	2.00
14	0.50	2.00	1.00	1.00	2.00	3.00	3.00	1.00	4.00	2.00	1.00	1.00	0.25	1	0.50
15	1.00	3.00	1.00	4.00	2.00	5.00	2.00	1.00	5.00	2.00	1.00	1.00	0.50	2.00	1

Notes: Number of comparisons = 105.

Consistency Ratio CR = 3.8%.

Principal eigenvalue = 15.853.

Eigen vector solution: 4 iterations, delta = 3.8E−8.

Table 4. Priority matrix of factors

Factors	Priorities (%)	Rank
1. Proactive employees	10.1	2
2. Accountability	5.1	9
3. Follow-up	4.7	11
4. Manage the work force but avoid micromanagement	3.6	13
5. Encourage, motivate, reward, and recognize	4.9	10
6. Reach out to employees by seeking them out	2.1	15
7. Demand realistic targets	4.1	12
8. Team work	6.4	7
9. Break the monotony and rotate	2.3	14
10. Courses and improvement options	5.4	8
11. Tools and equipment to raise productivity	9.2	4
12. Availability of water, power, and other input supplies	8.9	5
13. Positive attitude and involvement of management	16.5	1
14. Encouragement to effective communication	7.1	6
15. Good working conditions	9.6	3

5. Findings and discussion

AHP applied to ratings (in pair wise mode) enabled us to point out the following observations:

(1) Five top factors are "Positive attitude and involvement of management," "Proactive employees," "Good working conditions," "Tools and equipment to raise productivity," and "Availability of water, power and other input supplies."

(2) Bottom three factors are "Manage the work force but avoid micromanagement," "Break the monotony and rotate," and "Reach out to employees by seeking them out."

(3) The factor "Positive attitude and involvement of management" is most preferred in all these factors and "Reach out to employees by seeking them out" is least preferred.

(4) CR ratio has been found within acceptable limit indicating satisfactory consistency of experts' opinion.

5.1. Discussions on top five factors to enhance productivity

A brief analysis of five major factors as a perspective to enhance productivity is discussed in the following sub-sections.

5.1.1. Positive attitude and involvement of management

In an organization, efficient product has been produced by management skill (Carmeli, 2003). It has been shown that participative management positively affects employees' job satisfaction. Management can set up regular focus group meetings with employees and supervisors from different work units to discuss issues of strategic planning related to operational system changes, relationship among work units, and organizational effectiveness (Kim, 2002). If management can take appropriate steps to positively influence the belief structures that bring about positive attitude formation that will then lead to more acceptance of the technology by the organization's members (Amoako-Gyampah & Salam, 2004). Senior managers need to receive adequate training and support to provide effective vision and direction for the organization's "people management" strategies and

> there must be a congenial corporate culture, capable and assertive human resource leadership, and truly supportive top management. The importance of top management's role must not be underestimated (Patterson, West, Lawthom, & Nickell, 1997).

5.1.2. Proactive employee

Employee loyalty must be earned through a culture of respect and integrity and learning and development. Established policies and practices that promote a workplace culture stimulate employee engagement. Listening carefully to what employees want and need, and providing opportunities and challenges to leverage the respective talents of employees (Lockwood, 2007) are important factors. It has been suggested that customer requirements fulfill in a better way when employees are engaged and to focus on work, it is a must that an employee trusts his management and it is also beneficial to that organization (Mayer & Gavin, 2005). An organization listen to the ideas of employees and take them into action if they are profitable to the organization, and the organization gives response to the employees by reward system or any other way to achieve better performance (Trkman, 2010). Engagement of employees with positive attitude toward value system and interests of the firm/supply chain has been pointed out important to improve performance of employee, department, firm, and the whole supply chain (Kompaso & Sridevi, 2010).

5.1.3. Good working condition

It has been studied that women want to work in good working conditions. However, women working in various sectors sometimes have been found compromising with working conditions (Loutfi, 2001). It has been emphasized that "the average person can think up twice as many ideas when working with a group than when working alone" (Diehl & Stroebe, 1987). Effective communication may also help in developing mutual trust among people working in an organization towards providing healthy environment (Porter, 1998). Individual labor is more productive than group labor in very rare situations, but in an organization, it is a must to develop an environment where people work in groups (Johnson, 1991).

5.1.4. Tool and equipment to raise productivity

It has been observed that the quality of raw materials and equipment affects the quality performance of a project (Enshassi, Mohamed, & Abushaban, 2009). Any company wants to avoid the equipment failure to reduce its maintenance cost and it is done by performing predictive and preventive maintenance strategies (Swanson, 2001). In an organization, a proper tool increasing or decreasing the productivity depends upon how the employees use the tool (Drucker, 1995). It has been stated that for European manufacturers, Japanese and North Americans provide more information and education is the most important tool (De Meyer, Nakane, Miller, & Ferdows, 1989). Highest quality of a product at lower cost can be obtained by standardizing the tools and equipments (Drucker, 1999).

5.1.5. Availability of input supplies

Worldwide, the main concern of many scientists, researchers, and farmers is the low-input, self-sustaining, energy-efficient, and diversified agriculture system (Altieri, 1999). There are many firms which can access stationary inputs. Efficiency and quality must be improved for increasing the productivity and the requirements of the firm better meet the input supplies by local suppliers (Porter, 2000). Using improved, new, and less costly inputs, a firm becomes more productive (Javorcik, 2004). In a denser, larger, and urban environment, a plant is more productive with the same input (Puga, 2010).

5.2. Perspectives of productivity enhancement

Fifteen factors to enhance productivity grouped together under four perspectives (Table 5): Human Resource Perspective; Task Perspective; Motivation Perspective; and Top Management Perspective were observed to be important factors that influence productivity. Top management perspective has been identified as the top most important perspective, sharing 44.2% of overall priorities followed by positive attitude and involvement of management; good working conditions; tools and equipment to raise productivity; and availability of water, power, and other input supplies (Figure 1). A hierarchy has been shown of the perspectives of factors to enhance productivity in Figure 2. An action plan has also been presented in Figure 3 toward helping the practicing managers of the manufacturing sector of India for achieving enhanced productivity.

Table 5. Priority matrix of perspectives

Perspectives	Factors	Priorities of factors (%)	Priorities of perspectives (%)	Rank of perspectives
Human resource perspective	Team work	6.4	24.0	**II**
	Reach out to employees by seeking them out (Humane approach)	2.1		
	Proactive employees	10.1		
	Courses and improvement options (Training)	5.4		
Task perspective	Accountability	5.1	17.5	**III**
	Follow-up	4.7		
	Demand realistic targets	4.1		
	Manage the work force but avoid micromanagement	3.6		
Motivation perspective	Encourage, motivate, reward, and recognize	4.9	14.3	**IV**
	Break the monotony and rotate	2.3		
	Encouragement to effective communication	7.1		
Top management perspective	Tools and equipment to raise productivity	9.2	44.2	**I**
	Availability of water, power, and other input supplies	8.9		
	Positive attitude and involvement of management	16.5		
	Good working conditions	9.6		

Note: Bold values show rank of perspectives which has been discussed in Section 5.2.

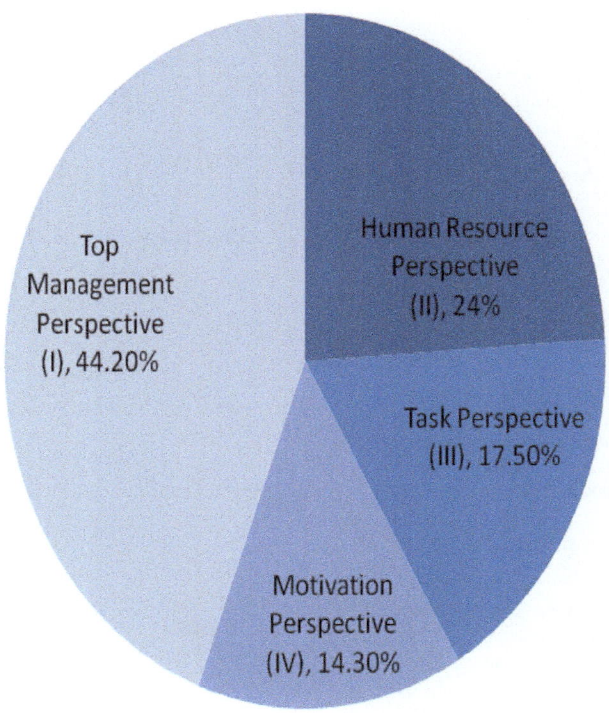

Figure 1. Perspectives' importance.

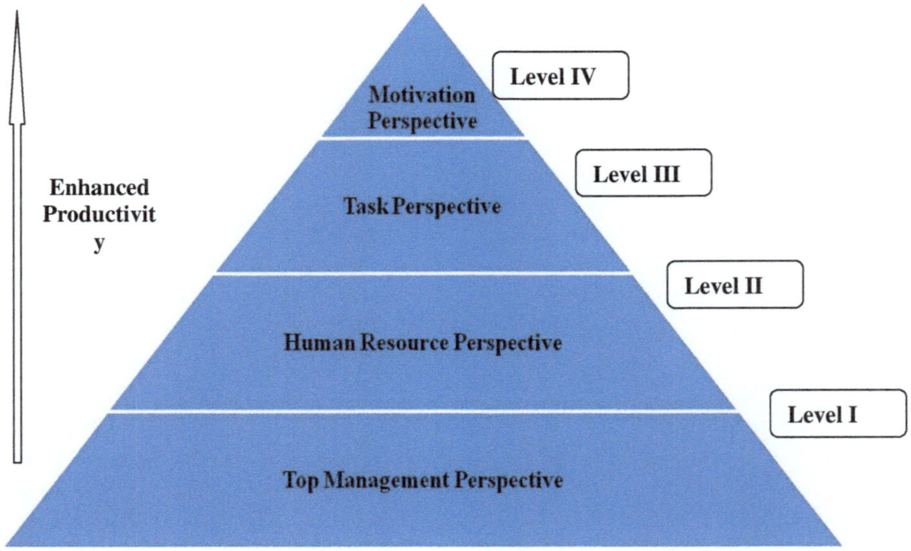

Figure 2. KUMADUHAL's hierarchy of productivity factors' perspectives.

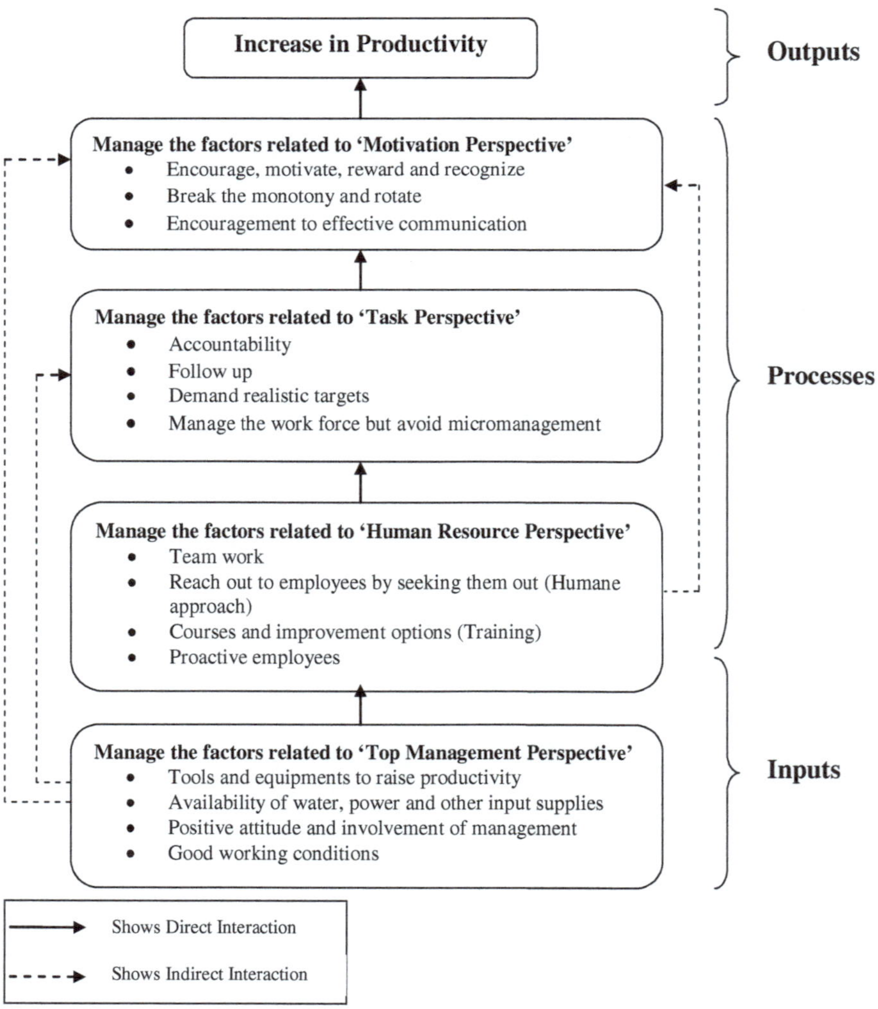

Figure 3. Action plan to enhance productivity.

6. Concluding remarks

With extensive literature review, we have identified 15 major factors that help in enhancing productivity. Further ranking of these factors has been carried out using AHP approach. "Positive attitude and involvement of management," "Proactive employees," "Good working conditions," "Tools and equipment to raise productivity," and "Availability of water, power and other input supplies" have been ranked as top five factors. The factors have been grouped into four perspectives and "Top management" perspective has been found to be the most important.

6.1. Limitations of study

The results as presented in the study are specific to the "Indian manufacturing sector;" further, there can be biasness in the experts' judgments.

6.2. Scope of future research

This research can be extended to other sectors/countries/region. The results may be evaluated utilizing the "Case Study" approach. Also, an extended application of AHP approach may be applied to categorize the factors using FCAHP technique (Kumar, 2014a) as future research. Further, the interactions among the categories/perspectives may be evaluated using CIAHP approach proposed by Kumar (2014b).

Funding
The authors received no direct funding for this research.

Author details
Sanjay Kumar[1]
E-mail: skbhardwaj19711971@gmail.com
Manish Duhan[1]
E-mail: manishduhan333@gmail.com
Abid Haleem[2]
E-mail: haleem.abid@gmail.com
[1] Department of Mechanical Engineering, International Institute of Technology and Management, Sonepat, Haryana, India.
[2] Faculty of Engineering and Technology, Department of Mechanical Engineering, Jamia Millia Islamia, New Delhi, India.

References

Alfes, K., Shantz, A. D., Truss, C., & Soane, E. C. (2013). The link between perceived human resource management practices, engagement and employee behaviour: A moderated mediation model. *The International Journal of Human Resource Management, 24*, 330–351. http://dx.doi.org/10.1080/09585192.2012.679950

Ali, A. Y. S., Ali, A. A., & Adan, A. A. (2013). Working conditions and employees' productivity in manufacturing companies in sub-Saharan African context: Case of Somalia. *Educational Research International, 2*, 67–78.

Allcott, H., Collard-Wexler, A., & O'Connell, S. D. (2014). *How do electricity shortages affect productivity? evidence from India (No. w19977)*. Cambridge, MA: National Bureau of Economic Research. http://dx.doi.org/10.3386/w19977

Altieri, M. A. (1999). The ecological role of biodiversity in agroecosystems. *Agriculture, Ecosystems & Environment, 74*, 19–31.

Amoako-Gyampah, K., & Salam, A. F. (2004). An extension of the technology acceptance model in an ERP implementation environment. *Information & Management, 41*, 731–745.

Aral, S., Brynjolfsson, E., & Van Alstyne, M. (2012). Information, technology, and information worker productivity. *Information Systems Research, 23*, 849–867. http://dx.doi.org/10.1287/isre.1110.0408

Bartel, A. P., Beaulieu, N. D., Phibbs, C. S., & Stone, P. W. (2014). Human capital and productivity in a team environment: evidence from the healthcare sector. *American Economic Journal: Applied Economics, 6*, 231–259.

Bartol, K. M., & Srivastava, A. (2002). Encouraging knowledge sharing: The role of organizational reward systems. *Journal of Leadership & Organizational Studies, 9*, 64–76.

Bernard, A. B., & Jensen, J. B. (2004). Exporting and productivity in the USA. *Oxford Review of Economic Policy, 20*, 343–357. http://dx.doi.org/10.1093/oxrep/grh020

Blackburn, J. D., Scudder, G. D., & Van Wassenhove, L. N. (1996). Improving speed and productivity of software development: a global survey of software developers. *IEEE Transactions on Software Engineering, 22*, 875–885. http://dx.doi.org/10.1109/32.553636

Broedling, L. A., Crawford, C. S., Kissler, G. D., Mohr, D. A., Newman, A. R., White, M. A., Williams, H., ... Koslowski, T. J. (1980). An examination of productivity impediments in the Navy industrial community (NPRDC-TR-81-2, AD-A092 80815). San Diego, CA: Navy Personnel Research and Development Center.

Brynjolfsson, E., & Hitt, L. M. (1998). Beyond the productivity paradox. *Communications of the ACM, 41*, 49–55. http://dx.doi.org/10.1145/280324.280332

Burton, W. N., Conti, D. J., Chen, C. Y., Schultz, A. B., & Edington, D. W. (1999). The role of health risk factors and disease on worker productivity. *Journal of Occupational and Environmental Medicine, 41*, 863–877. http://dx.doi.org/10.1097/00043764-199910000-00007

Caballero, R., Gómez, T., & Ruiz, F. (2009). Goal Programming: Realistic targets for the near future. *Journal of Multi-Criteria Decision Analysis, 16*, 79–110. http://dx.doi.org/10.1002/mcda.v16:3/4

Campbell, D. J. (2000). The proactive employee: Managing workplace initiative. *The Academy of Management Executive, 14*, 52–66.

Carmeli, A. (2003). The relationship between emotional intelligence and work attitudes, behavior and outcomes. *Journal of Managerial Psychology, 18*, 788–813. http://dx.doi.org/10.1108/02683940310511881

Chang, C. D., & Mills, J. C. (2013). Effects of a reward system on resident research productivity. *JAMA Otolaryngology-Head & Neck Surgery, 139*(12), 1285–1290.

Cohen, J. G., Sherman, A. E., Kiet, T. K., Kapp, D. S., Osann, K., Chen, L. M., & O'Sullivan, Patricia S. (2012). Characteristics of success in mentoring and research productivity—A case-control study of academic centers. *Gynecologic Oncology, 125*, 8–13. http://dx.doi.org/10.1016/j.ygyno.2012.01.005

Colak, M., Cetin, T., Yilmaz, S., Yildiz, T., & Korkmaz, Y. (2013). Furniture industry in Turkey due to product and process additional work or the scope of the method. *Asian Economic and Financial Review, 3*, 817–830.

Cooper, C. L., & Cartwright, S. (1994). Healthy mind; healthy organization—A proactive approach to occupational stress. *Human Relations, 47*(4), 455–471. http://dx.doi.org/10.1177/001872679404700405

Cusumano, M. A. (1988). Manufacturing innovation: Lessons from the Japanese auto industry. *MIT Sloan Management Review, 30*, 29–39.

Daim, T. U., Ha, A., Reutiman, S., Hughes, B., Pathak, U., Bynum, W., & Bhatla, A. (2012). Exploring the communication breakdown in global virtual teams. *International Journal of Project Management, 30*(2), 199–212. http://dx.doi.org/10.1016/j.ijproman.2011.06.004

De Grip, A., & Sauermann, J. (2013). The effect of training on productivity: The transfer of on-the-job training from the perspective of economics. *Educational Research Review, 8*, 28–36. http://dx.doi.org/10.1016/j.edurev.2012.05.005

De Meyer, A., Nakane, J., Miller, J. M., & Ferdows, K. (1989). Flexibility: The next competitive battle the manufacturing futures survey. *Strategic Management Journal, 10*(2), 135–144. http://dx.doi.org/10.1002/(ISSN)1097-0266

De Putter, C. E., Selles, R. W., Polinder, S., Panneman, M. J. M., Hovius, S. E. R., & van Beeck, E. F. (2012). Economic impact of hand and wrist injuries: health-care costs and productivity costs in a population-based study. *The Journal of Bone & Joint Surgery, 94*, e56.

Diego-Mas, J. A., Asensio-Cuesta, S., Sanchez-Romero, M. A., & Artacho-Ramirez, M. A. (2009). A multi-criteria genetic algorithm for the generation of job rotation schedules. *International Journal of Industrial Ergonomics, 39*(1), 23–33. http://dx.doi.org/10.1016/j.ergon.2008.07.009

Diehl, M., & Stroebe, W. (1987). Productivity loss in brainstorming groups: Toward the solution of a riddle. *Journal of Personality and Social Psychology, 53*, 497–509. http://dx.doi.org/10.1037/0022-3514.53.3.497

Dike, V. (2005). *Reward systems, worker Morale and productivity* (Nigeria society at the dawn of the 21st century). Retrieved from Africa Economic Analysis: www.AfricaEconomicAnalysis.org

Donnellan, T., & Hennessy, T. (2007). The effect of milk quota 'expansion' on EU/Ireland production. In *Exploiting the freedom to milk: Proceedings of the Teagasc National Dairy Conference 2007* (pp. 1-19).

Drucker, P. F. (1995). The new productivity challenge. *Quality in Higher Education, 37*, 45–53.

Drucker, P. F. (1999). Knowledge-worker productivity: The biggest challenge. *California Management Review, 41*, 79–94.

Eccles, R. G., Perkins, K. M., & Serafeim, G. (2012). How to become a sustainable company. *MIT Sloan Management Review, 53*, 43–50.

Enshassi, A., Mohamed, S., & Abushaban, S. (2009). Factors affecting the performance of construction projects in the Gaza strip. *Journal of Civil Engineering and Management, 15*, 269–280. http://dx.doi.org/10.3846/1392-3730.2009.15.269-280

Erdogan, B., & Bauer, T. N. (2005). Enhancing career benefits of employee proactive personality: The role of fit with jobs and organizations. *Personnel Psychology, 58*, 859–891. http://dx.doi.org/10.1111/peps.2005.58.issue-4

Eriksson, M., & Lindroos, O. (2014). Productivity of harvesters and forwarders in CTL operations in northern Sweden based on large follow-up datasets. *International Journal of Forest Engineering, 25*, 179–200. http://dx.doi.org/10.1080/14942119.2014.974309

Fairbank, J. F., & Williams, S. D. (2001). Motivating creativity and enhancing innovation through employee suggestion system technology. *Creativity and Innovation Management, 10*, 68–74. http://dx.doi.org/10.1111/caim.2001.10.issue-2

Gallie, D., Zhou, Y., Felstead, A., & Green, F. (2012). Teamwork, skill development and employee welfare. *British Journal of Industrial Relations, 50*, 23–46. http://dx.doi.org/10.1111/bjir.2012.50.issue-1

Gill, I., Koettl, J., & Packard, T. (2013). Full employment: A distant dream for Europe. *IZA Journal of European Labor Studies, 2*(1), 1–34.

Head, S. (2007). They're micromanaging your every move. *New York Review of Books, 54*, 42–44.

Holden, R. J. (2011). Lean thinking in emergency departments: A critical review. *Annals of Emergency Medicine, 57*, 265–278. http://dx.doi.org/10.1016/j.annemergmed.2010.08.001

Jaskiewicz, W., & Tulenko, K. (2012). Increasing community health worker productivity and effectiveness: A review of the influence of the work environment. *Human Resources for Health, 10*, 38. doi:10.1186/1478-4491-10-38 http://dx.doi.org/10.1186/1478-4491-10-38

Javorcik, B. S. (2004). Does foreign direct investment increase the productivity of domestic firms? In search of spillovers through backward linkages. *American Economic Review, 94*, 605–627. http://dx.doi.org/10.1257/0002828041464605

Johnson, D. W. (1991). *Cooperative Learning: Increasing College Faculty Instructional Productivity. ASHE-ERIC Higher Education Report No. 4, 1991*. ASHE-ERIC Higher Education Reports, George Washington University, One Dupont Circle, Suite 630, Washington, DC 20036-1183.

Karia, N., & Hasmi Abu Hassan Asaari, M. (2006). The effects of total quality management practices on employees' work-related attitudes. *The TQM Magazine, 18*, 30–43. http://dx.doi.org/10.1108/09544780610637677

Kim, S. (2002). Participative management and job satisfaction: Lessons for management leadership. *Public Administration Review, 62*, 231–241. http://dx.doi.org/10.1111/puar.2002.62.issue-2

Kirk, R. (1990). Using workload analysis and acuity systems to facilitate quality and productivity. *Journal of Nursing Administration, 20*, 21–30.

Kompaso, S. M., & Sridevi, M. S. (2010). Employee engagement: The key to improving performance. *International Journal of Business and Management, 5*, p89.

Kumar, S. (2014a). Some application of analytic hierarchy process. *Data Mining and Knowledge Engineering, 6*, 382–384.

Kumar, S. (2014b). Contextual interactions analytic hierarchy process (CIAHP): introduction and application to analyse interactions among knowledge management implementation capability (KMIC) factors. *International Journal of Knowledge and Learning, 9*, 242–263. http://dx.doi.org/10.1504/IJKL.2014.068926

Kumar, S., Parashar, N., & Haleem, A. (2009). Analytical hierarchy process applied to vendor selection problem: Small scale, medium scale and large scale industries. *Business Intelligence Journal, 2*, 355–362.

Kumar, N., Kumar, S., Haleem, A., & Gahlot, P. (2013). Implementing lean manufacturing system: ISM approach.

Journal of Industrial Engineering and Management, 6, 996–1012.
Kumar, S., Kumar, N., & Haleem, A. (2014). Contributions in research in the field of innovation management: Analysis of critical success factors, benefits and risks. *International Journal of Quality and Innovation, 2*, 310–337. http://dx.doi.org/10.1504/IJQI.2014.066386
Kumar, S., Luthra, S., Garg, D., & Haleem, A. (2014, December 18–21). Lean six sigma implementation: An analytic hierarchy process approach. In *NITIE-POMS Conference 2014 on Manufacturing Excellence: Imperatives for Emerging Economies at National Institute of Industrial Engineering* (pp. 18–21). Mumbai.
Kumar, S., Luthra, S., Haleem, A., & Garg, D., (2014). Significance of knowledge management in operations management. *XVIII Annual International Conference of the Society of Operations Management Theme: Operations Management in Digital Economy, At Department of Management Studies Indian Institute of Technology, Roorkee, India*, 393–398.
Kumar, S., Luthra, S., Haleem, A., Garg, D., & Singh, S. (2014, November). Understanding requisites of product innovation management: An ISM Approach. In *Journal Advances in Global Business Research-Proceedings of the AGBA 11th World Congress, New Delhi, India* (Vol. 11, pp. 340–350).
Kumar, S., Singh, V., & Haleem, A. (2014). Knowledge management-enablers and barriers: a questionnaire-based study. *International Journal of Knowledge Engineering and Data Mining, 3*, 31–57. http://dx.doi.org/10.1504/IJKEDM.2014.066233
Kumar, S., Luthra, S., & Haleem, A. (2015). Benchmarking supply chains by analyzing technology transfer critical barriers using AHP approach. *Benchmarking: An International Journal, 22*, 538–558. http://dx.doi.org/10.1108/BIJ-05-2014-0040
Kumar, S., Luthra, S., Haleem, A., Mangla, S. K., & Garg, D. (2015). Identification and evaluation of critical factors to technology transfer using AHP approach. *International Strategic Management Review, 3*, 24–42. http://dx.doi.org/10.1016/j.ism.2015.09.001
Kupperschmidt, B. R. (2000). Multigeneration employees: strategies for effective management. *The health care manager*, 19, 65-hyhen.
Leblebici, D. (2012). Impact of workplace quality on employee's productivity: case study of a bank in Turkey. *Journal of Business Economics and Finance, 1*, 38–49.
Liqian, Z., & Jianming, C. (2012). Eco-environmental quality assessment of Xining City based on GIS and AHP. *Modern Applied Science, 6*, p84.
Locke, R. M., & Romis, M. (2012). Improving work conditions in global supply chains. *MIT Sloan Management Review, 48*, 54–62.
Lockwood, N. R. (2007). Leveraging employee engagement for competitive advantage. *Society for Human Resource Management Research Quarterly, 1*, 1–12.
Loutfi, M. F. (2001). *Women, Gender and Work: What Is Equality and How Do We Get There?*. International Labour Office, 1828 L. Street, NW, Washington, DC 20036.
Luthra, S., Kumar, S., Kharb, R., Ansari, M. F., & Shimmi, S. L. (2014). Adoption of smart grid technologies: An analysis of interactions among barriers. *Renewable and Sustainable Energy Reviews, 33*, 554–565. http://dx.doi.org/10.1016/j.rser.2014.02.030
Malmir, A., Khalili, K., & Damirchi31, Q. V. (2012). Classifying the effective factors on productivity of human resources by using AHP and TOPSIS methods. *Kuwait Chapter of Arabian Journal of Business and Management Review, 1*, 68–73.
Mamiseishvili, K., & Rosser, V. J. (2011). Examining the relationship between faculty productivity and job satisfaction. *Journal of the Professoriate, 5*, 100–132.
Manzoor, Q. A. (2012). Impact of employees motivation on organizational effectiveness. *Business management and strategy, 3*(1), 1–12.
Martin, C. A. (2005). From high maintenance to high productivity. *Industrial and Commercial Training, 37*, 39–44. http://dx.doi.org/10.1108/00197850510699965
Mayer, R. C. ,& Gavin, M. B. (2005). Trust in management and PErFORMANCE: WHO minds the shop WHILE the employees WATCH the boss? *Academy of Management Journal, 48*, 874–888. http://dx.doi.org/10.5465/AMJ.2005.18803928
Nichols, E., Rosenberg, A., Majoni, A., & Mukanjari, S. (2014). Divestitures and shareholder wealth in the long-run–the South African case. *Corporate Ownership & Control, 569*, 569–578.
O'leary, M. B., Mortensen, M., & Woolley, A. W. (2011). Multiple team membership: A theoretical model of its effects on productivity and learning for individuals and teams. *Academy of Management Review, 36*, 461–478. http://dx.doi.org/10.5465/AMR.2011.61031807
Patterson, M. G., West, M. A., Lawthom, R., & Nickell, S. (1997). *Impact of people management practices on business performance* (pp. vii–viii). London: Institute of Personnel and Development.
Porter, M. (1998, November–December). Cluster and the new economics of competition. *Harvard Business Review*
Porter, M. E. (2000). Location, competition, and economic development: Local clusters in a global economy. *Economic Development Quarterly, 14*(1), 15–34. http://dx.doi.org/10.1177/089124240001400105
Puga, D. (2010). The magnitude and causes of agglomeration economies. *Journal of Regional Science, 50*, 203–219. http://dx.doi.org/10.1111/jors.2010.50.issue-1
Ruck, K., & Welch, M. (2012). Valuing internal communication; management and employee perspectives. *Public Relations Review, 38*, 294–302. http://dx.doi.org/10.1016/j.pubrev.2011.12.016
Saaty, T. L. (2000). *Fundamentals of decision making and priority theory with the analytic hierarchy process* (Vol. 6). Pittsburg: Rws Publications.
Saaty, T. L. (2008). Decision making with the analytic hierarchy process. *International Journal of Services Sciences, 1*, 83–98. http://dx.doi.org/10.1504/IJSSCI.2008.017590
Sandeep, K., Sanjay, K., Pardeep, G., & Abid, H. (2013). Analysis of interdependence among the enablers of green concept implementation in Indian automobile supply chain. *Journal of Engineering Research and Studies*, 4(2), 05–11.
Swaminathan, B., & Anand, S. (2014). A multi attribute Ranking mechanism for peer-to-peer networks. *Journal of Computer Science, 10*, 1458. http://dx.doi.org/10.3844/jcssp.2014.1458.1465
Swanson, L. (2001). Linking maintenance strategies to performance. *International Journal of Production Economics, 70*, 237–244. http://dx.doi.org/10.1016/S0925-5273(00)00067-0
Synwoldt, U., & Gellerstedt, S. (2003). Ergonomic initiatives for machine operators by the Swedish logging industry. *Applied Ergonomics, 34*(2), 149–156. http://dx.doi.org/10.1016/S0003-6870(03)00006-1
Tausky, C., & Chelte, A. F. (1983). Accountability and productivity some longitudinal data. *Work and Occupations, 10*, 207–220. http://dx.doi.org/10.1177/0730888483010002004
Tersoo, P. (2014). Agribusiness as a veritable tool for rural development in Nigeria. *International Letters of Social and Humanistic Sciences, 03*, 26–36.
Tohidi, H. (2011). Teamwork productivity & effectiveness in an organization base on rewards, leadership, training, goals, wage, size, motivation, measurement and information technology. *Procedia Computer Science, 3*, 1137–1146. http://dx.doi.org/10.1016/j.procs.2010.12.185

Trkman, P. (2010). The critical success factors of business process management. *International Journal of Information Management, 30*(2), 125–134. http://dx.doi.org/10.1016/j.ijinfomgt.2009.07.003

Vail, K. E., Juhl, J., Arndt, J., Vess, M., Routledge, C., & Rutjens, B. T. (2012). When death is good for life considering the positive trajectories of terror management. *Personality and Social Psychology Review, 16*, 303–329. http://dx.doi.org/10.1177/1088868312440046

Wheelan, S. A., Burchill, C. N., & Tilin, F. (2003). The link between teamwork and patients' outcomes in intensive care units. *American Journal of Critical Care, 12*(6), 527–534.

White, R. D. (2010). The micromanagement disease: Symptoms, diagnosis, and cure. *Public Personnel Management, 39*, 71–76. http://dx.doi.org/10.1177/009102601003900105

Wright, R. F. (2000). Strategies for avoiding the micro management trap. *Management Decision, 38*(5), 362–364. http://dx.doi.org/10.1108/00251740010340544

Zoltners, A. A., & Lorimer, S. E. (2000). Sales territory alignment: An overlooked productivity tool. *Journal of Personal selling & sales Management, 20*(3), 139–150.

9

Production inventory model for two-level trade credit financing under the effect of preservation technology and learning in supply chain

Sunil Kumar[1], Nidhi Handa[1], S.R. Singh[2] and Dharmendra Yadav[3*]

*Corresponding author: Dharmendra Yadav, Department of Mathematics, Vardhaman (P.G.) College, MJP Rohilkhand University, Bijnor, Uttar Pradesh, India

E-mail: dharmendrayadav3580@gmail. com

Reviewing editor: Zude Zhou, Wuhan University of Technology, China

Abstract: The present study investigated the inventory model for a retailer under two levels of trade credit to reflect the supply chain management. Supplier offers trade credit period of M to the retailer while in turn retailer provides a trade credit period of N to his/her customers. The supplier is willing to provide the retailer a full trade credit period for payments and the retailer offers the partial trade credit period to his/her customers. Here, selling items are considered as perishable items such as fruits, fresh fishes, gasoline, photographic films, etc. so that its potential worth decreases. It is assumed that decay in potential worth of items can be increased by using preservation technology. The demand is considered as the function of selling price and trade credit. Ordering cost can be reducing due to learning by doing phenomenon. By applying convex fractional programming results, we obtain necessary and sufficient conditions of an optimal solution. Some theorems are developed to determine retailer's optimal ordering policies and numerical examples are given to illustrate these theorems. In addition, some managerial insights from the numerical examples are also concluded.

Subjects: Mathematical Modeling; Operations Management; Operations Research

Keywords: production inventory model; two-level trade credit; preservation technology; learning phenomenon

1. Introduction

The managing of inventories is one of the most significant tasks that every manager must do efficiently and effectively in any organization so that their organization can grow. Now, all the organizations are involved in a global competitive market and then these organizations are taking seriously the activities

ABOUT THE AUTHOR

Dharmendra Yadav is an assistant professor in Vardhaman (P.G.) College, MJP Rohilkhand University, Bijnor, U.P. (India). He received his PhD in Inventory Control from the CCS University. His interests are in operations and supply chain management and optimization. He has had his articles published in journals, such as *Opsearch*, *Asia-Pacific Journal of Operational Research*, *International Journal of Systems Science*, *International Journal of Operational Research*, and others.

PUBLIC INTEREST STATEMENT

The aim of this paper is to develop production inventory model. We assume that supplier offers full trade credit period of M to the manufacturer whereas manufacturer provides partial trade credit period of N to his/her wholesaler. Here, concept of preservation technology has been used to increase the potential worth of the product. We have determined the optimal cycle time so that the total inventory cost is minimum. We found that Huang (2003), Shah (1993), and Goyal (1985) are special case of our proposed model. In addition, some managerial insights on the basis of numerical examples are also concluded.

related to manage inventories. Thus, recently the practitioners and researchers have been increasing their interest in optimizing the inventory decisions in a holistic way. Thus, the manager try to find out different way through which the cost associated with inventory can reduce and the profit of the organization can increase.

The economic order quantity (EOQ) model is widely used by practitioners as a decision-making tool for the control of inventory. The basic EOQ model is based on the implicit assumption that the retailer must pay for the items as soon as he receives them from a supplier. However, this may not be true. In today's business transactions, it is more and more common to see that the supplier will allow a certain fixed time period (say, 30 days) for settling the amount that the supplier owes to retailer for the items supplied. We term this period as trade credit period. Usually, interest is not charged for the outstanding amount if it is paid within the permissible delay period. This credit term in financial management is denoted as "net 30". Therefore, the retailer can sell the goods and earn the interest on the accumulated revenue received, and delay the payment up to the last moment of the permissible period allowed by the supplier. However, if the payment is not paid within the permissible delay period, then interest is charged on the outstanding amount under the previously agreed terms and conditions. This brings some economic advantage to the retailers as they may earn some interest from the revenue realized during the period of permissible delay. The trade credit financing produces two benefits to the supplier:

(1) It should attract new customers who consider it to be a type of price reduction, and

(2) It should cause a reduction in the sales outstanding, since some established customers will pay more promptly in order to take advantage of permissible delay more frequently.

In a real world, the supplier often makes use of this policy to promote his commodities. In India, gas stations adopted a price policy that charged less money per gallon to the customer who paid by cash, instead of by a credit card. Likewise, a store owner in many Chinatowns around the world usually charges a customer 5% more if the customer pays by a credit card, instead of by cash. As a result, the customer must decide which alternative to take when the supplier provides not only a cash discount but also a permissible delay. In this regard, a number of research papers appeared which deal with the EOQ problem under the condition of permissible delay in payments. Goyal (1985) is the first person to consider the EOQ inventory model under the condition of trade credit. Chand and Ward (1987) analyzed Goyal's model under assumptions of the classical EOQ model, obtaining different results. Shinn, Hwang, and Sung (1996) extended Goyal (1985) model and considered quantity discount for freight cost. Huang (2004) investigated that the unit selling price and the unit purchasing price are not necessarily equal within the EPQ framework under a supplier's trade credit policy. Teng, Chang, and Goyal (2005) developed the optimal pricing and lot sizing under permissible delay in payments by considering the difference between the selling price and the purchase cost and demand is a function of price. There are several interesting and relevant papers related to trade credit such as Chung, Goyal, and Huang (2005), Mahata and Goswami (2007), Bera, Kar, Chakraborti, and Sinha (2014), Yadav, Singh, and Kumari (2015) and Rao and Rao (2015), and their references. All the above inventory models implicitly assumed one-level trade credit financing, i.e. it is assumed that the supplier would offer the retailer a delay period and the retailer could sell the goods and accumulate revenue and earn interest within the trade credit period. They implicitly assumed that the customer would pay for the items as soon as the items are received from the retailer. That is, they assumed that the supplier would offer the retailer a delay period but the retailer would not offer any delay period to his/her customer. In most business transactions, this assumption is unrealistic. Usually the supplier offers a credit period to the retailer and the retailer, in turn, passes on this credit period to his/her customers. For example, in India, the TATA Company can delay the payment of purchasing cost until the end of the delay period offered by his supplier. The TATA Company also offers permissible delay payment period to his dealership. Huang (2003) presented an inventory model assuming that the retailer also permits a credit period to its customer which is shorter than the credit period offered by the supplier, in order to stimulate the demand. Huang (2006) extended Huang's (2003) model to investigate the retailers inventory policy under two levels of trade credit and limited storage space. Mahata and Goswami (2007) developed an inventory model to determine an optimal ordering policy for deteriorating items under two-level trade credit policy in the fuzzy sense. Huang (2007) incorporated Huang's (2003) model to investigate the two-level trade credit policy in the EPQ frame work. Ho, Ouyang, and Su (2008) developed an integrated supplier–buyer inventory model with the assumption that demand is sensitive to retail price and the supplier adopts a two-part trade credit policy. Liao (2008) developed an EOQ model with non-instantaneous receipt and exponentially deteriorating items under two-level trade credit financing. Tsao (2009) developed an EOQ model under advance sales discount and two-echelon trade credits. Teng and Chang (2009) extended the Huang's (2007) model by relaxing the assumption $N < M$. Kreng and Tan (2010) modify Huang's (2003) model by developing optimal wholesaler's replenishment decisions in the EOQ model under two levels of trade credit policy depending on the order quantity. Min, Zhou, and Zhao (2010) developed an inventory model for deteriorating items under stock-dependent demand and two-level trade credit. The inventory models under two levels of trade credit policy using the viewpoint of Huang (2003) can be found in several research papers such as Kreng and Tan (2010), Feng, Li, and Zhao (2013), Zhou, Zhong, and Wahab (2013), and Chung (2013), Shastri, Singh, Yadav, and Gupta (2014).

Deterioration of items present in inventory had been studied in the past decades (Dye, Ouyang, & Hsieh, 2007; Lin & Lin, 2004; Lin, Tan, & Lee, 2000; Mahata, 2012; Min et al., 2010; Soni, 2013; Wee, 1995; Yadav et al., 2012a). In that research, researchers mainly focused on: (1) considering different pattern of deterioration rate which may be time dependent or constant, (2) quantity discount, and (3) supply chain coordination. However, investigation on preservation technology to reduce deterioration rate has received little attention in the past years. The consideration of preservation technology is important due to the fact that preservation technology can reduce the deterioration rate significantly. The higher rate of deterioration would result in a higher total annual relevant cost and a lower demand rate. Murr and Morris (1975) showed that a lower temperature will increase the storage life and decrease decay. Moreover, drying or vacuum technology is introduced to reduce the deterioration rate of medicine and foodstuff. Zauberman, Ronen, Akerman, and Fuchs (1990) developed a method for color retention of Litchi fruits with SO_2 fumigation. The tradeoff between the increased cost of investment and the increased profit due to decreased deterioration rate is the focus of our study. Ouyang, Wu, and Yang (2006) found that if the retailer can reduce effectively the deteriorating rate of item by improving the storage facility, the total annual relevant inventory cost will be reduced. Wang, Wang, and Yang (2007) focused on deciding on resources portfolio and allocating resources to various orders in each production period. Many enterprises invest on equipment to reduce the deterioration rate and extend the product expiration date. For example, refrigeration equipment is used to reduce the deterioration rate of fruits, flowers, and sea foods in the supermarket. Shastri et al. (2014) developed an EOQ inventory model for a retailer under two-levels of trade credit to reflect the supply chain management (SCM) by using preservation technology to increase the potential worth of the deteriorated items.

Arrow (1962) and Rosen (1972) observed that the cost associated with production system especially for a new product declines by a factor 10–50% each time the accumulated production volumes doubles, due to learning by doing. Many researchers have applied this learning-by-doing phenomenon into production-marketing model to obtain optimal pricing, advertising, quality, and other strategies. Salameh and Jaber (2000) assumed that the defective items could be sold as a single batch at a discounted price prior to receiving the next shipment, and found that the economic lot size tends to increase as the average percentage of imperfect quality items decreases. Jaber, Goyal, and Imran (2008) extended the work of Salameh and Jaber (2000) by assuming that the defective percentage of goods within a shipment was reduced as the learning curve increased. Tsai (2012) presented two models for determining an optimal integrated EOQ and economic production quantity policy in a recoverable manufacturing environment. Tsai assumed that the unit production time of the recovery process decreases with the increase in total units produced as a result of learning. Yadav et al. (2012b) developed an EOQ model by incorporating the effect of learning in holding cost, ordering cost, and on the number of defective items present in each lot. Teng, Lou, and Wang (2013) proposed an economic production quantity model from seller's prospective to determine his/her optimal trade credit period and production lot size simultaneous in which (1) trade credit increases not only sales but also opportunity cost and default risk, and (2) production cost declines and obeys a learning curve phenomenon. Khan, Jaber, and Ahmad (2014) developed an integrated mathematical model for determining an optimal vendor–buyer inventory policy by accounting for quality inspection errors of type-I and type-II at the buyer's end and learning in production follows the Wright (1936) learning curve at the vendor's end. Goyal, Singh, and Yadav (2015) developed an EOQ model for imperfect lot with partial backordering. They also considered the effect of learning while obtaining the optimal solution.

The present study investigated the inventory model especially for a retailer under two levels of trade credit to reflect the SCM. Here, we are taking into account the following factors:

(1) the selling items are perishable such as fruits, fresh fishes, gasoline, photographic films, etc. It is assumed that decay in potential worth of items can be increased by using preservation technology;
(2) supplier offers trade credit period of M to the retailer while in turn retailer provides a trade credit period of N to his/her customers;
(3) the replenishment rate is finite;
(4) the supplier is willing to provide the retailer a full trade credit period for payments and the retailer offers the partial trade credit period to his/her customers;
(5) the demand is the function of selling price and trade credit;
(6) ordering cost reduces due to learning by doing phenomenon.

Under these conditions, we model inventory system as a cost minimization problem. By applying convex fractional programming results, we obtain necessary and sufficient conditions of an optimal solution. Some theorems are developed to determine retailer's optimal ordering policies and numerical examples are given to illustrate these theorems. In addition, some managerial insights from the numerical examples are also concluded.

2. Assumptions and notations

The following assumptions and notation are considered to develop the model.

2.1. Assumptions

(1) Replenishments rate, P, is known and constant.
(2) Credit period reduces the retailer's inventory holding cost, and hence a positive impact on demand. Similarly selling price of the items also produces positive impact on the demand. So, we consider demand as a function of trade credit and selling price.
(3) Preservation technology is used to reduce the decay rate of items and there is no replacement or repair of deteriorated items during a given cycle.
(4) It is a well-known learning by doing phenomenon that the total ordering cost declines by a factor of 10–50% each time the accumulative production run especially during the introduction phase of a new product. Mathematically, it can be represented as

$$A(n) = A_0/n^{\delta}$$

where A_0 is the ordering cost in first production run and $\delta > 0$.
(5) The supplier offers the full trade credit to the retailer, when $T \geq M$, the account settled at $T = M$, the retailers pay off all units sold and keep his/her profits, and start paying for the interest charges on the items in stock with rate I_c. When $T \leq M$, the account is settled at $T = M$ and the retailer no need to pay any interest on the stock.

(6) The retailer just offers the partial trade credit to his/her customers. Hence, the customer must make a partial payment to the retailer when the items are sold. Then the customer must pay off the remaining balance at the end of the trade credit period offered by the retailer. That is, the retailer can accumulate interest from his/her customer payment with rate I_e.

(7) Shortages are not allowed and lead time is zero.

(8) Time horizon is infinite.

2.2. Notations

P	production rate per year
$D(s, M)$	$(=as^{-b}e^{cM}, a, b, c > 0)$ demand rate of the customers
ϵ	preservation technology cost for reducing deterioration rate in order to preserve the products, $\epsilon \geq 0$
$m(\epsilon)$	reduce deterioration rate, a function of ϵ
k	original deterioration rate, $k > 0$
h	inventory holding cost per item per year excluding interest charges
$A(n)$	$(=A_0/n^{\delta})$ ordering cost per order
c	the unit purchase cost per unit item
s	the unit selling price of items of good quality where $s \geq c$
α	customer's fraction of the total amount owed payable at the time of placing an order offered by the retailer $0 \leq \alpha \leq 1$
M	retailer's trade credit period offered by the supplier in years
N	customer's trade credit period offered by the retailer in years
I_c	the interest charged per year per \$ in stocks by the supplier
I_e	the interest earned per \$ per year
T_1	the time at which the production stops in a cycle
T	cycle time in years

3. Mathematical modeling of supply chain under two level of trade credits

In this paper, we developed a retailer's EPQ-based inventory model under two levels of trade credit to reflect the SCM. It is assumed that the position of retailer is strong in the system and can obtain full trade credit offered by the supplier and the retailer just offered partial trade credit to his customers. We also considered that items deteriorate with deterioration rate k in the absence of preservation technology. However, reduced deterioration rate $m(x)$ is assumed when the retailer invests "\$$x$" on the preservation technology cost.

A production process starts at $t = 0$ at the rate P and continues up to $t = T_1$ where the inventory level reaches the maximum level. During this period, inventory level rises due to production and decline due to demand and deterioration. At $t = T_1$, production process stops and afterward inventory level declines only due to demand and deterioration. Inventory level reaches to zero at $t = T$. This process continues as the planning horizon is infinite. The graphical representation of this inventory system is clearly depicted in Figure 1. Let $I(t)$ be the inventory level at any time t $(0 \leq t \leq T)$. Initially, the stock level is Q. The inventory level decreases both due to demand and deterioration until it becomes zero at time $t = T$.

The differential equation governing the transition of the inventory system during the planning horizon is

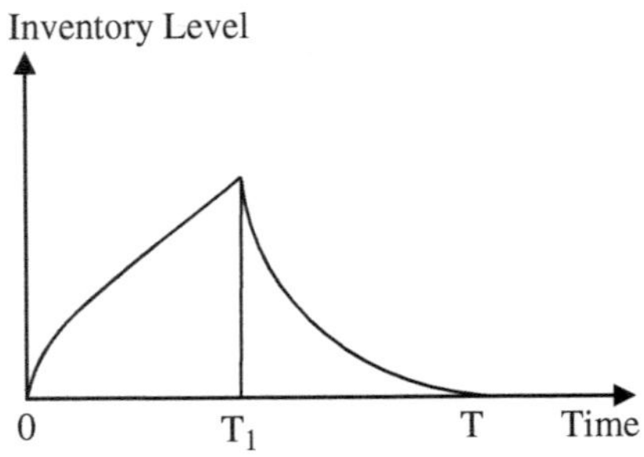

Figure 1. Pictorial representation of inventory level.

$$\frac{dI_1(t)}{dt} + (k - m(\epsilon))I(t) = P - D(s, M), \quad 0 \le t \le T_1 \tag{1}$$

With the initial conditions $I_1(0) = 0$. The solution of the differential Equation 1 with the initial condition $I_1(0) = 0$ is

$$I_1(t) = \frac{P - D(s, M)}{(k - m(\epsilon))} \left(1 - e^{-(k-m(\epsilon))t}\right), \quad 0 \le t \le T_1 \tag{2}$$

In the time interval $(T_1, T]$, the inventory declines due to demand and deterioration. Hence, the change in the inventory level is governed by the following differential equations:

$$\frac{dI_2(t)}{dt} + (k - m(\epsilon))I(t) = -D(s, M), \quad T_1 \le t \le T \tag{3}$$

With the boundary conditions $I_1(T) = 0$. The solution of the differential Equation 3 with the boundary condition $I_2(T) = 0$ is

$$I_2(t) = \frac{D(s, M)}{(k - m(\epsilon))} \left(e^{(k-m(\epsilon))(T-t)} - 1\right) \tag{4}$$

In addition, using the boundary condition

$$I_1(T_1) = I_2(T_1)$$

We obtain the following equations:

$$P\left(1 - e^{-(k-m(\epsilon))T_1}\right) = D(s, M)\left(e^{(k-m(\epsilon))(T-T_1)} - e^{-(k-m(\epsilon))T_1}\right) \tag{5}$$

$$T_1 = \frac{1}{(k - m(\epsilon))} \log\left\{1 + \frac{D(s, M)}{P}\left(e^{(k-m(\epsilon))T} - 1\right)\right\} \tag{6}$$

Now, we derive the total relevant cost for the seller which is comprised of the annual ordering cost, annual holding cost, annual interest payable, annual interest earned, and annual deterioration cost.

Annual ordering cost = $A(n)/T$

Annual holding cost (excluding interest charges) = $\frac{h}{T}\left[\int_0^{T_1} I_1(t)dt + \int_{T_1}^{T_2} I_2(t)dt\right] = \frac{h}{(k-m(\epsilon))^2 T}$ $\left((k - m(\epsilon))T_1 + e^{-(k-m(\epsilon))T_1} - 1\right) P + \frac{h}{(k-m(\epsilon))^2 T}\left(e^{(k-m(\epsilon))(T-T_1)} - (k - m(\epsilon))T - e^{-(k-m(\epsilon))T_1}\right) D(s, M)$ $= \frac{h}{(k-m(\epsilon))T}\left(PT_1 - D(s, M)T\right)$ using Equation 5.

Total annual cost due to deterioration of items during the cycle $= \frac{h}{T}\left(PT_1 - D(s,M)T\right)$

From the values of N and M, there are two possible cases:

Case-1: $M > N$ Case-2: $N \geq M$

Now, we discuss these two cases separately one by one.

Case-1: $M > N$

Based on the values of M, T, $T + N$ (i.e. the time at which the seller receives the payment from the last customer), three sub-cases can occur.

Sub-case-1.1: $M \leq T$

In this sub-case, the seller accumulates revenue and earns interest.

(i) from the portion of instant payment starting time 0 through M and

(ii) from the portion of delayed payment starting time N through M.

Hence, the interest earned per cycle is I_e times the total area of the triangle OMA and the triangle OMA′ as shown in Figure 2. Therefore, annual interest earned is given by

$$\frac{sI_eD(s,M)}{2T}\left(\alpha M^2 + (1-\alpha)(M-N)^2\right)$$

On the other hand, the seller grants its buyers a trade credit of N periods, and receives the money from its buyers from time N through $T + N$. Thus, at time M the seller receives $\alpha sD(s,M)M$ dollars from the instant payment and $(1-\alpha)sD(s,M)(M-N)$ dollars from the delayed payment, and pays its supplier $\alpha cD(s,M)M + (1-\alpha)cD(s,M)(M-N)$ dollars. The retailers must finance

(i) all items sold after M for the portion of instant payment, and

(ii) all items sold after $M - N$ for the portion of delayed payment at an interest charged I_c per dollar per year.

As a result, the interest payable per cycle is cI_c times the total area of the triangle ABC and the triangle A′B′C′ as shown in Figure 2.

Therefore, the annual interest payable is given by

$$\frac{cI_cD(s,M)}{2T}\left(\alpha(T-M)^2 + (1-\alpha)(T+N-M)^2\right)$$

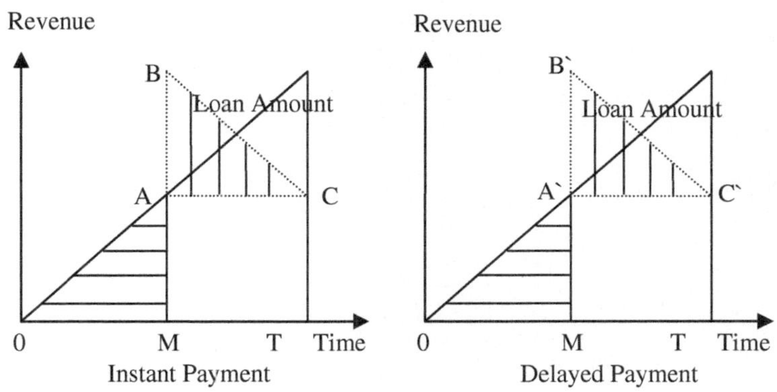

Figure 2. $M > N$ and $M \leq T$.

The seller's annual total relevant cost is

$$\begin{aligned}\mathrm{TRC}_{1.1}(T) = {} & \frac{A(n)}{T} + \frac{h + (k - m(\epsilon))c}{(k - m(\epsilon))T}\left(PT_1 - D(s,M)T\right) \\ & + \frac{cI_cD(s,M)}{2T}\left(\alpha\,(T-M)^2 + (1-\alpha)\,(T+N-M)^2\right) \\ & - \frac{sI_eD(s,M)}{2T}\left(\alpha M^2 + (1-\alpha)(M-N)^2\right) \end{aligned} \quad (7)$$

Sub-Case-1.2: $T \le M \le T + N$

Seller accumulates revenue and earns interest from two accounts:

(i) the portion of instant payment starting time 0 through M, and

(ii) the portion of delayed payment starting from N through M.

Hence, the seller's annual interest earned as shown in Figure 3 as

$$\frac{sI_eD(s,M)}{2T}\left(\alpha T^2 + 2\alpha T\,(M-T) + (1-\alpha)(M-N)^2\right)$$

The seller receives all instant payment by time T ($\le M$) so that there is no interest payable for the portion of instant payment. However, the seller must finance all items sold during time interval $[M - N, T]$. Therefore, the annual interest payable is

$$\frac{cI_cD(s,M)}{2T}(1-\alpha)\,(T+N-M)^2$$

The seller's annual total relevant cost by

$$\begin{aligned}\mathrm{TRC}_{1.2}(T) = {} & \frac{A(n)}{T} + \frac{h + (k - m(\epsilon))c}{(k - m(\epsilon))T}\left(PT_1 - D(s,M)T\right) \\ & + \frac{cI_cD(s,M)}{2T}(1-\alpha)\,(T+N-M)^2 - \frac{sI_eD(s,M)}{2T} \\ & \times \left(\alpha T^2 + 2\alpha T\,(M-T) + (1-\alpha)(M-N)^2\right)\end{aligned} \quad (8)$$

Sub-case-1.3: $T + N \le M$

The seller receives the total revenue before the trade credit period M, and hence there is no interest payable.

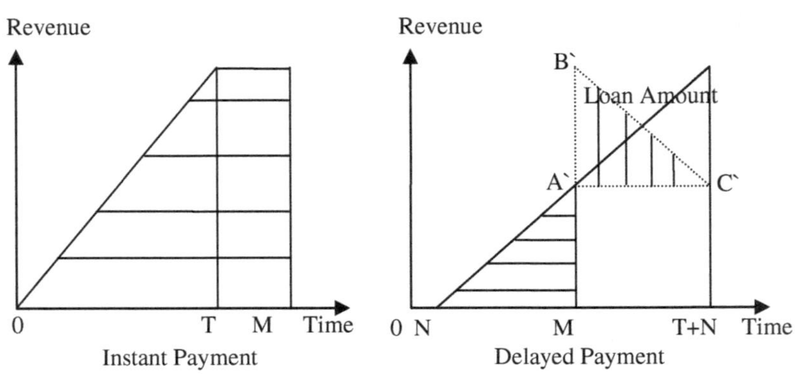

Figure 3. $M > N$ and $T \le M \le T + N$.

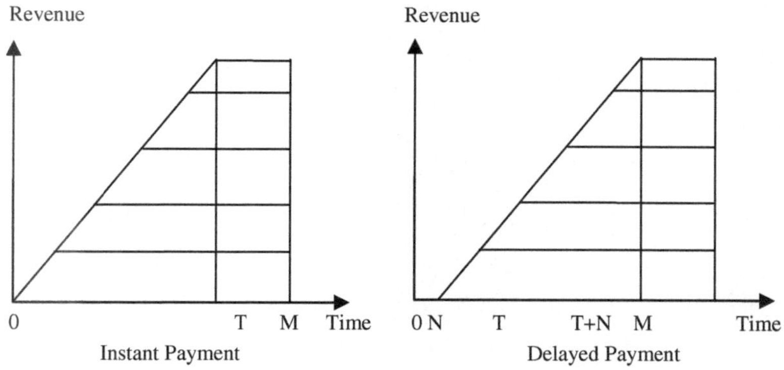

Figure 4. $M > N$ and $T + N \leq M$.

From Figure 4, the annual interest earned is

$$\frac{sI_eD(s,M)}{2T}\left(\alpha T^2 + 2\alpha T\,(M - T) + (1-\alpha)T^2 + 2(1-\alpha)T\,(M - T - N)\right)$$
$$= \frac{sI_eD(s,M)}{2}\left(2M - T - 2(1-\alpha)N\right)$$

The seller's annual total relevant cost by

$$\mathrm{TRC}_{1.3}(T) = \frac{A(n)}{T} + \frac{h + (k - m(\epsilon))c}{(k - m(\epsilon))T}\left(PT_1 - D(s,M)T\right) - \frac{sI_eD(s,M)}{2}\left(2M - T - 2(1-\alpha)N\right) \quad (9)$$

The seller's annual total relevant cost is given by

$$\mathrm{TRC}_1(T) = \begin{cases} \mathrm{TRC}_{1.1}(T) & \text{if } M \leq T \\ \mathrm{TRC}_{1.2}(T) & \text{if } T \leq M \leq T + N \\ \mathrm{TRC}_{1.3}(T) & \text{if } T + N \leq M \end{cases} \quad (10)$$

Hence, $\mathrm{TRC}_1(T)$ is continuous in T, and has the following properties

$$\mathrm{TRC}_{1.1}(M) = \mathrm{TRC}_{1.2}(M)$$

$$\mathrm{TRC}_{1.2}(M - N) = \mathrm{TRC}_{1.3}(M - N)$$

Case-2: $M \leq N$

Now, on the bases of the values of M and T, the following two sub-cases can occur: (i) $M \leq T$ and (ii) $M \geq T$. Now, we discuss them in length.

Sub-Case-2.1: $M \leq T$

From Figure 5, the annual interest earned from the instant payment is

$$\frac{sI_eD(s,M)}{2T}(\alpha M^2)$$

In this sub-case, for instant payment the seller must finance $\alpha cD(s,M)(T - M)$ at time M, and pay off the loan at time T. As to delayed payment, the seller must finance $(1-\alpha)cD(s,M)T$ for delayed payment at time M, and pay off the loan at time $T + N$. Therefore, the annual interest payable is

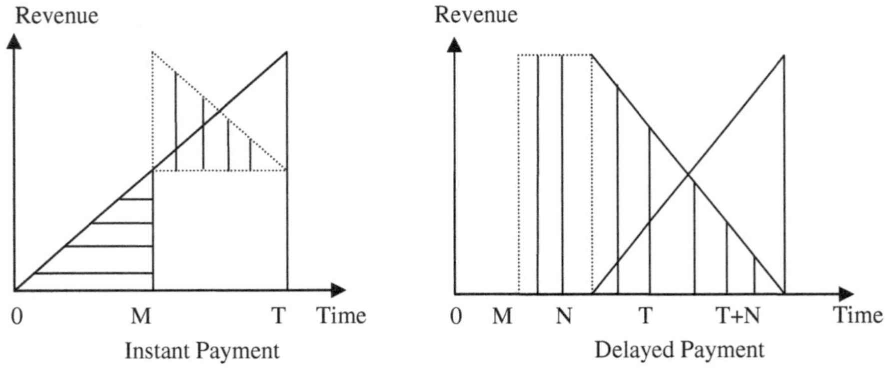

Figure 5. *M* ≤ *N* and *M* ≤ *T*.

$$\frac{cI_cD(s,M)}{2T}\left(\alpha(T-M)^2+(1-\alpha)T\left(T+2(N-M)\right)\right)$$

The annual total relevant cost of the seller is

$$\mathrm{TRC}_{2.1}(T)=\frac{A(n)}{T}+\frac{h+(k-m(\epsilon))c}{(k-m(\epsilon))T}\left(PT_1-D(s,M)T\right)+\frac{cI_cD(s,M)}{2T}\left(\alpha(T-M)^2+(1-\alpha)T\left(T+2(N-M)\right)\right)-\frac{sI_eD(s,M)}{2T}(\alpha M^2) \qquad (11)$$

Sub-Case-2.2: $M \geq T$

From Figure 6, the annual interest earned from the instant payment is

$$\frac{s\alpha I_eD(s,M)}{2}\left(T+2(M-T)\right)$$

In this sub-case, there is no interest payable for instant payment. However, the seller must finance $(1-\alpha)cD(s,M)T$ for delayed payment at time M, and pay off the loan at time $T+N$. Therefore, the annual interest payable is

$$\frac{c\alpha I_cD(s,M)}{2}(1-\alpha)\left(T+2(N-M)\right)$$

The seller's annual total relevant cost is

$$\mathrm{TRC}_{2.2}(T)=\frac{A(n)}{T}+\frac{h+(k-m(\epsilon))c}{(k-m(\epsilon))T}\left(PT_1-D(s,M)T\right)+\frac{c\alpha I_cD(s,M)}{2}(1-\alpha)\left(T+2(N-M)\right)$$

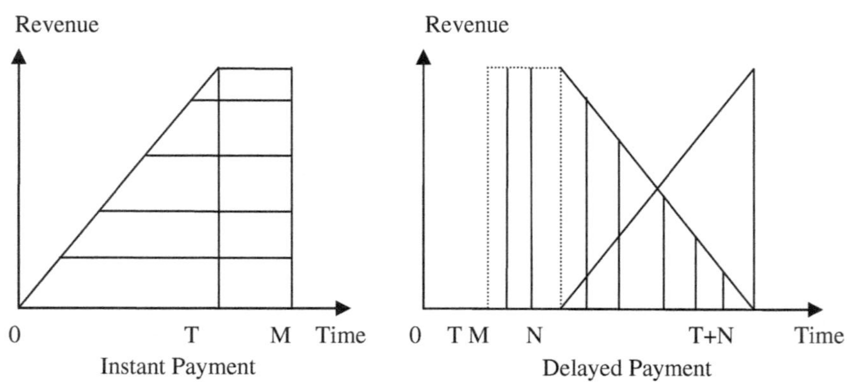

Figure 6. *M* ≤ *N* and *M* ≥ *T*.

$$\left(T + 2(M - T)\right) - \frac{s\alpha I_e D(s, M)}{2}\left(T + 2(M - T)\right) \quad (12)$$

The seller's annual total cost is

$$\mathrm{TRC}_2(T) = \begin{cases} \mathrm{TRC}_{2.1}(T) & \text{if } M \leq T \\ \mathrm{TRC}_{2.1}(T) & \text{if } M \geq T \end{cases} \quad (13)$$

It is clear that $\mathrm{TRC}_2(T)$ is continuous in T, and has the following properties

$$\mathrm{TRC}_{2.1}(M) = \mathrm{TRC}_{2.2}(M)$$

Now, we have to determine the optimal replenishment cycle T^{*} for both cases of $N < M$ and $N \geq M$.

4. Determination of the optimal replenishment cycle

In this section, the necessary and sufficient conditions for the determination of the optimal solution, say (T^*, $\mathrm{TRC}^*(T)$) are presented here for the case $N < M$ and $N \geq M$.

Intermediate Value Theorem: Let $f(x)$ be a continuous function on $[a, b]$ and $f(a)f(b) < 0$, then there exists a number $c \in (a, b)$ such that $f(c) = 0$.

Property-1: For a function f: $S \rightarrow R_1$ defined by $f(x) = g(x)/h(x)$ where g: $S \rightarrow R_1$, and S is a non-empty convex set in E_n. The following discuss convexity and concavity functions:

(a) g is convex on S, and $g(x) \geq 0$ for each $x \in S$.

(b) h is concave on S, and $h(x) \geq 0$ for each $x \in S$, and

(c) Both g and h are differentiable.

The $f(x)$ is convex if $g(x)$ is convex and $h(x)$ is concave.

4.1. Determination of optimal replenishment cycle for the case of N < M

By applying the above-mentioned results, we separately minimize each of $\mathrm{TRC}_{1.i}(T)$ for $i = 1, 2, 3$ and obtain the following theoretical results.

Theorem 1.

1. *$TRC_{1.1}(T)$ is a strictly pseudo-convex function in T, and hence there exists a unique minimum solution T_a^*.*
2. *If $M \leq T_a^*$ then $TRC_{1.1}(T)$ subjected to $M \leq T$ is minimized at T_a^*.*
3. *If $M \geq T_a^*$ then (T) subjected to $M \leq T$ is minimized at M.*

Proof. Let

$$g(T) = A(n) + \frac{h + (k - m(\epsilon))c}{(k - m(\epsilon))}\left(PT_1 - D(s, M)T\right) + \frac{cI_c D(s, M)}{2}\left(\alpha (T - M)^2 + (1 - \alpha)(T + N - M)^2\right)$$
$$- \frac{sI_e D(s, M)}{2}\left(\alpha M^2 + (1 - \alpha)(M - N)^2\right) \geq 0 \text{ for } T < 0$$

and

$h(T) = T > 0$

Taking first and second derivative of g(T), we get

$$\frac{dg(T)}{dT}=\frac{h+(k-m(\epsilon))c}{(k-m(\epsilon))}\left(\frac{PD(s,M)e^{(k-m(\epsilon))T}}{P+D(s,M)\left(e^{(k-m(\epsilon))T}-1\right)}-D(s,M)\right)+cI_cD(s,M)\left((T-M)+(1-\alpha)N\right)$$

$$\frac{d^2g(T)}{dT^2}=\frac{h+(k-m(\epsilon))c}{(k-m(\epsilon))}\left(\frac{P^2D(s,M)(k-m(\epsilon))e^{(k-m(\epsilon))T}}{\left(P+D(s,M)\left(e^{(k-m(\epsilon))T}-1\right)\right)^2}(1-D(s,M)/P)\right)+cI_cD(s,M)>0$$

Therefore, $TRC_{1.1}(T)=g(T)/h(T)$ is a strictly pseudo-convex function in T, which completes the proof of part (1).

Part (2) and (3) are the direct consequences of part (1).

To find T_a^*, we set $\frac{dTRC_{1.1}(T)}{dT}=0$

$$\frac{P\left(h+(k-m(\epsilon))c\right)}{(k-m(\epsilon))}\left(\frac{TD(s,M)e^{(k-m(\epsilon))T}}{P+D(s,M)\left(e^{(k-m(\epsilon))T}-1\right)}-\frac{1}{(k-m(\epsilon))}\log\left(1+\frac{D(s,M)}{P}\left(e^{(k-m(\epsilon))T}-1\right)\right)\right)$$
$$+\frac{D(s,M)\left[\alpha M^2+(1-\alpha)(M-N)^2\right]\left[sI_e-cI_c\right]}{2}+\frac{cI_cD(s,M)T^2}{2}-A(n)=0$$

THEOREM 2

1. *$TRC_{1.2}(T)$ is a strictly pseudo-convex function in T, and hence there exists a unique minimum solution T_b^*.*
2. *If $M-N\le T_b^*\le M$ then $TRC_{1.2}(T)$ subjected to $T\le M\le T+N$ is minimized at T_b^*.*
3. *If $T_b^*\le M-N$ then $TRC_{1.2}(T)$ subjected to $T\le M\le T+N$ is minimized at $M-N$.*
4. *If $T_b^*\ge M$ then $TRC_{1.2}(T)$ subjected to $T\le M\le T+N$ is minimized at M.*

Proof Let

$$g(T)=A(n)+\frac{h+(k-m(\epsilon))c}{(k-m(\epsilon))}\left(PT_1-D(s,M)T\right)+\frac{cI_cD(s,M)}{2}(1-\alpha)(T+N-M)^2$$
$$-\frac{sI_eD(s,M)}{2}\left(\alpha T^2+2\alpha T(M-T)+(1-\alpha)(M-N)^2\right)\ge 0\text{ for }T>0$$

and

$h(T)=T>0$

Taking first and second derivative of g(T), we get

$$\frac{dg(T)}{dT}=\frac{h+(k-m(\epsilon))c}{(k-m(\epsilon))}\left(\frac{PD(s,M)e^{(k-m(\epsilon))T}}{P+D(s,M)\left(e^{(k-m(\epsilon))T}-1\right)}-D(s,M)\right)$$
$$+cI_cD(s,M)(1-\alpha)(T-M+N)-sI_eD(s,M)\alpha(M-T)$$

$$\frac{d^2g(T)}{dT^2}=\frac{h+(k-m(\epsilon))c}{(k-m(\epsilon))}\left(\frac{P^2D(s,M)(k-m(\epsilon))e^{(k-m(\epsilon))T}}{\left(P+D(s,M)\left(e^{(k-m(\epsilon))T}-1\right)\right)^2}(1-D(s,M)/P)\right)$$
$$+D(s,M)\left(c(1-\alpha)I_c+\alpha sI_e\right)>0$$

Therefore, $TRC_{1.2}(T) = g(T)/h(T)$ is a strictly pseudo-convex function in T, which completes the proof of part (1).

Part (2), (3), and (4) are the direct consequences of part (1).

To find T_b^*, we set $\frac{dTRC_{1.2}(T)}{dT} = 0$

$$\frac{P\left(h+(k-m(\epsilon))c\right)}{(k-m(\epsilon))}\left(\frac{TD(s,M)e^{(k-m(\epsilon))T}}{P+D(s,M)\left(e^{(k-m(\epsilon))T}-1\right)}-\frac{1}{(k-m(\epsilon))}\log\left(1+\frac{D(s,M)}{P}\left(e^{(k-m(\epsilon))T}-1\right)\right)\right)$$
$$+\frac{D(s,M)(1-\alpha)(M-N)^2\left[sI_e-cI_c\right]}{2}+\frac{D(s,M)T^2\left[\alpha sI_e+(1-\alpha)cI_c\right]}{2}-A(n)=0$$

THEOREM 3

1. *$TRC_{1.3}(T)$ is a strictly pseudo-convex function in T, and hence there exists a unique minimum solution T_c^*.*
2. *If $T_c^* \le M-N$ then $TRC_{1.3}(T)$ subjected to $T+N \le M$ is minimized at T_c^*.*
3. *If $T_c^* \ge M-N$ then $TRC_{1.3}(T)$ subjected to $T+N \le M$ is minimized at $M-N$.*

Proof Let

$$g(T)=A(n)+\frac{h+(k-m(\epsilon))c}{(k-m(\epsilon))}\left(PT_1-D(s,M)T\right)-\frac{sI_eD(s,M)}{2}\left(2MT-T^2-2(1-\alpha)NT\right)\ge 0 \text{ for } T>0$$

and

$$h(T)=T>0$$

Taking first and second derivative of g(T), we get

$$\frac{dg(T)}{dT}=\frac{h+(k-m(\epsilon))c}{(k-m(\epsilon))}\left(\frac{PD(s,M)e^{(k-m(\epsilon))T}}{P+D(s,M)\left(e^{(k-m(\epsilon))T}-1\right)}-D(s,M)\right)-sI_eD(s,M)\left(M-T-(1-\alpha)N\right)$$

$$\frac{d^2g(T)}{dT^2}=\frac{h+(k-m(\epsilon))c}{(k-m(\epsilon))}\left(\frac{P^2D(s,M)(k-m(\epsilon))e^{(k-m(\epsilon))T}}{\left(P+D(s,M)\left(e^{(k-m(\epsilon))T}-1\right)\right)^2}(1-D(s,M)/P)\right)+sI_eD(s,M)>0$$

Therefore, $TRC_{1.3}(T) = g(T)/h(T)$ is a strictly pseudo-convex function in T, which completes the proof of part (1).

Part (2) and (3) are the direct consequences of part (1).

To find T_c^*, we set $\frac{dTRC_{1.3}(T)}{dT} = 0$

$$\frac{P\left(h+(k-m(\epsilon))c\right)}{(k-m(\epsilon))}\left(\frac{TD(s,M)e^{(k-m(\epsilon))T}}{P+D(s,M)\left(e^{(k-m(\epsilon))T}-1\right)}-\frac{1}{(k-m(\epsilon))}\log\left(1+\frac{D(s,M)}{P}\left(e^{(k-m(\epsilon))T}-1\right)\right)\right)$$
$$+\frac{sI_eD(s,M)T^2}{2}-A(n)=0$$

LEMMA 1 *Show that $\Delta_1 < \Delta_2$, where*

$$\Delta_1 = \frac{P\left(h+(k-m(\epsilon))c\right)}{(k-m(\epsilon))} \times \left(\frac{(M-N)D(s,M)e^{(k-m(\epsilon))(M-N)}}{P+D(s,M)\left(e^{(k-m(\epsilon))(M-N)}-1\right)} - \frac{1}{(k-m(\epsilon))}\log\left(1+\frac{D(s,M)}{P}\left(e^{(k-m(\epsilon))(M-N)}-1\right)\right)\right) + \frac{sI_eD(s,M)(M-N)^2}{2} - A(n)$$

and

$$\Delta_2 = \frac{P\left(h+(k-m(\epsilon))c\right)}{(k-m(\epsilon))}\left(\frac{MD(s,M)e^{(k-m(\epsilon))M}}{P+D(s,M)\left(e^{(k-m(\epsilon))M}-1\right)} - \frac{1}{(k-m(\epsilon))}\log\left(1+\frac{D(s,M)}{P}\left(e^{(k-m(\epsilon))M}-1\right)\right)\right) + \frac{D(s,M)\left[\alpha M^2+(1-\alpha)(M-N)^2\right]\left[sI_e-cI_c\right]}{2} + \frac{cI_cD(s,M)T^2}{2} - A(n)$$

Proof Since $TRC_{1.2}(T)$ is a strictly pseudo-convex function in T. We have

$$\frac{dTRC_{1.2}(T)}{dT} = \frac{1}{T^2}\left(\frac{P\left(h+(k-m(\epsilon))c\right)}{(k-m(\epsilon))}\left(\frac{TD(s,M)e^{(k-m(\epsilon))T}}{P+D(s,M)\left(e^{(k-m(\epsilon))T}-1\right)} - \frac{1}{(k-m(\epsilon))}\log\left(1+\frac{D(s,M)}{P}\left(e^{(k-m(\epsilon))T}-1\right)\right)\right) + \frac{D(s,M)(1-\alpha)(M-N)^2\left[sI_e-cI_c\right]}{2} + \frac{D(s,M)T^2\left[\alpha sI_e+(1-\alpha)cI_c\right]}{2} - A(n)\right)$$

is an increasing function in T, and hence we get

$$\frac{dTRC_{1.2}(M-N)}{dT} < \frac{dTRC_{1.2}(M)}{dT}$$

$$\Rightarrow \frac{\Delta_1}{(M-N)^2} < \frac{\Delta_2}{M^2}$$

$$\Rightarrow \Delta_1 < \Delta_2.$$

THEOREM 4

1. *If $\Delta_2 < 0$ then $TRC_1(T)$ is minimized at T_a^*.*
2. *If $\Delta_2 = 0$ then $TRC_1(T)$ is minimized at M.*
3. *If $\Delta_1 < 0$ then $TRC_1(T)$ is minimized at T_b^*.*
4. *If $\Delta_1 = 0$ then $TRC_1(T)$ is minimized at $M - N$.*
5. *If $\Delta_1 > 0$ then $TRC_1(T)$ is minimized at T_c^*.*

Proof We have

$$\frac{dTRC_{1.1}(T)}{dT} = \frac{1}{T^2}\left(\frac{P\left(h+(k-m(\epsilon))c\right)}{(k-m(\epsilon))}\left(\frac{TD(s,M)e^{(k-m(\epsilon))T}}{P+D(s,M)\left(e^{(k-m(\epsilon))T}-1\right)} - \frac{1}{(k-m(\epsilon))}\log\left(1+\frac{D(s,M)}{P}\left(e^{(k-m(\epsilon))T}-1\right)\right)\right) + \frac{D(s,M)\left[\alpha M^2+(1-\alpha)(M-N)^2\right]\left[sI_e-cI_c\right]}{2} + \frac{cI_cD(s,M)T^2}{2} - A(n)\right)$$

If $\Delta_2 < 0$, then

$$\lim_{T\to\infty} \frac{dTRC_{1.1}(T)}{dT} = \frac{cI_c D(s,M)}{2} > 0$$

and

$$\left[\frac{dTRC_{1.1}(T)}{dT}\right]_{T=M} = \frac{\Delta_2}{M^2} < 0$$

By applying the mean value theorem and Theorem 1, there exists a unique $T_a^* \in (M,\infty)$ such that $\frac{dTRC_{1.1}(T_a^*)}{dT} = 0$. Hence, $TRC_{1.1}(T)$ is minimizing at the unique point T_a^*. By using the analogous argument, we have

$$\frac{dTRC_{1.2}(M-N)}{dT} = \frac{\Delta_1}{(M-N)^2} < \frac{dTRC_{1.2}(M)}{dT} = \frac{\Delta_2}{M^2} < 0$$

which implies that $TRC_{1.2}(M)$ is minimizing at M. Likewise, we get

$$\lim_{\delta\to\infty} \frac{dTRC_{1.3}(\delta)}{dT} < \frac{dTRC_{1.3}(M-N)}{dT} = \frac{\Delta_1}{(M-N)^2} < 0$$

which implies that $TRC_{1.3}(T)$ is minimizing at M – N. Consequently, we obtain

$$TRC_{1.1}\,(T_a^*) < TRC_{1.1}(M) = TRC_{1.2}(M) < TRC_{1.2}(M-N) = TRC_{1.3}(M-N)$$

This shows the end of the first part of the theorem. Similarly, remaining part of the theorem can be obtained.

4.2. Determination of optimal replenishment cycle for the case of N ≥ M

By applying the above-mentioned results, we separately minimize each of $TRC_{2.i}(T)$ for $i = 1, 2$, and obtain the following theoretical results.

THEOREM 5

1. *$TRC_{2.1}(T)$ is a strictly pseudo-convex function in T, and hence there exists a unique minimum solution T_d^*.*
2. *If $M \le T_d^*$ then $TRC_{2.1}(T)$ subjected to $M \le T$ is minimized at T_d^*.*
3. *If $M \ge T_d^*$ then $TRC_{2.1}(T)$ subjected to $M \le T$ is minimized at M.*

To find T_d^*, we set $\frac{dTRC_{2.1}(T)}{dT} = 0$

$$\frac{P\,(h+(k-m(\epsilon))c)}{(k-m(\epsilon))}\left(\frac{TD(s,M)e^{(k-m(\epsilon))T}}{P+D(s,M)\left(e^{(k-m(\epsilon))T}-1\right)} - \frac{1}{(k-m(\epsilon))}\log\left(1+\frac{D(s,M)}{P}\left(e^{(k-m(\epsilon))T}-1\right)\right)\right)$$
$$+\frac{D(s,M)\alpha M^2\,[sI_e - cI_c]}{2} + \frac{cI_c D(s,M)T^2}{2} - A(n) = 0$$

It is obvious from the above theorem that above expression has a unique solution T_d^*. If $T_d^* \ge M$ then $TRC_{2.1}(T)$ is minimized at T_d^*. Otherwise, $TRC_{2.1}(T)$ is minimized at M.

THEOREM 6

1. *$TRC_{2.2}(T)$ is a strictly pseudo-convex function in T, and hence there exists a unique minimum solution T_e^*.*
2. *If M≥M $\leq T_e^*$ then $TRC_{2.2}(T)$ subjected to $M \geq T$ is minimized at T_e^*.*
3. *If $M \geq T_e^*$ then $TRC_{2.1}(T)$ subjected to $M \geq T$ is minimized at M.*

To find T_e^*, we set $\frac{dTRC_{2.2}(T)}{dT} = 0$

$$\frac{P\left(h+(k-m(\epsilon))c\right)}{(k-m(\epsilon))}\left(\frac{TD(s,M)e^{(k-m(\epsilon))T}}{P+D(s,M)\left(e^{(k-m(\epsilon))T}-1\right)}-\frac{1}{(k-m(\epsilon))}\log\left(1+\frac{D(s,M)}{P}\left(e^{(k-m(\epsilon))T}-1\right)\right)\right)$$
$$+\frac{\alpha sI_eD(s,M)T^2}{2}+\frac{cI_c(1-\alpha)D(s,M)T^2}{2}-A(n)=0$$

It is obvious from the above theorem that above expression has a unique solution T_e^*. If $T_e^* \leq M$ then $TRC_{2.2}(T)$ is minimized at T_e^*. Otherwise, $TRC_{2.2}(T)$ is minimized at M.

THEOREM 7

1. *If $\Delta_3 < 0$ then $TRC_2(T)$ is minimized at T_d^*.*
2. *If $\Delta_3 = 0$ then $TRC_2(T)$ is minimized at M.*
3. *If $\Delta_3 > 0$ then $TRC_2(T)$ is minimized at T_e^*.*

where

$$\Delta_3=\frac{P\left(h+(k-m(\epsilon))c\right)}{(k-m(\epsilon))}\left(\frac{MD(s,M)e^{(k-m(\epsilon))M}}{P+D(s,M)\left(e^{(k-m(\epsilon))T}-1\right)}-\frac{1}{(k-m(\epsilon))}\log\left(1+\frac{D(s,M)}{P}\left(e^{(k-m(\epsilon))M}-1\right)\right)\right)$$
$$+\frac{\alpha sI_eD(s,M)M^2}{2}+\frac{cI_c(1-\alpha)D(s,M)M^2}{2}-A(n)$$

5. Some special cases

5.1. Huang's model

If the potential value of the product does not change in infinite time then there is no need of preservation technology which we used here. So $k \to 0$, $m(\epsilon) \to 0$, $s = c$, $\alpha \to 0$ (it means that the retailer also offers the full trade credit to his/her customer), demand does not depend on selling price and trade credit period, and production rate is infinite then this inventory model is identical to that of Huang (2003).

5.2. Shah's model

If the potential value of the product does not change due to preservation technology then there is no need of preservation technology which we used here. So $m(\epsilon) \to 0$, $s = c$, $N \to 0$ (it means that the supplier would offer the retailer a delay period but the retailer would not offer the delay period to his/her customer) that is one level credit, $\alpha \to 0$, demand does not depend on selling price and trade credit and production rate is infinite then this inventory model is identical to that of Shah (1993).

5.3. Goyal's model

If the potential value of the product does not change in infinite time then there is no need of preservation technology which we used here. So $k \to 0$, $m(\epsilon) \to 0$, $s = c$, $N \to 0$ (it means that the supplier would offer the retailer a delay period but the retailer would not offer the delay period to his/her customer) that is one level credit, $\alpha \to 0$, demand does not depend on selling price and trade credit then this inventory model is identical to that of Goyal (1985).

6. Numerical analysis

The reduced deterioration rate, $m(\epsilon)$, is a function of the preservation technology cost "ϵ" such that $m(\epsilon) = k(1 - e^{k'})$, $k' > 0$, where k' is the simulation coefficient representing the percentage increase in $m(\epsilon)$ per dollar increase in ϵ, which means $m(\epsilon)$ is an increasing function bounded above by k.

Example 1 If a = 1,950, b = 0.09, c' = 0.08, c = \$50 per unit, h = \$15 per unit, I_c = 0.15, I_e = 0.10, N = 0.05, α = 0.5, k = 0.05, k' = 0.01, ϵ = \$40, it gives $m(\epsilon)$ = 0.016.

By using Theorem 4 we can easily obtain the optimal solution on three different sets of parameters as shown in Table 1.

Example 2 If a = 1,950, b = 0.09, c' = 0.08, c = \$50 per unit, h = \$15 per unit, I_c = 0.15, I_e = 0.10, s = 130, α = 0.5, k = 0.05, k' = 0.01, ϵ = \$40, it gives $m(\epsilon)$ = 0.016.

By using Theorem 5 we can easily obtain the optimal solution on two different sets of parameters as shown in Table 2.

6.1. Case study

A pharmaceutical company produces medicine. From the data it is calculated that the production cost per unit is \$50. Company observed that the demand pattern of the customer is $as^{-b}e^{c'M}$ i.e. it depends on the selling price and the trade credit period offered, where a = 1,950, b = 0.09, c' = 0.08. The holding cost per unit is \$15 whereas ordering cost place by retailer is \$145. Here, I_c = 0.15, I_e = 0.10. The trade credit offered by the company to his retailer is 0.1 years whereas retailer provides the trade credit period of 0.05 to his customers. Due to the strong position in the market, retailer can obtain full trade credit period whereas retailer provides partial trade credit to his customer where α = 0.5. Preservation technology is used here to decrease the rate of deterioration with ϵ = \$40.

In any decision-making situation, the change in the values of parameters may happen due to uncertainties. The decision-maker of the company observed that the deterioration rate, the selling price, and the production rate can changes as follows:

k = [0.03, 0.04, 0.05, 0.06, 0.07]

s = [65, 70, 75, 80, 85]

P = [2,300, 2,500, 2,700, 2,900, 3,100, 3,300, 3,500]

Now, in the front of the decision-maker it is a challenge to analyze how the optimal solution affected due to change in these parameters.

Table 1. Optimal policy in case of $M > N$

Values of parameters/cases	P	M	A	s	Δ_1	Δ_2	T_1	T^*	TRC (\$)
1.1	2,900	0.1	145	75	−141.93	−76.59	0.10	0.13(T_a)	1,998.78
1.2	3,900	0.1	95	75	−81.39	15.58	0.05	0.09(T_b)	1,637.73
1.3	4,500	0.15	150	125	42.57	252.01	0.05	0.08 (T_c)	631.28

Table 2. Optimal policy in case of $M \le N$

Values of parameters/cases	P	M	A	N	Δ_3	T_1	T^*	TRC (\$)
2.1	3,000	0.06	150	1.00	−136.06	0.10	0.13(T_d)	2,949.84
2.2	4,500	0.09	90	1.10	10.86	0.04	0.08(T_e)	20,468.00

In order to examine the implication of these changes, the sensitivity analysis will be of great help in decision-making.

6.1.1. Effect to deterioration rate

We now investigate the effects of varying rate of deteriorating in order to get more insight.

Figure 7 shows the deterioration rate at 0.03, 0.04, 0.05, 0.06, and 0.07 with other variables remain unchanged. It is shown that as the deterioration rate increases, the retailer's total inventory cost increases. A higher value of the deterioration rate results lower than the values for the optimal cycle time and the optimal order quantity, and higher value for the total inventory cost for the retailer.

6.1.2. Effect of selling price

To get the behavior of proposed model regarding selling, we investigate its effect on demand and the total inventory cost.

Figure 8 reflects the effect of selling price on demand and the total inventory cost of the retailer. It observes that as the selling price increases demand of the customer decreases and hence the total inventory cost of the retailer increases.

6.1.3. Effect of production rate

Sensitivity with respect to production rate has been performed here to get the effect on optimal policy of inventory management.

Figure 9 reflects the effect of production rate on production period, cycle time, and on the total inventory cost of the retailer. It is observed that as the production rate increases production period, cycle time decreases and the total inventory cost of the retailer decreases. So it is advisable to the decision-maker not to increase the production rate without the prior information about the customer's demand.

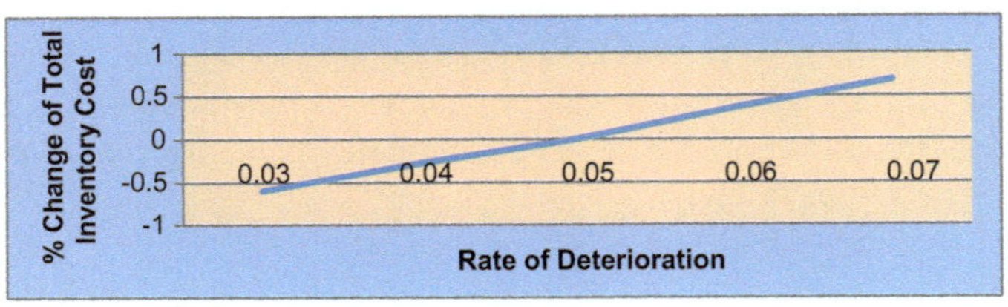

Figure 7. Effect of deterioration of total inventory cost.

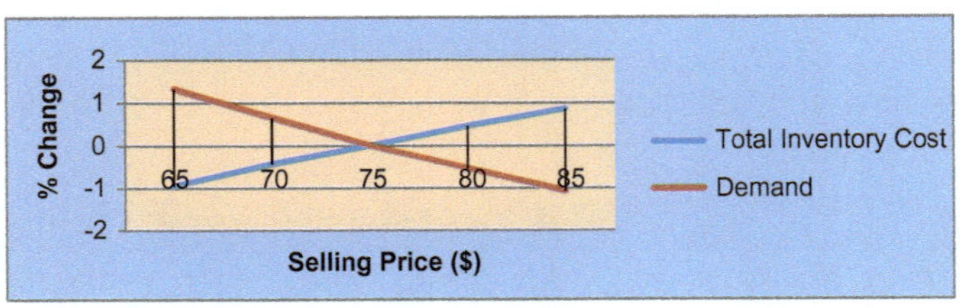

Figure 8. Effect of selling price on demand and total inventory cost.

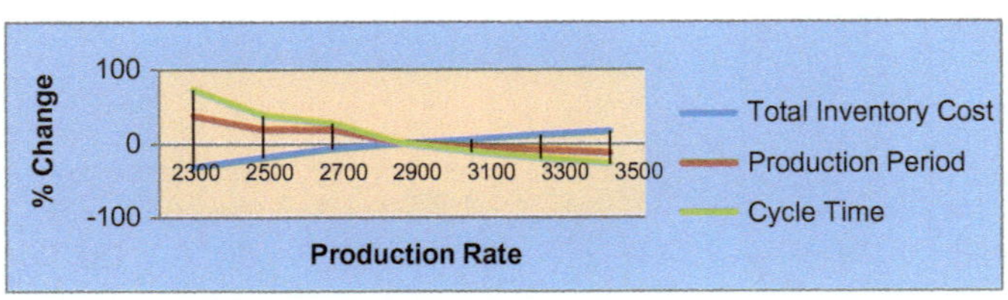

Figure 9. Effect of production rate on total inventory cost.

7. Discussion

Concept of preservation technology and the two-level trade credit has received a very little attention by the researchers while taking the decision regarding the manufacturing system. In this paper, we develop a production inventory model by assuming that the retailers hold the powerful position in the chain and can obtain the full trade credit offered by the supplier and retailer just offers partial trade credit to customers. We also assume that demand of the items not only depend upon the selling price but also credit period offered by the retailer. Here, retailer invests in preservation technology to control the deterioration rate of the product.

8. Conclusion

Our proposed production inventory model forms a general framework that includes many previous model as special cases such as Huang (2003), Shah (1993), and Goyal (1985). We also proposed theorems to find the optimal replenishment cycle time. Numeral examples are presented to illustrate the proposed theorems for finding the optimal cycle length. Sensitivity analysis with respect to deterioration rate, production rate, and selling price of items is also carried out to obtain a lot of managerial insights. The results of the paper not only provide a valuable reference for decision-makers in planning and controlling the inventory but also provide a useful model for many organizations that use the decision rule to improve their total operational cost.

9. Future work

There are several promising areas in which direction this model can be extended. This model can be extended by considering three different players such as supplier, manufacturer, and retailer to form integrated systems. One can also consider imprecise parameters in place of precise one.

Funding
The authors received no direct funding for this research.

Author details
Sunil Kumar[1]
E-mail: gkv.sunil@gmail.com
Nidhi Handa[1]
E-mail: nidhi_6744@yahoo.com
S.R. Singh[2]
E-mail: shivrajpundir@gmail.com
Dharmendra Yadav[3]
E-mail: dharmendrayadav3580@gmail.com
[1] Department of Mathematics, Gurukula Kangari Vishwavidyalaya, Hairdwar, Uttarakhand, India.
[2] Department of Mathematics, D.N. (P.G.) College, Meerut, Uttar Pradesh, India.
[3] Department of Mathematics, Vardhaman (P.G.) College, MJP Rohilkhand University, Bijnor, Uttar Pradesh, India.

References

Arrow, K. J. (1962). The economic implications of learning by doing. *The Review of Economic Studies, 29*, 155–173. http://dx.doi.org/10.2307/2295952

Bera, S., Kar, S., Chakraborti, T., & Sinha, B. (2014). An inventory model for deteriorating items under conditionally permissible delay in payments depending on the order quantity. *Applied Mathematics, 05*, 2675–2695. http://dx.doi.org/10.4236/am.2014.517256

Chand, S., & Ward, J. (1987). A note on economic order quantity under conditions of permissible delay in payments. *Journal of the Operational Research Society, 38*, 83–84.

Chung, K. J. (2013). The EPQ model under conditions of two levels of trade credit and limited storage capacity in supply chain management. *International Journal of Systems Science, 44*, 1675–1691. http://dx.doi.org/10.1080/00207721.2012.669864

Chung, K. J., Goyal, S. K., & Huang, Y. F. (2005). The optimal inventory policies under permissible delay in payments depending on the ordering quantity. *International Journal of Production Economics, 95*, 203–213. http://dx.doi.org/10.1016/j.ijpe.2003.12.006

Dye, C. Y., Ouyang, L. Y., & Hsieh, T. P. (2007). Inventory and pricing strategies for deteriorating items with shortages: A discounted cash flow approach. *Computers & Industrial Engineering, 52*, 29–40.

Feng, H., Li, J., & Zhao, D. (2013). Retailer's optimal replenishment and payment policies in the EPQ model under cash discount and two-level trade credit policy. *Applied Mathematical Modelling, 37*, 3322–3339. http://dx.doi.org/10.1016/j.apm.2012.07.012

Goyal, S. K. (1985). Economic order quantity under conditions of permissible delay in payments. *Journal of the Operational Research Society, 36*, 335–338. http://dx.doi.org/10.1057/jors.1985.56

Goyal, S. K., Singh, S. R., & Yadav, D. (2015). Economic order quantity model for imperfect lot with partial backordering under the effect of learning and advertisement dependent imprecise demand. *International Journal of Operational Research* (accepted manuscript).

Ho, C. H., Ouyang, L. Y., & Su, C. H. (2008). Optimal pricing, shipment and payment policy for an integrated supplier-buyer inventory model with two-part trade credit. *European Journal of Operational Research, 187*, 496–510. http://dx.doi.org/10.1016/j.ejor.2007.04.015

Huang, Y. F. (2003). Optimal retailer's ordering policies in the EOQ model under trade credit financing. *Journal of the*

Operational Research Society, 54, 1011–1015. http://dx.doi.org/10.1057/palgrave.jors.2601588

Huang, Y. F. (2004). Optimal retailer's replenishment policy for the EPQ model under the supplier's trade credit policy. *Production Planning and Control, 15*, 27–33. http://dx.doi.org/10.1080/09537280310001649372

Huang, Y. F. (2006). An inventory model under two-levels of trade credit and limited storage space derived without derivatives. *Applied Mathematical Modelling, 30*, 418–436. http://dx.doi.org/10.1016/j.apm.2005.05.009

Huang, Y. F. (2007). Optimal retailer's replenishment decisions in the EPQ model under two levels of trade credit policy. *European Journal of Operational Research, 176*, 1577–1591. http://dx.doi.org/10.1016/j.ejor.2005.10.035

Jaber, M. Y., Goyal, S. K., & Imran, M. (2008). Economic production quantity model for items with imperfect quality subject to learning effects. *International Journal of Production Economics, 115*, 143–150. http://dx.doi.org/10.1016/j.ijpe.2008.05.007

Khan, M., Jaber, M. Y., & Ahmad, A. R. (2014). An integrated supply chain model with errors in quality inspection and learning in production. *Omega, 42*, 16–24. http://dx.doi.org/10.1016/j.omega.2013.02.002

Kreng, V. B., & Tan, S. J. (2010). The optimal replenishment decisions under two levels of trade credit policy depending on the order quantity. *Expert Systems with Applications, 37*, 5514–5522. http://dx.doi.org/10.1016/j.eswa.2009.12.014

Liao, J. J. (2008). An EOQ model with noninstantaneous receipt and exponentially deteriorating items under two-level trade credit. *International Journal of Production Economics, 113*, 852–861. http://dx.doi.org/10.1016/j.ijpe.2007.09.006

Lin, C., & Lin, Y. S. (2004). A joint EOQ model for supplier and retailer with deteriorating items. *Asia-Pacific Journal of Operational Research, 21*, 163–178. http://dx.doi.org/10.1142/S0217595904000114

Lin, C., Tan, B., & Lee, W. C. (2000). An EOQ model for deteriorating items with time varying demand and shortages. *International Journal of Systems Science, 31*, 391–400. http://dx.doi.org/10.1080/002077200291235

Mahata, G. C. (2012). An EPQ-based inventory model for exponentially deteriorating items under retailer partial trade credit policy in supply chain. *Expert Systems with Applications, 39*, 3537–3550. http://dx.doi.org/10.1016/j.eswa.2011.09.044

Mahata, G. C., & Goswami, A. (2007). An EOQ model for deteriorating items under trade credit financing in the fuzzy sense. *Production Planning and Control, 18*, 681–692. http://dx.doi.org/10.1080/09537280701619117

Min, J., Zhou, Y. W., & Zhao, J. (2010). An inventory model for deteriorating items under stock-dependent demand and two-level trade credit. *Applied Mathematical Modelling, 34*, 3273–3285. http://dx.doi.org/10.1016/j.apm.2010.02.019

Murr, D. P., & Morris, L. L. (1975). Effect of storage temperature on post change in mushrooms. *Journal of the American Society for Horticultural Science, 100*, 16–19.

Ouyang, L. Y., Wu, K. S., & Yang, C. T. (2006). A study on an inventory model for non-instantaneous deteriorating items with permissible delay in payments. *Computers and Industrial Engineering, 51*, 637–651. http://dx.doi.org/10.1016/j.cie.2006.07.012

Rao, K. S., & Rao, S. E. (2015). Inventory model under permissible delay in payments and inflation with generalised Pareto lifetime. *International Journal of Procurement Management, 8*, 202–226. http://dx.doi.org/10.1504/IJPM.2015.066295

Rosen, S. (1972). Learning by experience as joint production. *The Quarterly Journal of Economics, 86*, 366–382. http://dx.doi.org/10.2307/1880798

Salameh, M. K., & Jaber, M. Y. (2000). Economic production quantity model for items with imperfect quality. *International Journal of Production Economics, 64*, 59–64. http://dx.doi.org/10.1016/S0925-5273(99)00044-4

Shah, N. H. (1993). A lot-size model for exponentially decaying inventory when delay in payments is permissible. *Cahiers du Centre d'études de recherche opérationnelle, 35*, 115–123. Retrieved from http://cat.inist.fr/?aModele=afficheN&cpsidt=4037390

Shastri, A., Singh, S. R., Yadav, D., & Gupta, S. (2014). Supply chain management for two-level trade credit financing with selling price dependent demand under the effect of preservation technology. *International Journal of Procurement Management, 7*, 695–717. http://dx.doi.org/10.1504/IJPM.2014.064978

Shinn, S. W., Hwang, H. P., & Sung, S. (1996). Joint price and lot size determination under conditions of permissible delay in payments and quantity discounts for freight cost. *European Journal of Operational Research, 91*, 528–542. http://dx.doi.org/10.1016/0377-2217(94)00357-2

Soni, H. (2013). Optimal replenishment policies for deteriorating items with stock sensitive demand under two-level trade credit and limited capacity. *Applied Mathematical Modelling, 37*, 5887–5895. http://dx.doi.org/10.1016/j.apm.2012.11.006

Teng, J. T., & Chang, C. T. (2009). Optimal manufacturer's replenishment policies in the EPQ model under two-levels of trade credit policy. *European Journal of Operational Research, 195*, 358–363. http://dx.doi.org/10.1016/j.ejor.2008.02.001

Teng, J. T., Chang, C. T., & Goyal, S. K. (2005). Optimal pricing and ordering policy under permissible delay in payments. *International Journal of Production Economics, 97*, 121–129. http://dx.doi.org/10.1016/j.ijpe.2004.04.010

Teng, J. T., Lou, K. R., & Wang, L. (2013). Optimal trade credit and lot size policies in economics production quantity models with learning curve production costs. *International Journal of Production Economics, 155*, 318–323. doi:10.1016/j.ijpe.2013.10.012

Tsai, D. M. (2012). Optimal ordering and production policy for a recoverable item inventory system with learning effect. *International Journal of Systems Science, 43*, 349–367. http://dx.doi.org/10.1080/00207721.2010.502261

Tsao, Y. C. (2009). Retailer's optimal ordering and discounting policies under advance sales discount and trade credits. *Computers and Industrial Engineering, 56*, 208–215. http://dx.doi.org/10.1016/j.cie.2008.05.006

Wang, K. J., Wang, S. M., & Yang, S. J. (2007). A resource portfolio model for equipment investment and allocation of semiconductor testing industry. *European Journal of Operational Research, 179*, 390–403. http://dx.doi.org/10.1016/j.ejor.2006.04.006

Wee, H. M. (1995). A deterministic lot-size inventory model for deteriorating items with shortages and a declining market. *Computers and Operations Research, 22*, 345–356.

Wright, T. (1936). Factors affecting the cost of airplanes. *Journal of the Aeronautical Sciences (Institute of the Aeronautical Sciences), 3*, 122–128. http://dx.doi.org/10.2514/8.155

Yadav, D., Singh, S. R., & Kumari, R. (2012a). Inventory model of deteriorating items with two-warehouse and stock dependent demand using genetic algorithm in fuzzy environment. *Yugoslav Journal of Operations Research, 22*, 51–78. http://dx.doi.org/10.2298/YJOR100219005Y

Yadav, D., Singh, S. R., & Kumari, R. (2012b). Effect of demand boosting policy on optimal inventory policy with backorder in fuzzy environment under the effect of learning. *International Journal of Procurement Management, 5*, 178–201. http://dx.doi.org/10.1504/IJPM.2012.045649

Yadav, D., Singh, S. R., & Kumari, R. (2015). Retailer's optimal policy under inflation in fuzzy environment with trade credit. *International Journal of Systems Science, 46*, 754–762. http://dx.doi.org/10.1080/00207721.2013.801094

Zauberman, G., Ronen, R., Akerman, M., & Fuchs, Y. (1990). Low PH treatment protects litchi fruit colour. *Acta Hortic, 269*, 309–314.

Zhou, Y. W., Zhong, Y. G., & Wahab, M. I. M. (2013). How to make the replenishment and payment strategy under flexible two-part trade credit. *Computers & Operations Research, 40*, 1328–1338. http://dx.doi.org/10.1016/j.cor.2012.12.013

10

Multi-objective optimization algorithms for mixed model assembly line balancing problem with parallel workstations

Masoud Rabbani[1]*, Reyhaneh Siadatian[1], Hamed Farrokhi-Asl[2] and Neda Manavizadeh[3]

*Corresponding author: Masoud Rabbani, School of Industrial Engineering, College of Engineering, University of Tehran, P.O. Box 11155- 4563 Tehran, Iran

E-mail: mrabbani@ut.ac.ir

Reviewing editor: Zude Zhou, Wuhan University of Technology, China

Abstract: This paper deals with mixed model assembly line (MMAL) balancing problem of type-I. In MMALs several products are made on an assembly line while the similarity of these products is so high. As a result, it is possible to assemble several types of products simultaneously without any additional setup times. The problem has some particular features such as parallel workstations and precedence constraints in dynamic periods in which each period also effects on its next period. The research intends to reduce the number of workstations and maximize the workload smoothness between workstations. Dynamic periods are used to determine all variables in different periods to achieve efficient solutions. A non-dominated sorting genetic algorithm (NSGA-II) and multi-objective particle swarm optimization (MOPSO) are used to solve the problem. The proposed model is validated with GAMS software for small size problem and the performance of the foregoing algorithms is compared with each other based on some comparison metrics. The NSGA-II outperforms MOPSO with respect to some comparison metrics used in this paper, but in other metrics MOPSO is better than NSGA-II. Finally, conclusion and future research is provided.

Subjects: Industrial Engineering & Manufacturing; Manufacturing Engineering; Production Engineering

Keywords: multi-objective optimization; mixed model assembly lines; balancing; sequencing; parallel workstations

ABOUT THE AUTHORS

Masoud Rabbani is a professor of Industrial Engineering in the School of Industrial and Systems Engineering at the University of Tehran.

Reyhaneh Siadatian is MSc degree student in school of Industrial Engineering at university of Tehran.

Hamed Farrokhi-Asl finished his Bachelor of Science in Industrial Engineering from University of Khaje-Nasir University of Technology, Iran at 2013. He continued his master's studies at University of Tehran and graduated at 2014.

Neda Manavizadeh obtained her PhD in Industrial Engineering from College of Engineering, University of Tehran (Start year: 2008). Currently, she is an assistant professor at Industrial Engineering Department, KHATAM University, Tehran, Iran.

PUBLIC INTEREST STATEMENT

In today's highly competitive industrial environment, increasing systems' flexibility to meet customers' demands is crucial. Assembly line is a flow of materials subjected to workstations. In mixed model assembly lines several products are made on an assembly line while the similarity of these products is so high. As a result, it is possible to assemble and produce several types of products simultaneously in rational time. So, the flexibility of this type of the assembly line is high to produce different product with high volume based on customers' demand. This paper deals with the mixed model assembly line balancing problem where the parallel workstations in dynamic periods are considered. Dynamic periods means that each period has effects on its next period, and this is used to increase efficiency of the line. Two algorithms are used to solve the proposed mathematical model and the results are compared with each other.

1. Introduction

Assembly line is a flow of materials and components subjected to workstations placed as a sequence. As parts move through this line of production, they are assembled together or to the main part in order to form the final products. One of the models used in this kind of line, is a hybrid model called mixed model assembly line (MMAL) (Soman, van Donk, & Gaalman, 2004). In MMALs several products are made on an assembly line while the similarity of these products is so high that the setup time for stations from one product to another one is assumed to be negligible. As a result, it is possible to assemble several types of products simultaneously with respect to orders entered to the line without any additional setup times. A variety of similar models in MMAL can meet the various demands of customers. In today's highly competitive manufacturing environment, increasing flexibility of systems according to customers ' desires is crucial. MMAL has several advantages, such as products manufacturing based on customers demand and increasing the flexibility of the line. As a result, it is so important to study and search about this topic (Manavizadeh, Hosseini, Rabbani, & Jolai, 2013).

According to the above mentioned, if several products with high workload come to a workstation consecutively, they may increase the cycle time. To prevent this happening, concepts of balance and sequence are used. Sequencing assembly line includes the order of products entering the assembly line and whenever several models come with high volume followed at a station, the cycle time may exceed the station (Boysen, Fliedner, & Scholl, 2009). Assembly line balancing emerges when designing an assembly line and it consists of finding a feasible assignment of tasks to workstations in such a path that the assembly costs are minimized, the demand is met and constraints of the assembly process are satisfied (Boysen, Fliedner, & Scholl, 2007). As a result, the aim of the assembly line balancing is optimal assignment of works to stations.

There are some models used in assembly line balancing problem. The first of them which is used in this study is called assembly balancing problem type-I. It consists of assignment of tasks to workstations where the number of workstations should be minimized for a given cycle time. The second types of assembly balancing problem is type-II and it occurs when there are certain numbers of workstations, but the cycle time is unknown (Akpınar & Bayhan, 2011).

The rest of the paper is organized as follows: in Section 2, some related literatures are reviewed. Mathematical model and the problem definition are presented in Section 3. The concept of a non-dominated solution is illustrated in Section 4. In Section 5, the methodology for tackling of this problem is described. Numerical results and comparison of two metaheuristics are presented in Section 6. Finally, the study is ended with the conclusion in Section 7.

2. Literature review

McMullen and Frazier (1998) found that using equal weight for each objective to solve a multi-objective assembly line balancing problem with simulated annealing algorithm when workstations are parallel has a better result. Simaria and Vilarinho (2004) solved MMAL balancing problem type-II with genetic algorithm to find the best balance in parallel workstations while the problem had constraints of zoning. McMullen and Tarasewich (2003) used ant techniques to solve the MMAL balancing problem with parallel workstations and probable time work. They compared ant techniques with several other heuristics, such as simulated annealing and found that this approach is competitive with the other methods.

Simaria and Vilarinho (2009) solved balancing two-sided assembly lines problem with an ant colony optimization algorithm to minimize number of workstations and balancing the workload with precedence, capacity, zoning, and synchronism constraints. They understood that their proposed procedure is performed well. Rabbani, Kazemi, and Manavizadeh (2012) proposed a new approach to minimize the number of workstations when these are intersecting in mixed model U-line balancing type-I problem and solved it with genetic algorithm. Hamzadayi and Yildiz (2012) used genetic algorithm and simulated annealing to solve MMAL balancing type-I problem with sequence in U-line with parallel workstations and zoning constraints. Boysen, Kiel, and Scholl (2011) focused on the added workload stations in sequencing MMAL and tried to minimize the number of work overload situations. They introduced different exact and heuristic algorithms such as branch and bound and tested their model in some instances.

Mamun, Khaled, Ali, and Chowdhury (2012) proposed a genetic algorithm to solve balancing mixed model assembly line of type-I with some features such as parallel workstations, zoning constraints, and sequence with limit resources. They defined reassignment process to increase flexibility in task allocation. Tiacci (2012) evaluated operational objectives of MMAL problem with parallel stations, stochastic task times, sequences, and buffers within workstations. They solved their model in Assembly Line Simulator (ASL) which is able to quickly model and simulate intricate assembly line. Akpınar, Bayhan, and Baykasoglu (2013) proposed a new hybrid algorithm which is combined ant colony optimization with genetic algorithm for MMAL balancing problem type-I with some features such as parallel workstations, zoning constraints, and sequence. Kellegöz and Toklu (2015) proposed a new method for balancing assembly line with parallel multi-manned workstations with zoning constraints and sequence. They used branch and bound algorithm for a comparative evaluation.

Tiacci (2015a) solved buffer allocation problem and assembly line balancing problem simultaneously for the first time. He considered stochastic task times, parallel workstations, and buffers between workstations for his problem and solves it with genetic algorithm. Yang and Gao (2016) presented MMAL problem with adjacent workforce cross-training where the skill of doing each task can be learned by two workers in adjacent stations and then they can reallocate the tasks occurred variation of demand. They solved this problem with a branch, bound, and remember algorithm. Sivasankaran and Shahabudeen (2014) had a review on assembly line balancing problems. They classified problem into single model or multi model, deterministic or probabilistic task times, types of assembly line (straight line or U- line) and with which algorithms these problem were solved. Finally, they found out which types of problems didn't work yet. According to the mention, an overview of approaches in the literature on MMAL balancing problem of type-I (MMALBP-I) is shown in Table 1.

According to the mentioned researches, this paper considers a bi-objective MMAL balancing problem with parallel workstations in dynamic situation. Two multi-objective metaheuristic algorithms

Table 1. An overview of approaches in the literature on MMALBP-I

Publication	Line configuration	Precedence constraint	Zoning constraint	Dynamic	Methodology
This study	Parallel workstations	√		√	NSGA-II, MOPSO
Manavizadeh et al. (2013)	U-line	√			SA
Ozbakir, Baykasoglu, Gorkemli, and Gorkemli (2011)	Parallel lines		√		ACO
Akpınar and Bayhan (2011)	Parallel workstations	√	√		GA, heuristics
Mamun et al. (2012)	Parallel workstations	√	√		GA
Akpınar et al. (2013)	Parallel workstations	√	√		GA, ACO
Gökçen et al. (2006)	Parallel lines	√			SA
Vilarinho and Simaria (2006)	Parallel workstations	√	√		ACO

are used to solve the problem. Finally, the performance of non-NSGA-II and multi-objective particle swarm optimization (MOPSO) algorithms is compared with each other for this problem. Details of this problem about objectives and assumptions are in the next section.

3. Problem description

In this paper, we are trying to consider the layout of a MMAL including parallel workstations in dynamic periods. Using parallel workstations has many features such as, two or more replicas of a workstation can perform the same set of tasks on different assemblies when required. Cycle times are allowed to be shorter than the longest task time in parallel workstations, thus it can increase the production rate and it provides better flexibility in designing the assembly line and the number of tasks performed by each worker increases (Vilarinho & Simaria, 2002).

Each work center (WC) comprises either one workstation for the instance of non-paralleling or multiple parallel workstations. A WC with multiple workstations is considered busy if every workstations inside are busy. An example of parallel workstations is shown in Figure 1. According to this figure, one operator is assigned to each workstation. There are seven workstations in the mentioned figure where two of them are parallel with each other. These two parallel workstations have created one WC. Another five workstations that there are no parallel workstations with them consisted five WCs where each of workstation is one WC. In parallel workstations, the tasks assigned to the WC are not shared among the workstations, while every workstation can perform each one of the tasks (Tiacci, 2015b).

In such a way, workstations are replicated only when processing time of a task is higher than the cycle time for at least one model, so the number of this replication does not exceeded by the tasks which have a task time higher than the cycle time (Akpınar & Bayhan, 2011; Vilarinho & Simaria, 2002). According to the literature, this research has following objectives.

3.1. Objectives

In this research, there are two objectives which should be achieved: First of them is minimizing number of workstations (Akpınar & Bayhan, 2011; Manavizadeh et al., 2013; Vilarinho & Simaria, 2002) and the second one is to maximize workload smoothness between workstations (Akpınar & Bayhan, 2011; Mamun et al., 2012; Vilarinho & Simaria, 2002, 2006). The second objective aims to balance the line and it means that for each model, the idle time is divided through the workstations equally (Mamun et al., 2012; Vilarinho & Simaria, 2006).

3.2. Assumptions

The common assumptions of the problem are listed below:

- Assembly line balancing problem type-I are used.
- Precedence constraints and precedence diagrams for each product are known. Precedence constraints indicate that an assignment should be allocated to a workstation, if it has no

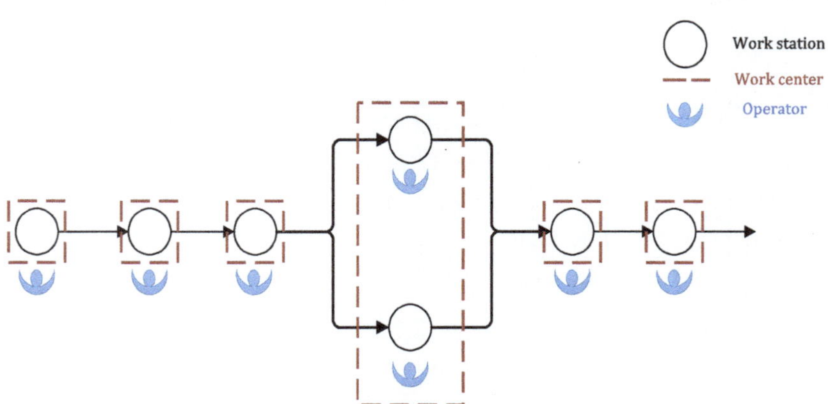

Figure 1. An assembly line with parallel workstations.

predecessors, or if all of its predecessors have already been assigned to that workstation or its previous workstations.

- Task performance's times of each product are known.
- Operators who are working in each workstation of the each line are multi-skilled and can be assigned to work at any station (flexible workers).
- Setup times between models are supposed as insignificant.
- All the operators are permanent and we do not have overtime periods.
- The number of workstations are variable.
- Cycle time is given and we have dynamic periods to determine all variables in different periods to achieve efficient solutions.
- We have a straight line with parallel workstations which can also be worked on each side of any line. Workstations along the line can be replicated to create parallel workstations.

3.3. Mathematical model

Every time one of the workstations has to wait for the completion of its predecessors, idle times will occurr. Minimizing the idle time which is related to the line is equivalent to minimizing the number of workstations (Manavizadeh et al., 2013). Minimizing the number of workstations can lead to reduced weighted line efficiency (WLE). So maximizing the WLE is another objective of this research to prevent this occurrence.

3.3.1. Notations and parameters

t = index of period $t = 1, 2, ..., T$

i = index of task $i = 1, 2, ..., N$

k = number of workstations $k = 1, 2, ..., K$

P = a specified planning horizon

M = number of similar models $m = 1, 2, ..., M$

H_i = the set of successors of task i

D_{mt} = the forecast demand of model m in period t

t_{imt} = the time required to perform task i on model m in period t

R_{kt} = the number of replicas of workstation k in period t

s_t = the real number of workstations needed to meet the demand in period t is: $s_t = \sum_{k=1}^{K} R_{kt}$

C_t = cycle time in period t. The forecast demand over the planning horizon, for model m in period t, is: $C_t = \frac{P}{\sum_{m=1}^{M} D_{mt}}$

q_{mt} = the overall proportion of the number of units of model m being assembled in period t is showed by: $q_{mt} = \frac{D_{mt}}{\sum_{p=1}^{M} D_{pt}}$

3.3.2. Decision variables

x_{ijt} = equal 1 if task i is assigned to workstation j in period t and equal 0 otherwise

r_{kt} = equal 1 if workstation k can be replicated in period t

IT_{kmt} = idle time of workstation k on model m in period t

I_{mt} = the stock of model m at the end of period t

NP_{mt} = total number of production of model m in period t

Objective function:

$$\text{MinZ} = \alpha \sum_{t=1}^{T} \sum_{k=1}^{K} k.x_{Nkt} + \beta \frac{\sum_{t=1}^{T} s_t}{\sum_{t=1}^{T} s_t - 1} \sum_{t=1}^{T} \sum_{m=1}^{M} \left(q_{mt} \sum_{k=1}^{s_t} \left(\frac{IT_{kmt}}{\sum_{l=1}^{s_t} IT_{lmt}} - \frac{1}{s_t} \right)^2 \right) \tag{1}$$

The objective of minimizing the number of the workstations is represented by the first term in objective function, and the objective of smoothing workload between the workstations is the second term (Akpınar & Bayhan, 2011).

In the objective function, α and β are coefficients indicating the weight of each objective which is determined by the management while the sum of these factor must be 1. Since minimizing the number of workstations is more important than the second objective so, its weight must be higher than second objective ($\alpha \succ \beta$) (Vilarinho & Simaria, 2006). Constraints of this model are formulated as follows:

Subject to:

$$\sum_{t=1}^{T}\sum_{k=1}^{K} x_{ikt} = 1 \quad i = 1, 2, ..., N \tag{2}$$

$$I_{m,t-1} - I_{mt} + NP_{mt} = D_{mt} \quad \forall m, \forall t > 1 \tag{3}$$

$$\sum_{k=1}^{K} x_{akt} - \sum_{t'=t}^{T}\sum_{k=1}^{K} x_{bkt'} \leq 0 \quad \forall t = 1...T \; a \in N, b \in H_a \tag{4}$$

$$\sum_{t=1}^{T}\sum_{i=1}^{N} t_{imt}.x_{ikt} + \sum_{t=1}^{T} IT_{kmt} = \sum_{t=1}^{T} c_t\left(1 + \sum_{t=1}^{T} r_{kt}.(R_{kt} - 1)\right) \quad \forall m, \forall k \tag{5a}$$

$$\sum_{t=1}^{T} r_{kt} \leq \sum_{t=1}^{T}\sum_{i=1}^{N} x_{ikt} \quad \forall k \tag{5b}$$

$$\lambda \sum_{t=1}^{T} r_{kt} \geq \sum_{t=1}^{T}\sum_{i=1}^{N} x_{ikt} \quad \forall k \tag{5c}$$

$$x_{ikt} \in \{0, 1\} \tag{6a}$$

$$r_{kt} \in \{0, 1\} \tag{6b}$$

$$IT_{kmt} \geq 0 \tag{6c}$$

$$I_{mt} \geq 0 \tag{6d}$$

$$NP_{mt} \geq 0 \tag{6e}$$

Constraint (2) ensures that all tasks are assigned to a station and each task is assigned only once (Gökçen, Ağpak, & Benzer, 2006).

Equality (3) indicates dynamic periods which shows the gap between two periods and describes available stocks at the end of period t which is used at the first of period $t + 1$. Dynamic environment is considered for objectives and other constraints.

Constraint (4) describes the precedence constraint. This means that before assigning task i to station k, all successors of its have been assigned to either station k or further stations. In this constraint, task b is a successor of task i.

Constraints (5a)–(5c) ensure that total time capacity of each workstation is up to task times assigned to each workstations and it will be equal to cycle time if there are no replication for that workstation. The maximum number of replicas of a workstation in a WC is up to the tasks which have a task time more than the cycle time for at least one model, and just the workstations where the time of the tasks assigned to it exceeds by the cycle time can be duplicated. In constraint (5c), λ is a very large positive integer (Akpınar & Bayhan, 2011).

Equalities (6a)–(6e) mean domain of the decision variables.

4. Non-dominated solution

Multi-objective problems (MOP) appear in the most orders in today's study. Attempting to solve a MOP seems to be more complete and sensible in many issues (Deb, Agrawal, Pratap, & Meyarivan, 2000). A general single-objective problem is defined as:

$$\text{Max or Min } f(x) \tag{7}$$

st:

$$g_i(x) \leq 0 \quad i = 1, 2, \ldots, m \tag{8}$$

$$h_j(x) = 0 \quad j = 1, 2, \ldots, p \tag{9}$$

$$x \in \Omega$$

where $g_i(x) \leq 0$ and $h_j(x) = 0$ are constraints of $f(x)$; Ω includes all possible x. While a general MOP is as follow:

$$\text{Max or Min } F(x) = [f_1(x), f_2(x), \ldots, f_k(x)] \tag{10}$$

st:

$$g_i(x) \leq 0 \quad i = 1, 2, \ldots, m \tag{11}$$

$$h_j(x) = 0 \quad j = 1, 2, \ldots, p \tag{12}$$

$$x \in \Omega$$

In multi-objective issues due to a few objectives, we attempt to discover great bargains rather than a single solution so the documentation of optimum solution is introduced as Pareto Optimum. A solution $x \in \Omega$ will be Pareto Optimal if and only if there is no $x' \in \Omega$ for which $v = F(x') = (f_1(x'), f_2(x'), \ldots, f_k(x'))$ that dominates $u = F(x) = (f_1(x), f_2(x), \ldots, f_k(x))$. In Pareto dominance, vector "u" dominates vector "v" if and only if "u" is partially worse than "v".

5. Methodology

Two solution's algorithms designed to discover logical Pareto solutions, the first one is based on non-dominated sorting genetic algorithm (NSGA-II) and the other one is based on multi-objective particle swarm optimization (MOPSO). In this section, metaheuristic approaches are described to solve the problem. The steps of the NSGA-II are schemed as follow (Akpınar & Bayhan, 2011):

5.1. Initial population

The initial population is created randomly; each individual is represented through three chromosomes. The first chromosome includes an order of different periods. The second chromosome contains an ordered sequence of all tasks which is used for each model according to the precedence diagram and its order is the same order that they are assigned to the WCs. The third chromosome consists workstations where the task is given to and if a workstation in a WC be replicated, the number of that workstation which is assigned to a gene will replicate in next gene for the number of parallel stations (Tiacci, 2015b). According to illustrations mentioned previously, tasks assigned to workstations required for each period are determined.

A simple example to explain creation of initial population is shown in Figure 2. According to this figure, each solution has three chromosomes to specify its properties which are period, task, and workstation, respectively. At first, the order of these chromosomes is generated, randomly. After that, they are transformed to meaningful chromosomes to show solutions of the problem. In this example, WC 1 has two workstations which are parallel, WC 2 has one workstation and WC 3 has two workstations which are parallel. According to Figure 2, the genes which are below each other are considered as an answer in a population. It means that each task in each period is assigned to a WC. According to this description, task number 1 in period "a" is assigned to WC 3, task number 3 in period "c" is assigned to WC 2 and task number 2 in period "b" is assigned to WC 1.

5.2. Crossover

The single-point crossover divides each of the parent into two parts (head and tail). This point is generated randomly. Recombination of parent one with parent two is generated new offsprings. The head of first offspring keeps the head part of the first parent and the tail part of the first offspring is filled with all missing tasks in the order in the second parent. According to the mentioned, the head of the second offspring is built based on the head part of the second parent and the tail part is filled with all missing tasks in the order in the first parent.

Each individual is represented through three chromosomes (i.e. tasks order, periods order, and workstations order). We should decide which chromosome should be selected for the crossover. For this reason, we generate one integer random number between one and seven, namely in range [1–7]. This number defines which chromosome(s) is (are) chosen for applying crossover operator. Note that "7" is obtained from $\binom{3}{1}+\binom{3}{2}+\binom{3}{3}=7$ (Rabbani, Farrokhi-Asl, & Ameli, 2016). State 1 occurs when the first chromosome of each parents is chosen for crossover and the other chromosomes remain unchanged which is shown in Figure 3. State 2 occurs when both of the first and the second chromosomes of each parents is chosen for crossover and the third chromosome is remained unchanged. State 3 occurs when the first and the third chromosomes of each parents is chosen for crossover and the second chromosome is remained unchanged and so on. Both of the generated offsprings become feasible as their head and the tail parts is also filled according to the precedence feasible order.

Figure 3 shows the state one when the first chromosome of each parent is chosen for crossover and the other chromosomes remain unchanged. According to this figure, period's chromosomes for both parents are changed in the offsprings and other chromosomes are remained unchanged. The figure shows that the head part of the first parent is located in the head part of the first offspring and the tail part of it is filled by unused genes from the second parent. According to the mentioned, *a, b,* and *c* genes of the head part of the first parent is located unchanged in the head part of the first offspring. Where *d, e,* and *f* genes from the second parent are not used in the first offspring. So, these three genes are located in the tail part of the first offspring, respectively.

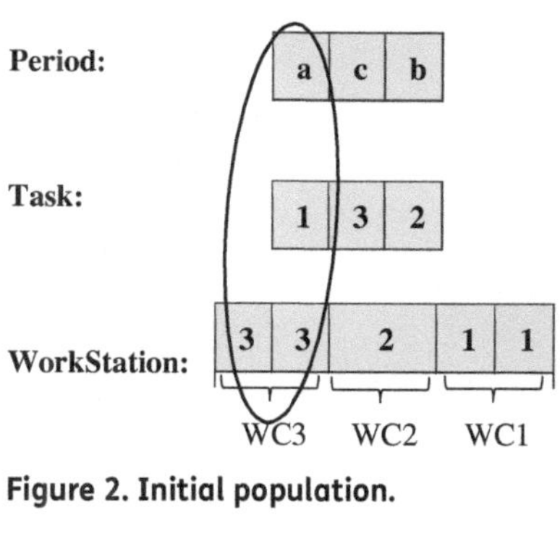

Figure 2. Initial population.

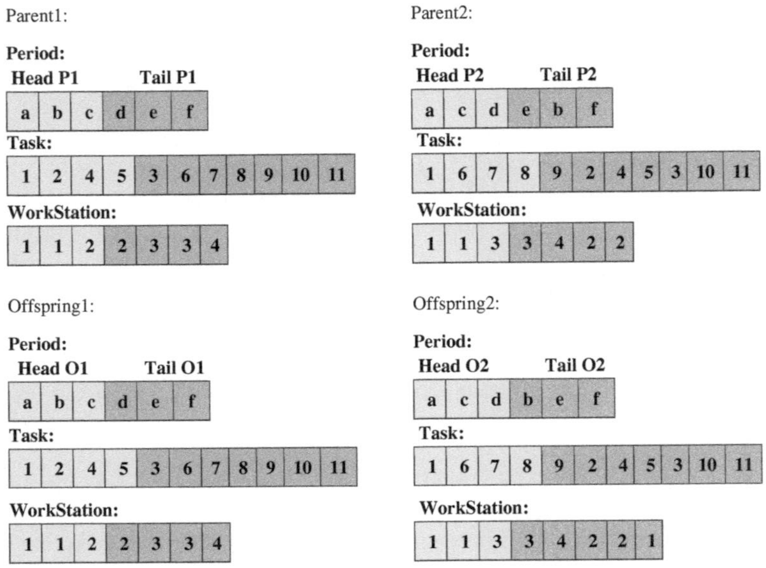

Figure 3. Single-point crossover.

5.3. Mutation

We used swap operation for mutation which two genes are selected randomly for new offspring and the place of these genes are replaced in the chromosome. Other genes of the parent without changing are placed in the offspring. Each individual is represented through three chromosomes, so for doing mutation [1–7] condition will be occur which are as same as crossover. In the first state which is shown in Figure 4, changes are occurred in period chromosome and the other two chromosomes are remained unchanged. As it is shown in the figure, the second gene from head part and the fifth gene from tail part of the period's chromosome from the parent are selected randomly and the palaces of them are changed. The other genes of this chromosome are remained unchanged. So, the arrangement of the genes for this chromosome is changed from *a-b-c-d-e-f* to *a-e-c-d-b-f*.

Crossover and mutation are used to define how the next generation is created. The strategy illustrative of which individuals will be stay in the population and which ones will be replaced. The individuals of new generation may be created by the previous generation and the offspring produced by crossover and mutation.

5.4. MOPSO algorithm

Particle swarm optimization is a metaheuristic algorithm which is designed for continuous problems. This algorithm is inspired by the behavior of birds who are in search of food. The movement' speed of each individual into the food is compared to the best behavior (Parsopoulos & Vrahatis, 2002). The multi-objective particle swarm optimization provides the Pareto front nearer to the required Pareto fronts and its performance is better than some other multi-objective algorithms (Coello & Lechuga, 2002). The steps of the MOPSO algorithm are as follows:

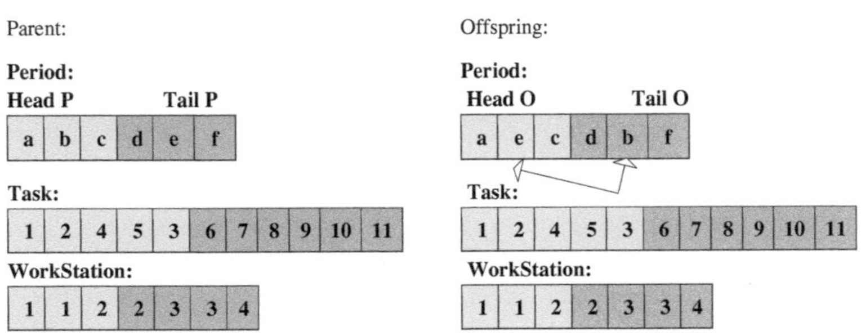

Figure 4. Swap mutation.

Step 1: *Generate the initial population of particles.*

Step 2: *Each particle' velocity is generated and saved in the velocity vector which determines the direction in which a particle should move to improve its new position.*

Step 3: *The new position of each particle is saved as memory of particle in the personal best, which shows the best position of each individual. The new position of each particle is calculated by sum of its previous position and its velocity vector. If the new position calculated for each particle is better than its previous position, the personal best should be equal to that position. The previous position in the memory should be kept when the new position is dominated by its.*

Step 4: *Leaders which is the position of non-dominated particles is saved in an external memory called global best. Leaders lead other particles towards better areas in search space and considered for all the individual.*

Step 5: *Until the number of iteration is same as the maximum number of iterations, we should repeat step 4.*

6. Numerical results

The performance of the NSGA-II and MOPSO algorithms is investigated and the related results are analyzed. The algorithms are coded in MATLAB R2013b and run on Intel CORE i7 2.30 GHz on personal computer with 6 GB RAM.

A numerical example is used to validate model with GAMS 23.5 software and the results of NSGA-II and MOPSO algorithms are compared with each other. In this example, precedence diagram can be shown as Akpınar and Bayhan (2011) in Figure 5 which is known for us in the problem with 30 tasks. For small size problem some of these tasks are considered. The initial experiment is performed by small-sized problems, which includes 10 sample test problems of different sizes, and 10 sample test problems of different sizes were used for large-sized problem. In this example, tasks are done over a planning horizon of 400 time units. A workstation can be replicated if a processing time of its task is more than the cycle time. The maximium number of replicas of workstations in a WC in this problem is supposed to be three (Akpınar & Bayhan, 2011).

The exact results of GAMS software for the proposed model are shown in Tables 2 and 3. Obtained results for variables are summarized in these two tables. In Table 4, the obtained variables by solving an each objective function separately is placed in another objective function to calculate its value.

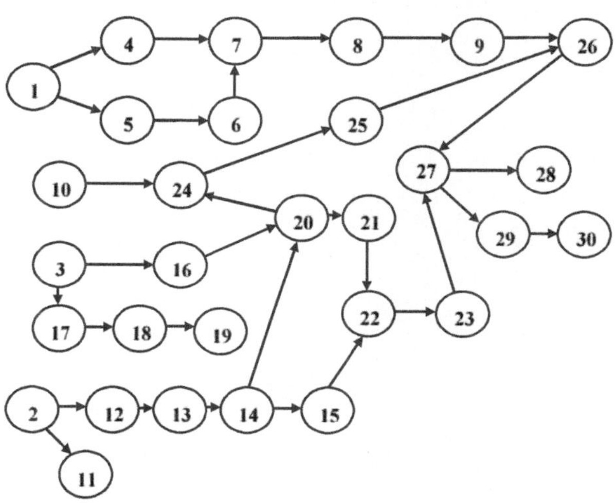

Figure 5. Precedence diagram.

Table 2. GAMS results for first objective

Index	1.1.1	1.1.2	1.2.1	1.2.2	1.3.1	1.3.2	2.1.1	2.1.2	2.2.1	2.2.2	2.3.1	2.3.2	3.1.1	3.1.2	3.2.1	3.2.2	3.3.1	3.3.2	4.1.1	4.1.2	4.2.1	4.2.2	4.3.1	4.3.2
Variable																								
x_{ijt}	1	0	0	0	0	0	1	0	0	0	0	0	0	0	1	0	0	0	1	0	0	0	0	0
r_{kt}	1		0				1		0				1		0									
IT_{kmt}	1	0	3	0			6	0	8	0			6	0	6	0								
I_{mt}	6		0				4		0															
NP_{mt}	10		0				10		0															

Table 3. GAMS results for second objective

Index	1.1.1	1.1.2	1.2.1	1.2.2	1.3.1	1.3.2	2.1.1	2.1.2	2.2.1	2.2.2	2.3.1	2.3.2	3.1.1	3.1.2	3.2.1	3.2.2	3.3.1	3.3.2	4.1.1	4.1.2	4.2.1	4.2.2	4.3.1	4.3.2
Variable																								
x_{ijt}	1	0	0	0	0	0	0	0	0	0	1	0	1	0	0	0	0	0	1	0	0	0	0	0
r_{kt}	0		0				0		0				0		0									
IT_{kmt}	1	0	0	4			6	0	6	0			3	0	0	4								
I_{mt}	6		0				0		0															
NP_{mt}	10		0				6		4															

Table 4. Competition of objectives in GAMS

	Z_1	Z_2
Z_1	$Z_1 = 1$	$Z_2 = 0.3357$
Z_2	$Z_1 = 1$	$Z_2 = 0.147$

6.1. Meta-heuristic methods

In NSGA-II and MOPSO algorithms, we classified test problems as small-sized and large-sized test problems with respect to the number of periods, tasks, workstations, and models including in them. The results of comparison of NSGA-II and MOPSO algorithms are shown in Tables 5–14. Problem characters in Tables 5–14 are in order of the number of periods, tasks, workstations, and models they include. Each parameter of algorithms is used as follows.

General parameters:

- Number of population size for both algorithm is set to 100.
- Experiments for both algorithm are repeated 20 times.
- Maximum number of iteration for both algorithms in each run of algorithm is considered as 50.

NSGA-II parameters:

- Crossover rate and mutation rate are in order 0.8 and 0.3.

MOPSO parameters:

- Number of repository size is noted 100.
- Values of C_1 (personal learning coefficient) and C_2 (global learning coefficient) are set to 2.
- Grid inflation parameter (α) is set to 0.1.
- Leader selection pressure parameter (β) is noted as 4.
- Gamma which is defined as extra repository member selection pressure parameter is considered 2.

In order to compare the efficiency of these algorithms five comparison metrics such as number of Pareto solutions, diversification metric (DM), spacing metrics (SM), mean ideal distance (MID), and mean square index (MSI) were used. DM recognizes the extension of solution set and calculated as follows:

$$DM = \sqrt{\sum_{i=1}^{n} \max(\| x_t^i - y_t^i \|)} \tag{13}$$

where $\max(\| x_t^i - y_t^i \|)$ is distance between the non-dominated solutions x_t^i and y_t^i. The results of our computations for this metric are shown in Tables 5 and 6.

Tables 7 and 8 described the computation results of SM wh ich provides us to know the Euclidean distance between two adjacent solutions of Pareto front. This kind of metric provides us to know details about the uniformity of the distribution of the solutions which is obtained by each of the algorithm. This metric is computed as follows:

$$SM = \sqrt{\frac{1}{N-1} \times \sum_{i=1}^{n} (d_i - \bar{d})^2} \tag{14}$$

where d_i is the Euclidean distance between solution i and the nearest solution belonged to Pareto set of solutions and $\bar{d}$ is the average value of all d_i.

Table 5. Diversity metric (DM) for small size		
Average	**3.58527**	**5.023423**
Problem characters	NSGA-II	MOPSO
2-3-4-2	2.8284271247461	2.45015319037307
2-4-6-2	2.828427124746	2.82878899852692
2-5-7-2	2	5.65733205717679
2-6-8-2	2.2360679774	10.8167288058419
2-7-10-2	5.74458549317	5.91607978309962
3-8-8-3	4.0003136569	5.65688750321298
3-8-9-3	8.7178286821	6.0000109716944
3-9-12-3	1.4142646273	7.74596669241483
3-10-12-3	6.082784615	0
3-10-14-3	0	3.16227766016838

Table 6. DM for large size		
Average	**4.030011**	**11.50504**
Problem characters	NSGA-II	MOPSO
4-19-18-5	1.41929671309859	9.16515138991168
4-20-20-5	1.41464100574115	0
4-21-18-5	2.82853860	17.9443584449264
5-22-19-6	2.82866812540513	0.883720930232558
5-25-20-6	1.4145532275	16.6733320005331
5-26-23-6	6.55758735584786	16.431676725155
5-27-25-7	6.00007446626679	10.4880884817015
6-28-27-7	7.9373566656	9.9498743710662
6-28-28-8	5.0003502062	25.7681974534503
6-30-28-8	4.89904845846309	7.74596669241483

Table 7. SM for small size		
Average	**0.337056**	**0.183864**
Problem characters	NSGA-II	MOPSO
2-3-4-2	0.5	0
2-4-6-2	0.5	0
2-5-7-2	0	0.932091039049302
2-6-8-2	0	0.639882423582033
2-7-10-2	0.153873333	0.266666666666667
3-8-8-3	0.4996576884	0
3-8-9-3	0.4443394946	0
3-9-12-3	0	0
3-10-12-3	1.2726878292	0
3-10-14-3		0

MID metric shows sum of the distance between obtained solutions (f^i; where i is the number of Pareto) and the ideal answer (f^{best}) for each objective function. The results of this metric are shown in Tables 9 and 10. Ideal answer is the point in population which has the best answer in all of

Table 8. SM for large size

Average	**1.083484**	**0.629952**
Problem characters	NSGA-II	MOPSO
4-19-18-5	0	0.363636363636364
4-20-20-5	0	
4-21-18-5	0.49953363	0.6875
5-22-19-6	0.498589097127162	0.883720930232558
5-25-20-6	0	0.989247311827957
5-26-23-6	0.399877644727188	1.32075471698113
5-27-25-7	0.499959408866703	0
6-28-27-7	7.9373566656	1.1219512195122
6-28-28-8	0.4281549607	0.932707355242567
6-30-28-8	0.571371224816095	0

Table 9. MID for small size

Average	**0.981219**	**0.984215**
Problem characters	NSGA-II	MOPSO
2-3-4-2	1.000223930499	0.91378458764677
2-4-6-2	0.976587442255	0.98944701349708
2-5-7-2	0.959888719832	1.00633917800085
2-6-8-2	0.98153139073	1.01666355204049
2-7-10-2	1.0027624299	1.00472327236632
3-8-8-3	1.007753485	1.01230488121194
3-8-9-3	1.0007072821	0.98930468444202
3-9-12-3	0.97399547051	1.00368664165091
3-10-12-3	0.99634459732	0.90553851381
3-10-14-3	0.91239464725	1.0003560058896

Table 10. MID for large size

Average	**0.998784**	**0.998749**
Problem characters	NSGA-II	MOPSO
4-19-18-5	0.99658145956181	1.0013596627309
4-20-20-5	0.99096373896803	0.98165616894063
4-21-18-5	0.99736741523	1.00100576896847
5-22-19-6	0.99969522164064	1.00149981451826
5-25-20-6	0.99321634447	1.00088307431177
5-26-23-6	1.00230794372585	1.00035328221594
5-27-25-7	1.00119200055204	0.99972863946362
6-28-27-7	1.003963293	1.00081737175658
6-28-28-8	1.0019985629	1.0003934050078
6-30-28-8	1.0005499319529	0.99978886673097

Table 11. MSI for small size

Average	0.02849	0.061832
Problem characters	NSGA-II	MOPSO
2-3-4-2	1.11022302462516e-16	0.0987654320987654
2-4-6-2	1.80411241501588e-16	0.0904977375565611
2-5-7-2	5.55111512312578e-17	0.176470588235294
2-6-8-2	1.11022302462516e-16	0.126315789473684
2-7-10-2	0.0391304347826	2.775557561562e-16
3-8-8-3	0.114942528735	0.0775862068965516
3-8-9-3	0.0732758620689	0.0486815415821498
3-9-12-3	0.012018381053	4.440892098500e-16
3-10-12-3	0.045528455284	0
3-10-14-3	0	1.110223024625e-16

Table 12. MSI for large size

Average	0.068898	6.72e-16
Problem characters	NSGA-II	MOPSO
4-19-18-5	0.120229042105453	5.55111512312578e-16
4-20-20-5	0.034778450949676	0
4-21-18-5	0.0430062498	4.9960036108132e-16
5-22-19-6	0.0639204545454549	6.10622663543836e-16
5-25-20-6	0.03100103762	3.88578058618805e-16
5-26-23-6	0.0886902165524723	3.33066907387547e-16
5-27-25-7	0.059846154767623	2.77555756156289e-16
6-28-27-7	0.1508811481	1.77635683940025e-15
6-28-28-8	0.0519081911	8.32667268468867e-16
6-30-28-8	0.044718805316506	1.4432899320127e-15

Table 13. Number of Pareto solutions and computational time for small size

	NPS		Computational time(s)	
Average	3	2.9	74.41495	9.71821
Problem characters	NSGA-II	MOPSO	NSGA-II	MOPSO
2-3-4-2	3	2	44.71	3.304637
2-4-6-2	3	2	48.195905	3.930728
2-5-7-2	2	4	46.378222	4.814735
2-6-8-2	3	6	56.205583	5.498103
2-7-10-2	4	4	57.127584	5.665040
3-8-8-3	3	2	85.788409	12.088588
3-8-9-3	5	2	94.436604	14.718414
3-9-12-3	2	4	94.434902	14.325840
3-10-12-3	4	1	110.221109	14.119085
3-10-14-3	1	2	106.651202	18.716929

Table 14. Number of Pareto solutions and computational time for large size

	NPS		Computational time(s)	
Average	**3.8**	**3.2**	**641.545**	**277.2629**
Problem characters	NSGA-II	MOPSO	NSGA-II	MOPSO
4-19-18-5	2	3	351.614671	70.098852
4-20-20-5	2	1	366.901229	110.831746
4-21-18-5	3	5	378.450320	153.406588
5-22-19-6	3	5	508.753348	250.982827
5-25-20-6	2	4	583.892324	201.820911
5-26-23-6	6	3	591.855149	312.749703
5-27-25-7	5	2	792.193048	404.199652
6-28-27-7	5	3	920.721841	525.944696
6-28-28-8	5	4	1,063.405027	521.565098
6-30-28-8	5	2	857.667723	221.028653

objective functions in the problem. The ideal answer in this problem is considered as the point of (0,0) because all objectives should be minimized. $f_{1,total}^{max}$ is the maximum value of the first objective function and $f_{1,total}^{min}$ is the minimum value of it. It is better to have less value of this metric. This metric is computed as follows:

$$\mathrm{MID} = \frac{\sum_{i=1}^{N} \sqrt{\left(\frac{f_1^i - f_1^{best}}{f_{1,total}^{max} - f_{1,total}^{min}}\right)^2 + \left(\frac{f_2^i - f_2^{best}}{f_{2,total}^{max} - f_{2,total}^{min}}\right)^2}}{n} \quad (15)$$

MSI which is shown in Tables 11 and 12 is the mean square of the maximum value of objective function minus the minimum value of the objective function.

Number of Pareto solutions measures ability of the algorithms to find effective point and computational time, which are summarized in Table 13 for small size and Table 14 for large-sized problems, respectively.

DM measure shows that in the most examples MOPSO operate better than NSGA-II in both sizes of the problems, because it has a higher value than NSGA-II. In SM measure, the average value of NSGA-II is more than MOPSO in both small and large sizes of the problem. So, we can say that MOPSO is superior to NSGA-II. MID measure does not show superiority for any of them. In this metric, the average value for NSGA-II in the small sizes is less than MOPSO, so it performs better than MOPSO in small sizes and in the large-scale MOPSO is better than NSGA-II. MOPSO is superior to NSGA-II in MSI measure. The computational time for MOPSO is almost better than NSGA-II. In most examples, NSGA-II is better than MOPSO in producing Pareto numbers.

7. Conclusion

This research deals with MMALBP-I which had some particular features such as parallel workstations and precedence constraints to reduce the number of workstations and maximize the workload smoothness between workstations for a given cycle time. This model has been implemented in a dynamic situation.

In the first step GAMS software was used to solve small-scale problem. For the second step, NSGA-II and multi-objective swarm optimization were used to solve the problem for the numerical example in small and large scales. The results obtained from those algorithms were compared with each other. The results which are displayed in most examples of DM and MSI measure, MOPSO is superior to NSGA-II, but in the SM NSGA-II operate better than MOPSO. The value of MID measure shows that in small scale, NSGA-II is better than MOPSO; and in large scale MOPSO is better than NSGA-II, so this measure doesn't show superiority for any of them. The computational time for MOPSO is almost better than NSGA-II, but in most examples NSGA-II is better than MOPSO in producing Pareto numbers.

For future research, MMAL balancing problem of type-II with U- line can be used in a dynamic situation when cycle time is unknown and number of workstations is known using the constraints of number of operators and skill level of them.

Funding
The authors received no direct funding for this research.

Author details
Masoud Rabbani[1]
E-mail: mrabbani@ut.ac.ir
Reyhaneh Siadatian[1]
E-mail: r_siadatian@ut.ac.ir
Hamed Farrokhi-Asl[2]
E-mail: hamed.farrokhi@ut.ac.ir
Neda Manavizadeh[3]
E-mail: n.manavi@khatam.ac.ir
[1] School of Industrial Engineering, College of Engineering, University of Tehran, P.O. Box 11155-4563 Tehran, Iran.
[2] School of Industrial Engineering, Iran University of Science & Technology, Tehran, Iran.
[3] Department of Industrial Engineering, KHATAM University, Tehran, Iran.

References

Akpınar, S., & Bayhan, G. M. (2011). A hybrid genetic algorithm for mixed model assembly line balancing problem with parallel workstations and zoning constraints. *Engineering Applications of Artificial Intelligence, 24*, 449–457. http://dx.doi.org/10.1016/j.engappai.2010.08.006

Akpınar, S., Bayhan, G. M., & Baykasoglu, A. (2013). Hybridizing ant colony optimization via genetic algorithm for mixed-model assembly line balancing problem with sequence dependent setup times between tasks. *Applied Soft Computing, 13*, 574–589. http://dx.doi.org/10.1016/j.asoc.2012.07.024

Boysen, N., Fliedner, M., & Scholl, A. (2007). A classification of assembly line balancing problems. *European Journal of Operational Research, 183*, 674–693. http://dx.doi.org/10.1016/j.ejor.2006.10.010

Boysen, N., Fliedner, M., & Scholl, A. (2009). Sequencing mixed-model assembly lines: Survey, classification and model critique. *European Journal of Operational Research, 192*, 349–373. http://dx.doi.org/10.1016/j.ejor.2007.09.013

Boysen, N., Kiel, M., & Scholl, A. (2011). Sequencing mixed-model assembly lines to minimise the number of work overload situations. *International Journal of Production Research, 16*, 4735–4760. http://dx.doi.org/10.1080/00207543.2010.507607

Coello, C. A., & Lechuga, M. S. (2002). MOPSO: A proposal for multiple objective particle swarm optimization. *Proceedings of the IEEE Congress on Evolutionary Computation, 2*, 1051–1056.

Deb, K., Agrawal, S., Pratap, A., & Meyarivan, T. (2000). A fast elitist non-dominated sorting genetic algorithm for multi-objective optimization: NSGA-II. *Lecture notes in computer science, 1917*, 849–858. http://dx.doi.org/10.1007/3-540-45356-3

Gökçen, H., Ağpak, K., & Benzer, R. (2006). Balancing of parallel assembly lines. *International Journal of Production Economics, 103*, 600–609.

Hamzadayi, A., & Yildiz, G. (2012). A genetic algorithm based approach for simultaneously balancing and sequencing of mixed-model U-lines with parallel workstations and zoning constraints. *Computers & Industrial Engineering, 62*, 206–215.

Kellegöz, T., & Toklu, B. (2015). A priority rule-based constructive heuristic and an improvement method for balancing assembly lines with parallel multi-manned workstations. *International Journal of Production Research, 53*, 736–756. http://dx.doi.org/10.1080/00207543.2014.920548

Mamun, A. A., Khaled, A. A., Ali, S. M., & Chowdhury, M. M. (2012). A heuristic approach for balancing mixed-model assembly line of type I using genetic algorithm. *International Journal of Production Research, 50*, 5106–5116. http://dx.doi.org/10.1080/00207543.2011.643830

Manavizadeh, N., Hosseini, N. S., Rabbani, M., & Jolai, F. (2013). A simulated annealing algorithm for a mixed model assembly U-line balancing type-I problem considering human efficiency and Just-In-Time approach. *Computers & Industrial Engineering, 64*, 669–685.

McMullen, P. R., & Frazier, G. V. (1998). Using simulated annealing to solve a multiobjective assembly line balancing problem with parallel workstations. *International Journal of Production Research, 36*, 2717–2741. http://dx.doi.org/10.1080/002075498192454

McMullen, P. R., & Tarasewich, P. (2003). Using ant techniques to solve the assembly line balancing problem. *IIE Transactions, 35*, 605–617. http://dx.doi.org/10.1080/07408170304354

Ozbakir, L., Baykasoglu, A., Gorkemli, B., & Gorkemli, L. (2011). Multiple-colony ant algorithm for parallel assembly line balancing problem. *Applied Soft Computing, 11*, 3186–3198. http://dx.doi.org/10.1016/j.asoc.2010.12.021

Parsopoulos, K. E., & Vrahatis, M. N. (2002). Particle swarm optimization method in multiobjective problems. In *Proceedings of the 2002 ACM Symposium on Applied Computing, ACM* (pp. 603–607). Madrid. http://dx.doi.org/10.1145/508791

Rabbani, M., Farrokhi-Asl, H., & Ameli, M. (2016). Solving a fuzzy multi-objective products and time planning using hybrid meta-heuristic algorithm: Gas refinery case study. *Uncertain Supply Chain Management, 4*, 93–106. http://dx.doi.org/10.5267/j.uscm.2015.12.002

Rabbani, M., Kazemi, S. M., & Manavizadeh, N. (2012). Mixed model U-line balancing type-1 problem: A new approach. *Journal of Manufacturing Systems, 31*, 131–138. http://dx.doi.org/10.1016/j.jmsy.2012.02.002

Simaria, A. S., & Vilarinho, P. M. (2004). A genetic algorithm based approach to the mixed- model assembly line balancing problem of type II. *Computers & Industrial Engineering, 47*, 391–407.

Simaria, A. S., & Vilarinho, P. M. (2009). 2-ANTBAL: An ant colony optimization algorithm for balancing two-sided assembly lines. *Computers & Industrial Engineering, 56*, 489–506.

Soman, C. A., van Donk, D. P., & Gaalman, G. (2004). Combined make-to-order and make-to-stock in a food production system. *International Journal of Production Economics, 90*, 223–235. http://dx.doi.org/10.1016/S0925-5273(02)00376-6

Sivasankaran, P., & Shahabudeen, P. (2014). Literature review of assembly line balancing problems. *The International Journal of Advanced Manufacturing Technology, 73*, 1665–1694. http://dx.doi.org/10.1007/s00170-014-5944-y

Tiacci, L. (2012). Event and object oriented simulation to fast evaluate operational objectives of mixed model assembly lines problems. *Simulation Modelling Practice and Theory, 24*, 35–48. http://dx.doi.org/10.1016/j.simpat.2012.01.004

Tiacci, L. (2015a). Simultaneous balancing and buffer allocation decisions for the design of mixed-model assembly lines with parallel workstations and stochastic task times. *International Journal of Production Economics, 162*, 201–215. http://dx.doi.org/10.1016/j.ijpe.2015.01.022

Tiacci, L. (2015b). Coupling a genetic algorithm approach and a discrete event simulator to design mixed-model un-paced assembly lines with parallel workstations and stochastic task times. *International Journal of Production Economics, 159*, 319–333. http://dx.doi.org/10.1016/j.ijpe.2014.05.005

Vilarinho, P. M., & Simaria, A. S. (2002). A two-stage heuristic method for balancing mixed-model assembly lines with parallel workstations. *International Journal of Production Research, 40*, 1405–1420. http://dx.doi.org/10.1080/00207540110116273

Vilarinho, P. M., & Simaria, A. S. (2006). ANTBAL: An ant colony optimization algorithm for balancing mixed-model assembly lines with parallel workstations. *International Journal of Production Research, 44*, 291–303. http://dx.doi.org/10.1080/00207540500227612

Yang, C., & Gao, J. (2016). Balancing mixed-model assembly lines using adjacent cross-training in a demand variation environment. *Computers & Operations Research, 65*, 139–148.

11

An optimal inventory policy under certainty distributed demand for cutting tools with stochastically distributed lifespan

Cun Rong Li[1]* and Jiadong Cheng[1]

*Corresponding author: Cun Rong Li, Department of Industrial Engineering, School of Mechanical & Electronics Engineering, Wuhan University of Technology, 122 Luoshi Road, Hongshan District, Wuhan City 430070, P.R. China
E-mail: cunrong_li@163.com

Reviewing editor: Zude Zhou, Wuhan University of Technology, China

Abstract: Traditional inventory policy was deeply investigated for various kinds of demand in different industrial sectors. More extensive explorations on inventory policy, including the combination with manufacturing process, detailed attributes of the purchased products, etc. was conducted by many researchers. During manufacturing process, lifespan of cutting tools have significant impact on both the quantity of inventory and production cost. In this paper, the impact of maximum allowable stopping time for cutting tools on production-inventory policy under general production demands was investigated. An optimal inventory policy with general demand (OIPGD) was developed with which the allowable stopping time for tools, order-up-to-level inventory, and order cycle can be optimally determined by an exhaustive searching algorithm. Examples with different distributions on tool lifespan and production demand is presented to show the implementation of the OIPGD model. The results and the sensitivity analysis about the parameters show that optimized combination of selection for tool allowable stopping time, order-up-to-level, and order cycle time can dramatically minimize the total cost of the whole inventory activity.

Subjects: Engineering Economics; Industrial Engineering & Manufacturing; Production Engineering

Keywords: inventory policy; general demand; lifespan assessment; cutting tools

1. Introduction

Under most conditions with inventory policy, more extensive explorations on inventory, including the combination with manufacturing process, detailed attributes of the purchased products, etc.

ABOUT THE AUTHOR

Cun Rong Li received his PhD degree from Wuhan University of Technology (WHUT) majored in Mechanical Design & Theory. He is now an associate professor in Industrial Engineering Department in WHUT. His research interests are supply chain, enterprise information system, and manufacturing automation. He visited the Advanced Manufacturing Center, Liverpool University in 2005, Mechanical and Industrial Engineering Department, Louisiana University in 2013 as a visiting scholar. He published more than 20 research papers and engaged in more than 10 research projects from government and enterprise. He is also a reviewer of several international journals.

PUBLIC INTEREST STATEMENT

This paper aims at putting up with an optimizing inventory policy, and its purpose is to determine the lowest inventory cost model. In this cost model, three most important parameters order-up-to-level *R*, the length of time between placement of replenishment orders, and maximum allowable usage time of cutting tools are taken into consideration and the cutting tools' lifespan is stochastically distributed. Before deriving such a cost model, the authors present a lifespan analysis of cutting tools and various compound cost related to the total inventory cost and then a few examples are showed to illustrate the model. By employing this theory, enterprise can minimize their total supply chain cost and cutting tools' usage cost.

were conducted by many researchers for a better manipulation of the supply chain system. If the tools inventory is not enough when the failure of cutting tools occurs, it will result in loss of production which will also influence the production schedule. On the other hand, if the tools inventory is too much, it will cause higher holding cost, which is also not permitted. However, some factors in manufacturing process may also be taken more into consideration, since they may have impact on inventory policy because of their influences on cutting tool lifespan. In manufacturing system, the demand of production usually follows some distributions. The tool lifespan and general distributed demand sometimes convoluted together and have concurrent impact on the total cost of the supply chain system for a company. As a result, research on the convolution problem of inventory policy for cutting tools has a great significance accordingly. The goal of this paper is to develop an optimal inventory policy with general demand (OIPGD) as the lifespan of cutting tools is random under general distributed production demand.

1.1. Procurement and inventory policy

Moritz and Roelof (2003) had analyzed the impact of this inbound flow on inventory and presented the usage of results on Markvo decision process which showed average cost optimality of an order policy. Mungan, Yu, and Sarker (2010) studied the manufacturing lot-sizing, procurement and delivery schedules for technology-connected companies with their products facing continuous price decrease problem during the life cycle. Timothy (2005) researched two types of inventory control model through the use of a simple periodic review model and also an alternative approach to sensitivity analysis of the model. Unlike traditional inventory research, Ben-Daya and Raouf (1994) and Hariga and Ben-Daya (1999) dealt lead time as a decision variable which gave a new model about inventory. Sarkar, Ghosh, and Chaudhuri (2012) and Chung and Lin (2001) had done a research about optimal inventory replenishment policy for a deteriorating item and discussed it with numerical analysis. Banerjee and Sharma considered a deterministic inventory model that incorporated price and time and conditions of the net profit function and then illustrated it with examples and sensitivity analysis. Yu, Mungan, and Sarker (2011) and Sarker, Duan, and Wu (2012) discussed a highly improved approach on consignment and inventory policy with generalized demand distributions of retailers. The above literatures have done extensive researches on procurement and inventory; however, when considering the lifespan of a product in a whole supply chain, only a few of those researches could be involved.

1.2. Inventory policy and demands

About the optimization of supply chain, various kinds of researches have been done with different aspects taken into consideration. Kamath and Pakkala (2002) proposed an optimal policy for an inventory model which allows stock disposal and shows a solution of a dynamic programming backward recursion. A commonly used inventory policy for dealing with demands with different penalty costs is the critical-level inventory policy. Hwang and Hahn (2000) investigated an optimal procurement policy for items with an inventory-level-developed demand rate and fixed lifetime based on the demand rate of a function of current inventory level, and each unit having its fixed expiry date. Toktay and Wein (2001) show a production stage that produces a single item in make-to-stock and demand for finished goods in detail, such as accurate condition in a simulation study, safety stock, stochastic correlated demand, etc. in forecasting–production–inventory systems. Zequeira, Valdes, and Berenguer (2007) study the optimal buffer inventory to satisfy the demand during the interruption period due to maintenance action considering the possibility of imperfect production and opportunities for the fabrication of the buffer inventory.

1.3. Lifespan assessment of cutting tools

From the well-known F. W. Taylor's formula (see Shaw, 2005) $V = A/T_m$ (where V represent the cutting speed, T is the lifespan of the cutting tools, and m means different influence of T and A in different cutting condition), is described the relationship of cutting tools' lifespan, which is related to the cutting parameters; however, when the penalty of cutting tools' failure is taken into consideration, the way to determine the maximum allowable lifespan will be different.

Traditional research on lifespan of cutting tools is concentrated only on the materials' performance, while it is not enough. Axinte, Belluco, and De Chiffre (2001) proposed a method to acquire reliable measurements of tool lifespan in turning and provided a basis for the quantification of tool lifespan measurement uncertainty. Arsecula經ratne (2003) used Oxley's machining theory to predict cutting forces and cutting temperature which can be used to estimate the tool lifespan under RC conditions and further predicted tool deformation by accounting for the rake face temperature. Non-linear regression analysis techniques are used by Oraby and Hayhurst (2004) to establish models for wear and tool life determination in terms of the variation of a ratio of force, rather than absolute values of force components acting at the tool tip. Serra, Ouahabi, and Rmili (2012) presented a new criterion for the detection of cutting tool end of life from cutting tool vibration, the study aiming at the development of a cutting tool wear monitoring strategy applicable to the particular issues of turning process. Li and Sarker (2013) discussed how the parameters influence the lifespan and put forward an evaluation method; however, they have not taken the inventory policy into consideration.

As discussed above, there are very fewer researches concentrated on the lifespan distribution of cutting tools in supply chain. But in fact, the lifespan has great impact on the inventory policy. At the same time, the optimal order-up-to-level can minimize the total cost of the manufacturing process and also lacks investigation. There are so many factors that have influence on cutting tools, such as the tool type, the tool size, the tool materials, the working condition and even the machining angle. The tools' inventory has much influence on the holding cost and the penalty cost that have impact on the total cost. As a result, it is quite important to find an appropriate order-up-to-level which is determined by the lifespan of cutting tools to minimize the cost. Thus, this paper has formulated an inventory model which takes the lifespan distribution, order-up-to-level, and order cycle into consideration to make optimal decision about minimizing the total cost and give examples to show its application.

2. Model formulation

Lifespan and order-up-to-level of cutting tools have significant impacts on the minimization of inventory total cost. In this section, an attempt is made to formulate the objective function (total cost) which needs to be minimized to find the optimal lifespan, order cycle, and order-up-to-level for the manufacturing process.

2.1. Assumptions and notations

As the inventory issue of cutting tools is quite complex, in order to realize the aim of this paper, a few assumptions will be taken into consideration during the formulation of the process.

2.1.1. Assumptions

To simplify the research, the following assumptions are necessary to model in this paper:

(1) In order to avoid complication, only one type of machining tool is considered for model formulation.

(2) Total products for machining in a period are fixed.

(3) A penalty cost will be existed when the failure of a machining tool causes potential damage of work piece.

(4) Tool vendors are located nearby and result in negligible tool supplying time.

(5) Tool shortage results in downtime with machine and is not back ordered.

(6) The tool demand rate for production follow a general distribution, $g(x)$.

2.1.2. Notations and definitions summary

The definitions used in this paper are list as below:

Decision variables

R: order-up-to-level (tools)

T_c: Cycle length, the length of time between placement (or receipt) of replenishment orders

T_m: Maximum allowable usage time of cutting tools (in time unit)

Parameters

T_u: Average unit processing time per product (time unit/product)

D_p: Demand rate of products (products/year)

T: Lifespan of cutting tools (in time units)

C_f: Fixed cost for every order (\$/order)

c_x: Unit replacement cost of a tool (\$/tool)

c_h: Unit holding cost of a tool (\$/tool/year)

c_p: Unit penalty cost of tool failure (\$/tool)

D_p: Demand rate of product (products/year)

π: Unit shortage cost (\$/tool)

Cost function components and intermediate variables

C_o: Annual ordering cost

C_p: Expected total penalty cost

Cs: Expected annual total shortage cost

C_h: Expected annual holding cost

x_d: Random demand of production per month

x_t: Random demand of cutting tools depending on x_d and T_m

D_T: Demand of tools required per year

T_w: Total working time needed per year

TC: Total cost (\$/year)

2.2. Expected lifespan estimation for pre-empt cutting tools

The lifespan of a machining tool, T, is discussed by Li and Sarker (2013), which follows a distribution $f(T)$ that is mainly affected by three main parameters V (machining speed), f (cutting feed), and d (machining depth), other factors(such as cutting angle) which are insignificant for practical usage are ignored. As depicted in Figure 1, the horizontal axis T means the usage time distribution and the vertical axis means the usage frequency of the cutting tools, from Figure 1 it is obvious that most of the cutting tools' lifespan is in the maximum allowable usage time area and only a few cutting tools failure outside the range. So the optimal lifespan of a machining tool is absolutely in that area and through discussion it can be defined as T_m. When the lifespan is shorter than T_m, it means lower penalty but more tools will be used, thus causes higher purchasing cost and higher holding cost. On the other hand, if the lifespan is longer than that, the penalty will be higher, while the purchasing and holding cost lower. Under this condition, it is important to find an optimal lifespan of cutting tools to make decisions about the inventory policy and then reach the purpose of minimizing the total cost.

If a stopping policy is implemented, the distribution, $f(T)$ with $-\infty \leq T \leq \infty$ is modified to $f(T)$ with $-\infty \leq T \leq T_m$. It is obvious that if the cutting tools lasted less than T_m, the order-up-to-level will increase and the order cycle will decrease. Now the key point is to find this most optimal stopping time T_m of tool lifespan which will minimize the total cost TC. So in order to make a good decision about procurement policy for the revised situation, the optimal stopping time T_m needs to be based on the expected lifespan $E(T)$ of the revised distribution.

For the part with $T \leq T_m$, marked as area A_1, as shown in Figure 1, the expected lifespan of all tools can be defined as

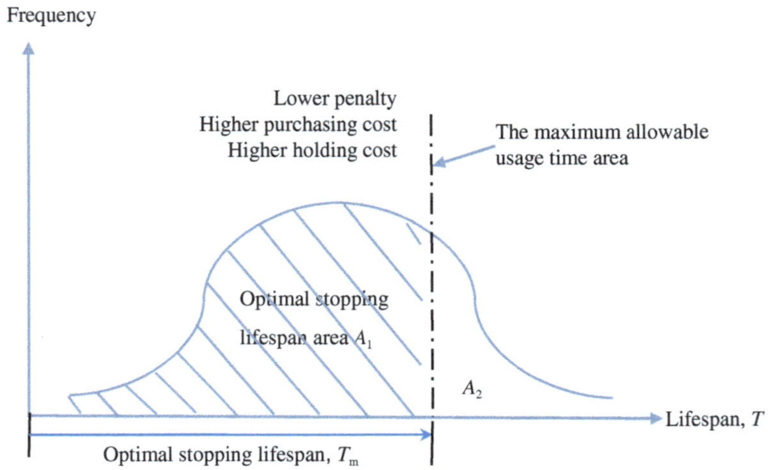

Figure 1. General distribution of cutting tools' lifespan.

$$E(T)_{T \le T_m} = \int_0^{T_m} Tf(T)dT \quad (1)$$

For the other part marked as area A_2 in Figure 1, where $T > T_m$, which means the lifespan of the tools exceeds the maximum allowable usage time; thus the lifespan of all these tools is practically T_m. So, the expected lifespan for these pre-empted tools (in area A_2) is *defined* as

$$E(T_m)_{T > T_m} = T_m \int_{T_m}^{\infty} f(T)dT = T_m \left[1 - F(T_m)\right] \quad (2)$$

where $f(T)$ follows the same generic distribution.

Thus, $F(T_m) = P(T \le T_m)$ is the probability of a tool failure earlier than T_m and $1 - F(T_m) = P(T > T_m)$ is the probability of a tool exceeding the maximum allowable usage time T_m when the tool can be replenished before any natural failure which result in no cost to the whole system.

Since the total distribution of the tools contains two different groups, A_1 and A_2 as represented above, each group has different weight in the expected lifespan. Now the total expected lifespan of both groups can be defined as

$$E(T)_{A_1 + A_2} = E(T)_{T \le T_m} + E(T_m)_{T > T_m} \quad (3)$$

2.3. Demand of cutting tools with generally distributed demand of production

In practical manufacturing process, the total demand for a production within a relative long period is steady, which makes an approximately constant demand. However, the demand usually varies due to the requirements of terminal user and sometimes follows a kind of distribution.

It is assumed that the product has a demand of D_p products/year. The average unit processing time is T_u per product (time unit/product). Thus, the total manufacturing working time T_w required per year for the product can be expressed as $T_w = T_u D_p$ time unit/year. For an expected pre-empt lifespan (in time units/tool) of a machining tool, $E(T)_{A_1 + A_2}$, the total demand of cutting tools per year, D_T (tools/year), can be computed as

$$D_T = T_u D_p / E(T)_{A_1 + A_2} \quad (4)$$

where the tool lifespan T follows a generic distribution $f(T)$.

It is assumed that the demand within an order cycle, T_c, is a continuous variable with stationary probability distribution. Units demanded out of stock are at high expense and the average shortages are not negligible relative to the on-hand inventory.

Assume that random variable x_d has a density function $g(x_d)$ with a mean μ_d and standard deviation σ_d, and this is for the demand of production. Considering the density function $g(x_d)$ depends on the ordering cycle T_c, and the demand of tools for the period x_d is also a function of T_m, so for the period of T_c, the expected demand of tools, x_t, can be expressed as $x_t = T_u x_d / E(T)_{A_1+A_2}$, combined with Equation 3, x_t can also be rewritten as

$$x_t = \frac{T_u x_d}{E(T)|_{T=T_m} + E(T_m)} \tag{5}$$

where T follows a generic distribution of $f(T)$, x_d follows a distribution $g(x_d)$.

2.4. Analysis of OIPGD model

In this paper, the cost components of OIPGD consists of annual ordering cost C_o, expected total penalty cost C_p, expected annual total replacement cost C_r, expected annual total shortage cost C_s, and expected annual holding cost C_h. This section will discuss each of the components, respectively, because the selection of optimal lifespan with the allowable maximum stopping time for cutting tools incurs change of total tool requirements, D_T, and eventually has impact on the order-up-to-level quantity, R, and review interval (i.e. ordering cycle), T_c. This relation is illustrated in Figure 2.

(a) Annual ordering cost, C_o

Since the demand has a continuous density function, the inventory level will always fall from the maximum possible inventory level, R, to zero. The practical inventory will decide another order with every ordering review. As a result, there are $1/T_c$ orders per year and the expected annual cost of ordering, C_o, (only for purchasing) can be expressed as

$$C_o = C_f / T_c \tag{6}$$

(b) Expected total penalty cost, C_p

Penalty cost depends on the allowable maximum stopping time T_m. As shown in Figure 1, the tool fails in group A_1 will result in penalty cost. For a single machining tool, the probability of fail before T_m is $F(T_m)$, which is the cumulative probability while the tool life $T \le T_m$. The expected penalty cost can be obviously expressed as $c_p F(T_m) = c_p \int_{-\infty}^{T_m} f(T)dT$. Equation 4 gives the expected total demand of cutting tools, thus the yearly total penalty cost is

$$C_p = D_T c_p F(T_m) = \frac{T_u D_p c_p}{E(T)_{A_1+A_2}} \int_{-\infty}^{T_m} f(T)dT \tag{7}$$

(c) Expected annual total replacement cost, C_r

Since a lot of cutting tools will be replaced each year and replacement takes time and effort, the replacement cost should be taken into consideration. As the lifespan of cutting tools gets longer, the replacement of cutting tools gets less. c_x is the cost of replacement per unit and Equation 4 gives the expected total demand of cutting tools, So the expected replacement cost can be easily expressed as

$$C_r = c_x T_u D_p / E(T)_{A_1+A_2} \tag{8}$$

(d) Expected annual total shortage cost, C_s

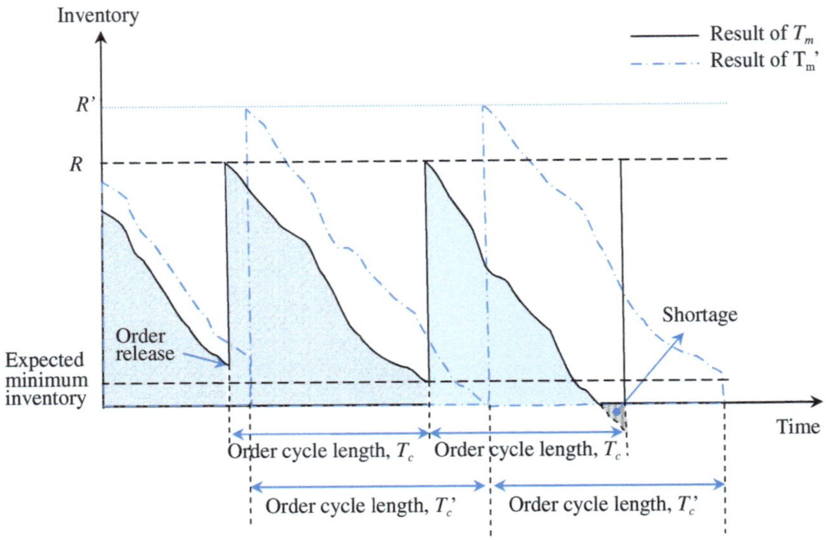

Figure 2. Illustration of influence of stopping time T_m on review cycle T_c and order-up-to-level R.

From the distribution of production demand x_d in the ordering cycle, $x_d \sim g(x_d)$, the distribution of tools demand, x_t can be deducted as (see Appendix A)

$$x_t \sim g_{x_t}(x_t) = \frac{E(T)_{A_1+A_2}}{T_u} g\left(\frac{E(T)_{A_1+A_2}}{T_u} x_t\right) \tag{9}$$

Obviously, x_t is a stochastic demand of cutting tools for a review period T_c, if $x_t > R$, a shortage of tools occurs (which is shown in Figure 2). This will result in downtime of production and consequently shortage cost. Under this circumstance, for an ordering (review) period, the expected shortage of tools, Q_s, is given by

$$Q_s = \int_R^{\infty} [(x_t - R) g_{x_t}(x_t)] dx_t = \int_R^{\infty} \left[(x_t - R)\frac{E(T)_{A_1+A_2}}{T_u} g\left(\frac{E(T)_{A_1+A_2}}{T_u} x_t\right)\right] dx_t \tag{10}$$

Thus, the expected annual total shortage cost C_s can be expressed as

$$C_s = \frac{\pi}{T_c} \int_R^{\infty} \left[(x_t - R)\frac{E(T)_{A_1+A_2}}{T_u} g\left(\frac{E(T)_{A_1+A_2}}{T_u} x_t\right)\right] dx_t \tag{11}$$

(e) Expected annual holding cost, C_h

With expected annual holding cost C_h, it compromises to two sub components, (1) regular expected inventory, $\bar{I}$, and (2) the non-back ordered inventory (there is no need to fill the shortage of previous review period when a shortage occurs in previous period).

Once the allowable maximum stopping time, T_m, is determined, the demand for cutting tools has a constant mean and the inventory varies on the average linearly between the minimum and maximum inventory. Here the expected minimum inventory is $R - u_t$ combined with Equation 5, that is,

$$\bar{I}_{\min} = R - \frac{T_u \mu_d}{E(T)_{A_1+A_2}} \tag{12}$$

The expected maximum inventory, that occurs at the arrival of the replenishment, is given by $R - \mu_t + D_T T_c$. Combined with Equations 4 and 5, this sub component can be rewritten as

$$\bar{I}_{max} = R + \frac{T_c T_u D_p - T_u \mu_d}{E(T)_{A_1+A_2}} \tag{13}$$

Because the inventory varies approximately in a linear way, the average inventory for sub component $\bar{I}$ is halfway between expected maximum inventory, $\bar{I}_{max}$ and expected minimum inventory, $\bar{I}_{min}$. Thus, the average inventory for this sub component is easily computed as

$$\bar{I} = \frac{\bar{I}_{min} + \bar{I}_{max}}{2} = R + \frac{T_c T_u D_p - 2T_u \mu_d}{2E(T)_{A_1+A_2}} \tag{14}$$

Adding up the inventories expressed in Equations 10 and 14, the expected annual holding cost, C_h, can be expressed as

$$C_h = c_h \left\{ R + \frac{T_c T_u D_p - 2T_u \mu_d}{2E(T)_{A_1+A_2}} + \int_R^{\infty} \left[(x_t - R) \frac{E(T)_{A_1+A_2}}{T_u} g\left(\frac{E(T)_{A_1+A_2}}{T_u} x_t \right) \right] dx_t \right\} \tag{15}$$

The total cost objective function $TC(T_m, T_c, R)$ consists of five components of C_o, C_p, C_r, C_s and C_h, and is given by $TC(T_m, T_c, R) = C_o + C_p + C_s + C_h + C_r$. With Equations 6, 7, 11, and 15, the total cost objective function can be concluded as

$$\begin{aligned} TC(T_m, T_c, R) &= \frac{\pi}{T_c} \int_R^{\infty} \left[(x_t - R) \frac{E(T)_{A_1+A_2}}{T_u} g\left(\frac{E(T)_{A_1+A_2}}{T_u} x_t \right) \right] dx_t + \frac{c_f}{T_c} + \frac{T_u D_p c_p}{E(T)_{A_1+A_2}} \int_{-\infty}^{T_m} f(T) dT \\ &+ c_h \left\{ R + \frac{T_c T_u D_p - 2T_u \mu_d}{2E(T)_{A_1+A_2}} + \int_R^{\infty} \left[(x_t - R) \frac{E(T)_{A_1+A_2}}{T_u} g\left(\frac{E(T)_{A_1+A_2}}{T_u} x_t \right) \right] dx_t \right\} + c_x T_u D_p / E(T)_{A_1+A_2} \end{aligned} \tag{16}$$

By merging the similar items, function (16) can be rewritten as

$$\begin{aligned} TC(T_m, T_c, R) = \frac{c_f}{T_c} + \frac{T_u D_p c_p}{E(T)_{A_1+A_2}} \int_{-\infty}^{T_m} f(T) dT + c_h \left(R + \frac{T_c T_u D_p - 2T_u \mu_d}{2E(T)_{A_1+A_2}} \right) \\ + \left(\frac{\pi}{T_c} + c_h \right) \int_R^{\infty} \left[(x_t - R) \frac{E(T)_{A_1+A_2}}{T_u} g\left(\frac{E(T)_{A_1+A_2}}{T_u} x_t \right) \right] dx_t + c_x T_u D_p / E(T)_{A_1+A_2} \end{aligned} \tag{17}$$

where tool life, $T \sim f(T)$, $E(T)_{A_1+A_2} = E(T)|_{T=T_m} + E(T_m)$, $E(T)|_{T=T_m} = \int_0^{T_m} T f(T) dT$, $E(T_m) = T_m \left[1 - F(T_m) \right]$.

3. Solution methodology and computation results

Total cost objective function $TC(T_m, T_c, R)$ in Equation 17 for the replenishment policy consists of three decision variables. T_u, D_p, c_f, c_u, c_p, c_x, and c_h are the system characteristic parameters. Variables T_m (optimal stopping time of cutting tools), ordering cycle time T_c, and order-up-to-level R are r equired to be solved to satisfy the minimum total cost.

3.1. Exhaustive search algorithm for OIPGD model

In order to find the optimal order-up-to quantity, R, review time, T_c, and tool allowable maximum stopping time, T_m, partial derivative approach can be employed to calculate the stationary points.

With decision variable T_c, by differentiation, $\partial TC / \partial T_c = 0$ leads to

$$\frac{c_h T_u D_p}{2E(T)_{A_1+A_2}} - \frac{c_f + \pi \int_R^{\infty} \left[(x_t - R) \frac{E(T)_{A_1+A_2}}{T_u} g\left(\frac{E(T)_{A_1+A_2}}{T_u} x_t \right) \right] dx_t}{T_c^2} = 0,$$

which yields

$$T_c = \sqrt{\frac{2E(T)_{A_1+A_2}\left\{C_+\pi\int_R^\infty\left[(x_t-R)\frac{E(T)_{A_1+A_2}}{T_u}g\left(\frac{E(T)_{A_1+A_2}}{T_u}x_t\right)\right]dx_t\right\}}{c_hT_uD_p}} \quad (18)$$

For decision variable T_m, by differentiation, $\partial TC/\partial T_m = 0$ leads to

$$\frac{T_uD_pc_pf(T_m)}{E(T)_{A_1+A_2}} - \frac{T_uD_pc_p[1-F(T_m)]}{[E(T)_{A_1+A_2}]^2}\int_{-\infty}^{T_m} f(T)dT - \frac{c_h(T_cT_uD_p-2T_u\mu_d)[1-F(T_m)]}{2[E(T)_{A_1+A_2}]^2}$$
$$+\left(\frac{\pi}{T_c}+c_h\right)\int_R^\infty\left\{(x_t-R)\left[\left(\frac{1-F(T_m)}{T_u}\right)g\left(\frac{E(T)_{A_1+A_2}}{T_u}x_t\right)\right.\right.$$
$$\left.\left.+\left(\frac{E(T)_{A_1+A_2}}{T_u}\right)g'\left(\frac{E(T)_{A_1+A_2}}{T_u}x_t\right)\left(\frac{1-F(T_m)}{T_u}\right)x_t\right]\right\}dx_t$$
$$-c_xT_uD_p[1-F(T_m)]/2\left[E(T)_{A_1+A_2}\right]^2=0 \quad (19)$$

which is too complicated to yield a mathematical expression for T_m (Appendix B).

Similar to the above differential method, $\partial TC/\partial R = 0$ leads to the following expression (see Equation C4 in Appendix C):

$$c_h+\left(\frac{\pi}{T_c}+c_h\right)\frac{E(T)_{A_1+A_2}}{T_u}\int_R^\infty g\left(\frac{E(T)_{A_1+A_2}}{T_u}x_t\right)dx_t=0 \quad (20)$$

which is also too complex to be simplified further.

3.2. Exhaustive search algorithm method for OIPGD model

Equations 19 and 20 cannot be transformed into a simplified expression, and it is also too complicated to delaminate two of the three decision variables R, T_c, and T_m through Equations 18, 19, and 20. However, with Equation 18, by using exhaustive search method, only two dimensions R and T_m can be searched in their feasible domain, then the result of T_c computed. With R, T_m, and dependent variable T_c, a result of total cost can be computed. In this section, an exhaustive search algorithm (ESA) method is developed to find the optimal solution of R, T_c, and T_m by satisfying the minimum total cost TC.

Equation 19 is a mathematical expression of T_m. And numerical algorithms can be used to solve the value of T_m and then the variable R can be solved by Equation 20. Eventually, total cost TC can be worked out by using Equation 17. This computational algorithm is given as follows. In this algorithm, $g(T_m)$ will be defined to indicate the left-hand side of Equation 19 in terms of T_m. If $g(T_m)$ equals or is close to 0, the corresponding value of T_m is determined to be the optimal solution.

3.2.1. ESA method

Step 1: Initialize C_f, c_p, c_u, c_h, c_x, T_u, D_p, Δ_T (step-size lifespan of T), Δ_{T_c} (step-size lifespan of T_c), Δ_R (an arbitrary value of R) and define the distribution of T, the length of time between placement T_c, and the range of R. Specify the parametric values for distribution of T and express $f(T)$ and $E(T)$.

Step 2: Set the starting values Δ_R, Δ_T, Δ_{T_c}, and evaluate the cost function $TC^*(T_c,T_m,R) \leftarrow \infty$.

Step 3: (a) Calculate cumulative distribution probability $F(T)$,

(b) Compute expected lifespan $E(T)_{A_1+A_2}$ with stopping time $T_m \leftarrow T$ using Equation 3.

(c) Compute the cycle length T_c with Equations 3 and 18.

Step 4: (a) Compute $TC(T_c,T_m,R)$ using Equation 17.

(b) If $TC(T_c, T_m, R) < TC^*(T_c, T_m, R) \leftarrow \infty$,

Set $TC^*(T_c, T_m, R) \leftarrow TC(T_c, T_m, R), R^* = R, T_m{}^* = T_m$, set . $T_c^* = T_c + \Delta T_c$

If $TC(T_c, T_m, R) \leftarrow TC^*(T_c, T_m, R)$, go to Step 4(a)

Else set $T_m = T_m + \Delta T$,

If $TC(T_c, T_m, R) < TC^*(T_c, T_m, R)$, go to Step 4(a)

Else set $R = R + \Delta R$,

If $TC(T_c, T_m, R) < TC^*(T_c, T_m, R)$, go to Step 4(a)

Else set $R^* = R, T_m{}^* = T_m, T_c^* = T_c$.

Step 6: Stop and current values of T_m^*, T_c^* and R^* are the best solution and $TC^*(R^*, T_m^*, T_c)$ is the minimum total cost.

3.3. Numerical analysis

The following example illustrates the simulation results obtained from the algorithm of ESA. In real machining process for a mechanical manufacturing enterprise, a milling cutter named as "S-NA2T030" is used to proceed with their own product. When the product is with material hardness value of HRC45~50, the lifespan of the cutter follows a normal distributed with mean $\mu_T = 4.5$ h and standard deviation of $\sigma = 0.2$ h. This company processes approximately $D_p = 72{,}800$ products per year by using this tool. For every product, an average of $T_u = 0.1$ h machining time is required. The fixed purchasing cost for each order $C_f = \$30.00$/order, the penalty cost for each failure of the tool is $c_p = \$0.50$/tool, the replacement cost per tool is $c_x = \$6.00$/tool, the holding cost per tool per year is $c_h = \$6$/tool/year, and the tool unit shortage cost $\pi = \$20.00$/tool. Under this situation, the objective of this practical problem is to minimize the total cost *TC* by determining the optimal values of order cycle T_c, optimal stopping time T_m, and order-up-to-level *R*.

By running the ESA program, the optimal decision variables, search result can be concluded of this vulnerable parts changing and procurement problem. The optimal decision variables, $T_m = 3.70$ h, an order cycle $T_c = 0.12$ years ≈ 44 days (about one and a half months), an order-up-to-level $R = 1{,}020$ tools, and the minimized total cost $TC = \$19{,}322$.

Above is the example of when the cutting tools' lifespan follows normal distribution and the theoretical results with the ESA method. For further discussion, two different lifespan distributions will be used here to demonstrate the generic solution methodology with the ESA method.

Assume that the lifespan of the cutting tools is uniform (*a*, *b*), the distribution of which is shown below in Table 1.

With *f*(*T*) following the uniform distribution, from the data above, the range of the lifespan of cutting tools can be assumed as $a = 4$ and $b = 5$, by running the ESA method, the optimal result is

Table 1. Expected lifespan $E(T)\,|_{T=T_m}$ of tools for uniform distribution

Distribution	$f(T)$	$F(T_m)$	$E(T)\|_{T=T_m}$
Uniform (*a*, *b*)	$\frac{1}{b-a}$	$\frac{T_m-a}{b-a}$	$\left(\frac{T_m-a}{b-a}\right)\left(\frac{T_m+a}{2}\right)$

Table 2. Optimal solution for different distributions

Distributions	$f(T)$ ($a = 4, b = 5$)	T_m (h)	T_c (years)	R (tools)	TC ($)
Uniform (*a*, *b*)	$\frac{1}{b-a}$	4.7	0.08	2,000	23, 143
Normal (μ, σ^2)	$\frac{1}{\sqrt{2\pi}\sigma}e^{-(T-\mu)^2/2\sigma^2}$	3.7	0.12	1,020	19, 322

T_m = 4.7 h, an order cycle T_c = 0.08 years ≈ 29 days (about one-and-a-half months), an order-up-to-level R = 2,000 tools, and the minimized total cost TC = \$23,143.

A detailed comparison of the two distributions about the function and the result through the usage of ESA method are listed in Table 2. Using the ESA method, all the possibilities are compared and the most optimal solution is obtained which is shown below, and as the distribution changes the optimal solution will be different. In the table below, a generic solution methodology is demonstrated that can be applied accordingly for other types of demand distributions.

4. Sensitivity analysis

In this research, the maximum allowable usage time of cutting tools T_m, the length of time between placement (or receipt) of replenishment orders T_c, and the order-up-to-level of tools R are interacted. If one of the parameters changes, the other two are influenced, and the total cost TC is absolutely changed. Since there are three variables, it is hard to clearly depict all the variables as to how they influence the object function. In order to investigate which one is the most influential, one of the variables is fixed and then analysed.

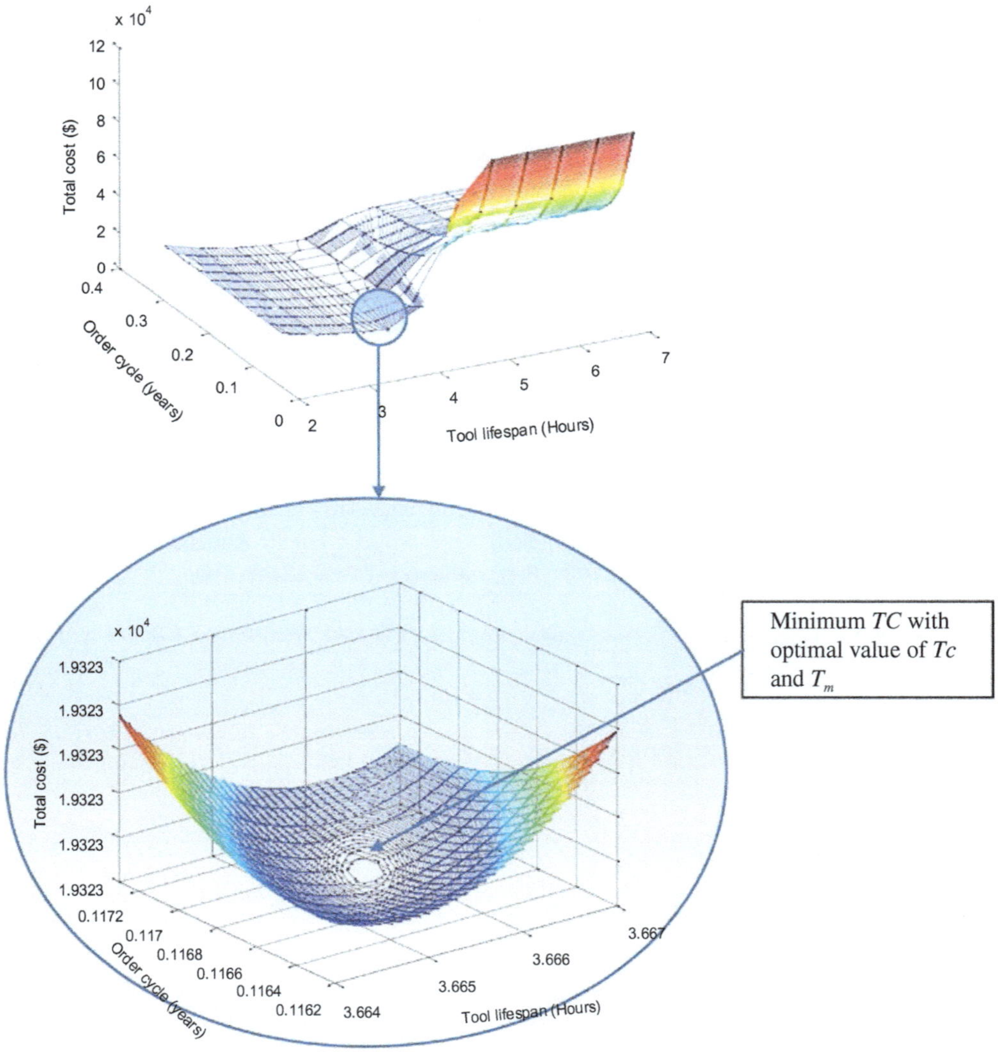

Figure 3. Fluctuation of TC according to changing of T_m and T_c.

4.1. Impact analysis of T_m and T_c on TC while R is fixed

This paper has established the model of *TC*, therefore, when individually inspecting after T_m, T_c, *R* effects on the cutting tools reliability lifespan, reviewed parameter changes in a reasonable range, and the other parameters respectively take typical values.

While $R = 1{,}000$, the relationship between T_m, T_c, and *TC* is shown in Figure 3. As the unit of time is hour, accuracy to two decimal places is sufficient, while for the unit of T_c, three decimal places is sufficient. The result is limited to the accuracy and it can be seen from figure that the total cost $TC = \$19{,}323$, $T_c = 0.117$ years ≈ 44 days, $T_m = 3.66$ h.

It is easy to see that T_m is more influential than T_c from Figure 3. When T_m and T_c are smaller than the optimal solution, the total cost changes slightly and when T_m is bigger than 3.67, it starts to change sharply.

4.2. Influence analysis of T_m and R on TC while T_c is fixed

While $T_c = 0.1$, the relationship between T_m, *R*, and *TC* is shown in Figure 4. And with the accuracy fixed, it can be seen from Figure 4 that the total cost $TC = 19{,}332\$$, $R = 1{,}015$, $T_m = 3.67$ h.

From Figure 4, it is easy to see the optimal solution. As the total machining time $T_w = T_u \times D_p = 7{,}280$, while the average lifespan is around 4.5, the cutting tools needed per year is around 1,617 products, so when the order-to-quality is bigger than that, the total cost will not change much.

In addition, through analysis of the two pictures above, it can be concluded that the lifespan of cutting tools T_m has the greatest influence on the total cost, the length of time between placement (or receipt) of replenishment orders T_c and order-up-to-level *R* have relatively small effects on total cost.

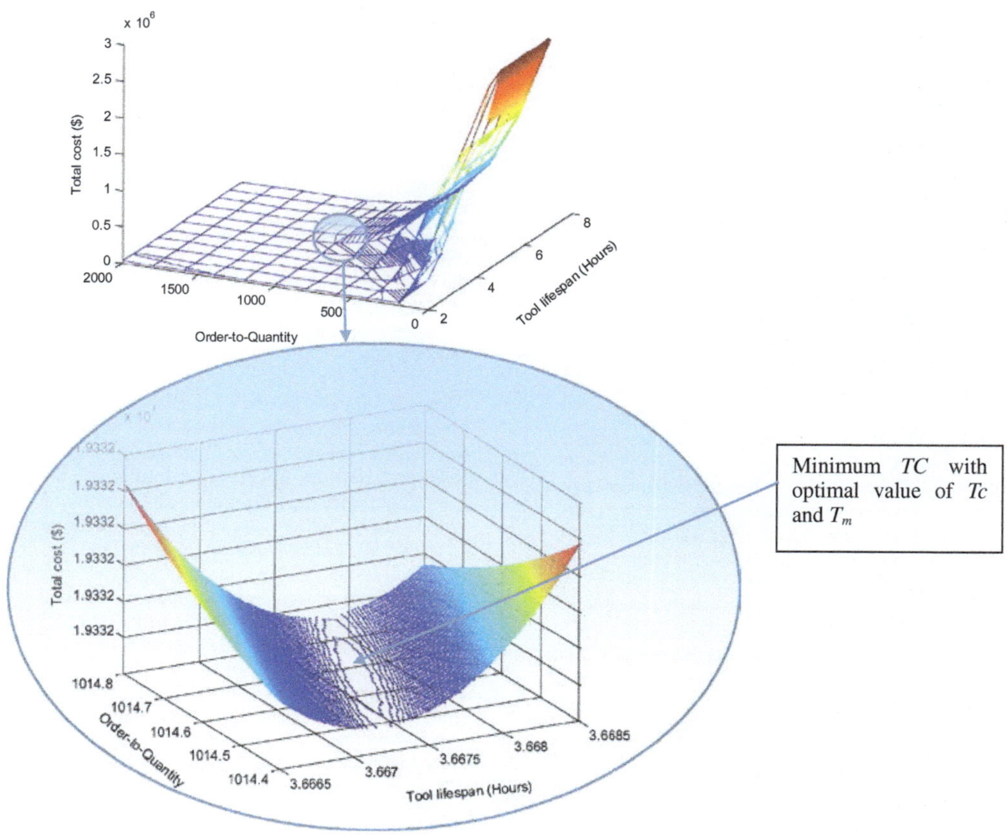

Figure 4. Fluctuation of *TC* according to changing of T_m and *R*.

5. Conclusions

In this paper, the author has depicted an interactive relationship among three important factors which influence the inventory policy under certainty distribution demand for cutting tools with stochastically distributed lifespan. They are the maximum allowable usage time of cutting tools, the length of time between placement (or receipt) of replenishment orders, and order-up-to quantity. An optimal ESA method was found to solve this problem, with two commonly used lifespan distribution of cutting tools to make a comparison and come to a conclusion that the ESA method is useful for general distribution. Because with three variables, the analysis is four-dimensional and is too complicated to express, the research fixed one variable and depicted the object function with the other two variables to see the relationship and compare the different results. From the research above it is easy to find that "T_m" is the most influential factor of the total cost. But in this paper, the author only considered one kind of machining tool and ignored the compound influence of the manufacturing system. For future work, it is necessary to study different cutting tools, which follow various lifespan distribution and compound influence.

Funding
The authors received no direct funding for this research.

Author details
Cun Rong Li[1]
E-mail: cunrong_li@163.com
Jiadong Cheng[1]
E-mail: jeffery900821@gmail.com
[1] Department of Industrial Engineering, School of Mechanical & Electronics Engineering, Wuhan University of Technology, 122 Luoshi Road, Hongshan District, Wuhan City 430070, P.R. China.

References

Arseculoratne, J. A. (2003). On prediction of tool life and tool deformation conditions in machining with restricted contact tools. *Machine Tools & Manufacture, 43*, 657–669.

Axinte, D. A., Belluco, W., & De Chiffre, L. (2001). Reliable tool life measurements in turning—An application to cutting fluid efficiency evaluation. *Machine Tools & Manufacture, 41*, 1003–1014.

Ben-Daya, M., & Raouf, A. (1994). Inventory models involving lead time as a decision variable. *Journal of the Operational Research Society, 45*, 579–582. http://dx.doi.org/10.1057/jors.1994.85

Chung, K. J., & Lin, C. N. (2001). Optimal inventory replenishment models for deteriorating items taking account of time discounting. *Computers & Operations Research, 28*, 67–83.

Hariga, M., Ben-Daya, M. (1999). Some stochastic inventory models with deterministic variable lead time. *European Journal of Operational Research, 113*, 42–51.

Hwang, H., & Hahn, K. H. (2000). An optimal procurement policy for items with an inventory level-dependent demand rate and fixed lifetime. *European Journal of Operational Research, 127*, 537–545. http://dx.doi.org/10.1016/S0377-2217(99)00337-9

Kamath, K. R., & Pakkala, T. P. M. (2002). A Bayesian approach to a dynamic inventory model under an unknown demand distribution. *Computers & Operations Research, 29*, 403–422.

Li, C. R., & Sarker, B. R. (2013, November). Lifespan prediction of cutting tools for high-value-added products. *International Journal of Advanced Manufacturing Technology, 69*, 1887–1894. doi:10.1007/s00170-013-5160-1

Moritz, F., & Roelof, K. (2003). On optimal inventory control with independent stochastic item returns. *European Journal of Operational Research, 151*, 25–37.

Mungan, D., Yu, J., & Sarker, B. R. (2010). Manufacturing lot-sizing, procurement and delivery schedules over a finite planning horizon. *International Journal of Production Research, 48*, 3619–3636. http://dx.doi.org/10.1080/00207540902878228

Oraby, S. E., & Hayhurst, D. R. (2004). Tool life determination based on the measurement of wear and tool force ratio variation. *Machine Tools & Manufacture, 44*, 1261–1269.

Sarkar, T., Ghosh, S. K., & Chaudhuri, K. S. (2012). An optimal inventory replenishment policy for a deteriorating item with time-quadratic demand and time-dependent partial backlogging with shortages in all cycles. *Applied Mathematics and Computation, 218*, 9147–9155. http://dx.doi.org/10.1016/j.amc.2012.02.072

Serra, R., Ouahabi, A., & Rmili, W. (2012, July 22–27). An end-of-life criterion from cutting tool vibrations. In *15th International Conference on Experimental Mechanics*, Porto.

Shaw, M. C. (2005). *Metal cutting principles*. Oxford: Oxford University Press.

Timothy, L. U. (2005). Inventory models with inventory-level-dependent demand: A comprehensive review and unifying theory. *European Journal of Operational Research, 162*, 792–804.

Toktay, L. B., & Wein, L. M. (2001). Analysis of a forecasting-production-inventory system with stationary demand. *Management Science, 47*, 1268–1281. http://dx.doi.org/10.1287/mnsc.47.9.1268.9787

Yu, J., Mungan, D., & Sarker, B. R. (2011). An integrated multi-stage supply chain inventory model under an infinite planning horizon and continuous price decrease. *Computers & Industrial Engineering, 61*, 118–130.

Yu, J., Sarker, B. R., Duan, Q., & Wu, B. (2012). Single-manufacturer, multi-retailer consignment policy for retailers' generalized demand distributions. *Journal of the Operational Research Society, 63*, 1708–1719. http://dx.doi.org/10.1057/jors.2012.13

Zequeira, R. I., Valdes, J. E., & Berenguer, C. (2007). Optimal buffer inventory and opportunistic preventive maintenance under random production capacity availability. *International Journal of Production Economics, 111*, 686–696.

Appendix A. Proof of distributed demand for tools

Demand of production in a review period x_d follows a generic distribution of $g(x_d)$. There exists a relation expressed in Equation 5. Thus, for demand of cutting tools, let us note that x_t follows a distribution $g_{x_t}(y)$, from the definition of cumulative probability, $G_{x_t}(y)$ is given by

$$G_{x_t}(y)=P(x_t \le y) \tag{A1}$$

Combined with Equation 5, Equation A1 is given by

$$G_{x_t}(y)=P\left(\frac{T_u x_d}{E(T)|_{T=T_m}+E(T_m)}\le y\right) \tag{A2}$$

Because $T_u x_d/[E(T)|_{T=T_m}+E(T_m)]>0$, Equation A2 can be rewritten as

$$G_{x_t}(y)=P\left(x_d\le \frac{E(T)|_{T=T_m}+E(T_m)}{T_u}y\right)=G\left(\frac{E(T)|_{T=T_m}+E(T_m)}{T_u}y\right) \tag{A3}$$

A differentiation on both sides of Equation A3 yields

$$g_{x_t}(y)=\frac{E(T)|_{T=T_m}+E(T_m)}{T_u}f\left(\frac{E(T)|_{T=T_m}+E(T_m)}{T_u}y\right) \tag{A4}$$

Appendix B. Differential process on $\partial TC/\partial T_m=0$

The total cost objective function is expressed as follows:

$$\begin{aligned}TC(T_m,T_c,R)=&\frac{C_f}{T_c}+\frac{T_u D_p c_p}{E(T)_{A_1+A_2}}\int_{-\infty}^{T_m} f(T)dT+c_h\left(R+\frac{T_c T_u D_p-2T_u\mu_d}{2E(T)_{A_1+A_2}}\right)\\&+\left(\frac{\pi}{T_c}+c_h\right)\int_R^{\infty}\left[(x_t-R)\frac{E(T)_{A_1+A_2}}{T_u}g\left(\frac{E(T)_{A_1+A_2}}{T_u}x_t\right)\right]dx_t\end{aligned} \tag{B1}$$

By differentiation on T_m, $\partial TC/\partial T_m=0$ yields

$$\begin{aligned}&\frac{T_u D_p c_p f(T_m)}{E(T)_{A_1+A_2}}-\frac{T_u D_p c_p[1-F(T_m)]}{[E(T)_{A_1+A_2}]^2}\int_{-\infty}^{T_m} f(T)dT-\frac{c_h(T_c T_u D_p-2T_u\mu_d)[1-F(T_m)]}{2[E(T)_{A_1+A_2}]^2}\\&+\left(\frac{\pi}{T_c}+c_h\right)\frac{\partial}{\partial T_m}\left\{\int_R^{\infty}\left[(x_t-R)\left(\frac{E(T)_{A_1+A_2}}{T_u}\right)g\left(\frac{E(T)_{A_1+A_2}}{T_u}x_t\right)\right]dx_t\right\}=0\end{aligned} \tag{B2}$$

because

$$\partial E(T)_{A_1+A_2}/\partial T_m=\partial E(T)_{A_1+A_2}/\partial T_m=\frac{\partial}{\partial T_m}\left(\int_0^{T_m} Tf(T)dT+T_m[1-F(T_m)]\right)=1-F(T_m) \tag{B3}$$

the differential part within Equation B2

$$\frac{\partial}{\partial T_m}\left\{\int_R^{\infty}\left[(x_t-R)\left(\frac{E(T)_{A_1+A_2}}{T_u}\right)g\left(\frac{E(T)_{A_1+A_2}}{T_u}x_t\right)\right]dx_t\right\}$$
$$=\int_R^{\infty}\left\{(x_t-R)\left[\left(\frac{1-F(T_m)}{T_u}\right)g\left(\frac{E(T)_{A_1+A_2}}{T_u}x_t\right)+\left(\frac{E(T)_{A_1+A_2}}{T_u}\right)g'\left(\frac{E(T)_{A_1+A_2}}{T_u}x_t\right)\left(\frac{1-F(T_m)}{T_u}\right)x_t\right]\right\}dx_t \quad \text{(B4)}$$

Appendix C. Differential process on $\partial TC/\partial R = 0$

The total cost objective function is expressed as follows:

$$TC(T_m, T_c, R)=\frac{C_f}{T_c}+\frac{T_u D_p c_p}{E(T)_{A_1+A_2}}\int_{-\infty}^{T_m} f(T)dT+c_h\left(R+\frac{T_c T_u D_p-2T_u\mu_d}{2E(T)_{A_1+A_2}}\right)$$
$$+\left(\frac{\pi}{T_c}+c_h\right)\int_R^{\infty}\left[(x_t-R)\frac{E(T)_{A_1+A_2}}{T_u}g\left(\frac{E(T)_{A_1+A_2}}{T_u}x_t\right)\right]dx_t \quad \text{(C1)}$$

Let us differentiate on R, $\partial TC/\partial R = 0$ yields

$$\partial TC/\partial R=c_h+\partial\left\{\left(\frac{\pi}{T_c}+c_h\right)\int_R^{\infty}\left[x_t\frac{E(T)_{A_1+A_2}}{T_u}g\left(\frac{E(T)_{A_1+A_2}}{T_u}x_t\right)\right]dx_t\right\}\Big/\partial R-\partial$$
$$\times\left\{\left(\frac{\pi}{T_c}+c_h\right)\int_R^{\infty}\left[R\frac{E(T)_{A_1+A_2}}{T_u}g\left(\frac{E(T)_{A_1+A_2}}{T_u}x_t\right)\right]dx_t\right\}\Big/\partial R=0 \quad \text{(C2)}$$

which leads to

$$c_h-\left(\frac{\pi}{T_c}+c_h\right)\frac{E(T)_{A_1+A_2}}{T_u}g\left(\frac{E(T)_{A_1+A_2}}{T_u}R\right)R+\left(\frac{\pi}{T_c}+c_h\right)\frac{E(T)_{A_1+A_2}}{T_u}$$
$$\times\left[\int_R^{\infty}g\left(\frac{E(T)_{A_1+A_2}}{T_u}x_t\right)dx_t+g\left(\frac{E(T)_{A_1+A_2}}{T_u}R\right)R\right]=0 \quad \text{(C3)}$$

Thus,

$$c_h+\left(\frac{\pi}{T_c}+c_h\right)\frac{E(T)_{A_1+A_2}}{T_u}\int_R^{\infty}g\left(\frac{E(T)_{A_1+A_2}}{T_u}x_t\right)dx_t=0 \quad \text{(C4)}$$

12

Using CNOs in international marketing and outbound logistics

Kim Jansson[1*] and Iris Karvonen[1]

*Corresponding author. Kim Jansson, Business Ecosystems, Value Chains and Foresight, VTT Technical Research Centre of Finland, Tekniikantie 2, P.O. Box 1000, Espoo FI-02044 VTT, Finland

E-mail: Kim.Jansson@vtt.fi

Reviewing editor: Zude Zhou, Wuhan University of Technology, China

Abstract: The paper analyses if the collaborative networked organisations (CNO) concept can bring advantages in organising the international outbound logistics for SMEs. In the manufacturing domain, the European CNO research has identified benefits from using the concept in traditional supply chains, collaboration in various inbound networks and business ecosystems. Less focus has been on outbound logistics for delivering products and related service to customers at remote locations. The analysis is based on conducted company interviews. The interviewed companies have a good record of successful international operations. The used international delivery models are mapped into taxonomy of well-known outbound logistics models. The paper proposes a customer interface network model, based on the CNO concept to tackle problems encountered.

Subjects: Logistics, Manufacturing Engineering, Manufacturing Technology, Operations Management, Operations Research, Production Engineering

Keywords: outbound logistics, collaborative networked organisations, international distribution, SME, delivery models

ABOUT THE AUTHORS

Kim Jansson received a M.Sc. degree in Industrial Engineering and Computer Science. He is currently working as a senior scientist at VTT Technical Research Centre of Finland in a team focused on "Business Eco-system Development". His main research areas are collaboration networks in manufacturing and engineering, product lifecycle services, sustainable manufacturing remanufacturing, information systems and marine engineering. He is currently involved in several national and international research projects in collaboration with industrial companies and organizations. Iris Karvonen works as a Principal Scientist at VTT Technical Research Centre of Finland in a team focused on "Business renewal in organizations and networks". Her main research areas are collaboration networks in manufacturing and engineering, product lifecycle services, sustainable manufacturing, information systems and risk management in enterprise networks. She has participated in several national research and company projects and Nordic, European and global research projects and clusters.

PUBLIC INTEREST STATEMENT

The delivery of products and services to remote customers is an area of big concern for companies expanding from local to international markets. For a small- and medium-size company having limited resources, this is a real problem. Both finance resources and the availability of competent personnel, are lower than those of large enterprises. These small- and medium-size companies extending to geographically new markets need to consider how and through which stages the entrance to remote markets is made. They need to decide which operations are transferred to the remote countries and which are kept in the home country, how to build the distribution network, what kind of collaborative partners are needed and how to identify them? The paper describes how collaboration and networking could be used in international logistics. The long-term goal is to support European small- and medium-size companies to produce goods and deploy efficient outbound logistics for international markets.

1. Introduction

The outbound logistics is an area of big concern for organisations seeking to expand from local to international markets. For a SME type company having limited resources, this is a real problem. The resources of SMEs, both as finance and as competent personnel, are lower than those of large enterprises. There are examples of companies that have made substantial efforts to sell to new market areas but have not succeeded and have finally withdrawn from the location. SMEs willing to extend to geographically new markets need to consider how and through which stages the entrance is made. For example, which operations are transferred to the target areas and which are kept in the home country or in the neighbourhood, how to build the distribution network, what kind of collaborative partners are needed and how to identify them?

The objective of this paper is to discuss the organisational forms of SMEs in global operations. The paper first presents the results from two qualitative research studies, based on interviews in SME companies. Both studies are motivated by the growth opportunities on the remote markets at the same time as the local European market is in change. The interviewed companies all have a successful history of internationalisation. A summary of used outbound logistics delivery models has been collected and interview observations are presented.

Furthermore, the paper describes how the collaborative networked organisations (CNO) concepts could be used to develop delivery models in international logistics. The long-term goal is to support European SMEs to produce goods and deploy efficient outbound logistics for international overseas markets.

The first set of interviews was conducted within the marine industries. A large-scale FIMECC (Finnish Metals and Engineering Competence Cluster); Innovations and Network (I&N) research programme was launched in the beginning of 2009 with special focus on marine industries (FIMECC, 2009). The "Net" project, being a minor project and a part of the I&N programme, focuses on SME companies' direct international businesses and customer segments. The research work carried out generates explorative data, based on questionnaires and interviews. The Finnish marine industry branch is to a high degree networked, involving a large number of companies, mostly SMEs, having the local shipyards as the main customer. Due to the dramatically decreasing number of new building orders received by the shipyards, the companies now have to search for direct overseas international business, develop their international business processes and international delivery logistics operations.

In parallel the "Industrial Global Logistics Value Networks" Tarvo Project (Tarvo, 2013) was started by VTT in 2012 with support from the national research funding organisation Tekes (2013). In the project, the focus is on Finnish manufacturing SME companies who are either entering the global markets or increasing their volume of export. The objective is to create modern delivery logistics networks for SME companies to increase sales in the Asian markets.

In this context, the meaning of outbound logistics or international delivery logistics is broader than just transportation for order fulfilment to the customer. International delivery logistics cover also the identification and creation of sales channels, product configuration and assembly, inventories, customer pick-up points, spare parts, repair and maintenance service, training, etc. at remote locations.

In both mentioned projects, a number of industrial interviews have been carried out with representatives of the company management. The interviews give real-world information on used models for customer value creation. This paper is based on the work carried out by VTT in the Tarvo project and in the Net projects together with Turku School of Economics at the University of Turku.

The paper is organised in the following way: Section 2 describes used international delivery network models. A summary of observations from the conducted interviews is included in Section 3. Section 4 reviews previous research results in the area of CNO. In Section 5, the concept of customer interface network is introduced. Section 6 gives the conclusions and proposes ways forward.

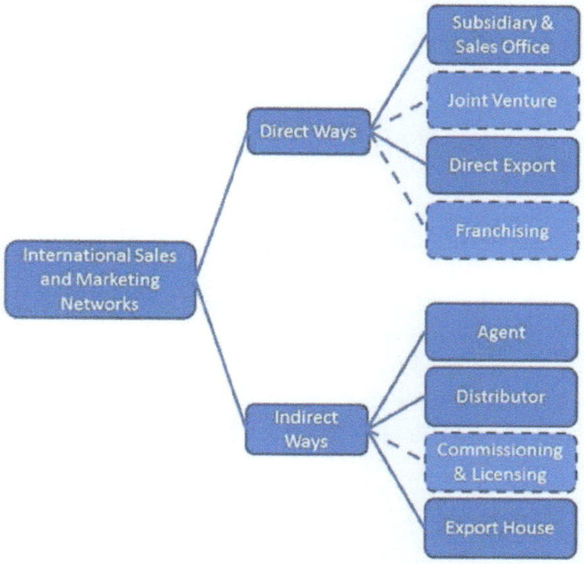

Figure 1. Taxonomy of organisation models in international distribution networks.

2. International distribution network models

There are several possibilities for entry into new markets, like using export company, agent, distributor or creating alliances with local companies. Often SMEs need to build their entrance to the market not only as sales channels but also as operational support and maintenance. To achieve sufficient volume in the market also, some manufacturing or engineering activities may need to be moved to the target area. Typically, the form of the presence on the distant market area does not remain the same. The partners and the organisational arrangement of the distribution network may evolve along the time.

In Figure 1, a taxonomy for international delivery networks is presented, based on a literature review. The network types have two main types: direct or indirect ways, depending on if there are intermediate organisations involved between the company and the customer. *Joint ventures*, *Franchising* and *Commissioning/Licensing* are drawn with dashed lines as they were not seen as common for the industry branches in the interviews.

The different options have different challenges regarding needed resources, identification and commitment of partners, contracting, pricing, and controlling the market and contacts to the customers. The following four graphs in Figure 2 give an overview of the roles, information flows and contract relations of the most used organisation models in the taxonomy. Table 1 gives an overview of the different features in organisation models.

3. Results from conducted interviews

The objectives of the Net project interviews are to get deep understanding of direct international business model characteristics in the SME marine sector. The purposes of the Tarvo project interviews are to review paths for setting up international delivery networks with special emphasis on the Chinese markets. Although the objectives of the two interviews groups were slightly different, the used interview questions were quite the same. The selection of the companies to be interviewed was done by the researchers. The candidate companies were on forehand known to be involved in international business. Thus, only qualitative results are expected, while the sampling is biased for quantitative results.

Altogether 17 companies were interviewed, of which, one was interviewed twice (once in both projects). Companies are kept anonymous in this paper. Together they represent 10 equipment providers for the marine and process industry, two turnkey marine outfitting providers, one software

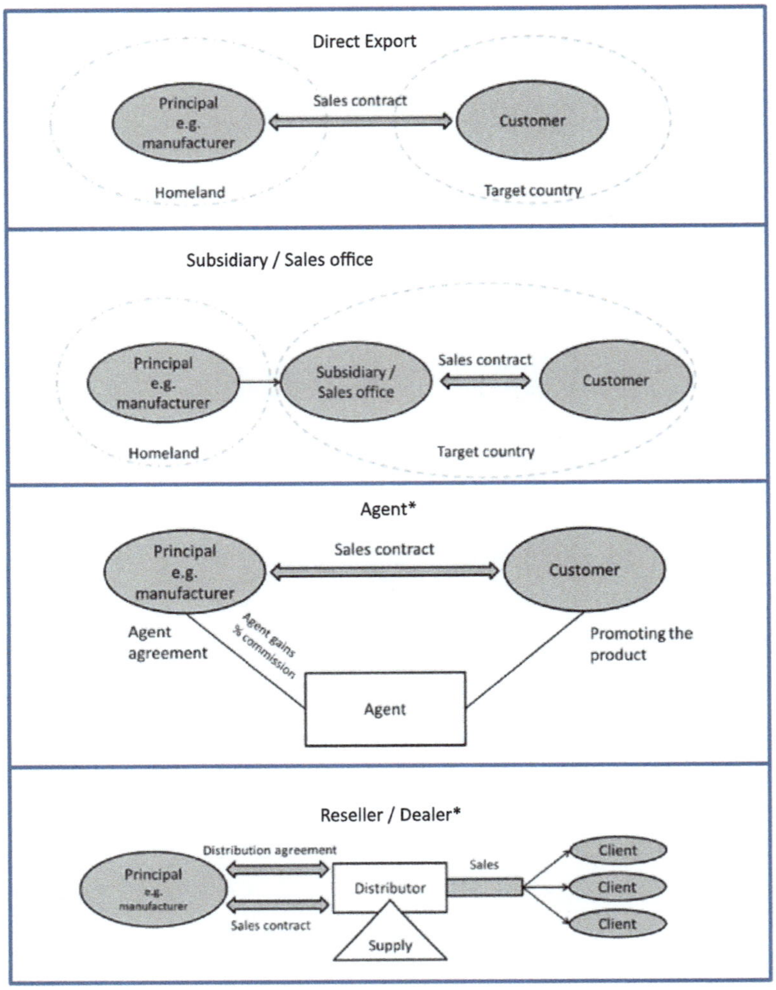

Figure 2. Overview of the used organisation models, adapted from (Caraiani & Potecea, 2010).

Table 1. Summary of the features of the used models

	Description	Benefits	Challenges	When to apply?
	Agent opens doors in export; sales contract between manufacturer and customer	Contact to customers. Knowledge of the agent area	Contractually difficult. Commitment of the agent. Takes more resources than direct export	Useful way to create contacts to customers when there are no contacts
Dealer	Dealer buys the products from the manufacturer and sells them to customers	The dealer takes the risk of sales. The dealer knows the area	No direct contact to the customers. Commitment of the dealer. Price setting. Requires more resources than direct export	For standard products and other situations when the contact to customers is not necessary
Direct export	Company has direct contact to customers	Direct contact to customers. No difficult contracts. Requires less resources than agent or dealer	No continuous presence at the market. Travelling costs. It may be difficult to reach the customers	In cases when the number of customers is restricted and does not require too much travelling (short distances/not very often)
Subsidiary company/ own sales office	The sales office or subsidiary company is in the same country or area as customers	Direct contact to customers. Commitment of employees. Using a subsidiary company, some taxes can be avoided	Costs of creating the remote office or company. Recruiting the employees. Collecting needed knowledge	Not in the first phases of globalisation
Partnership with a foreign company	Collaboration with a foreign company to market and sell the products together	Increased credibility and more business potential	How to build the relationship and identify proper partners?	The products or services of the partners complement each other
Export ring/export house	Companies with the same origin country collaborate for common export	The export house may support in export and market knowledge	No direct contacts to customers. Difficult to monitor the market	The company has low knowledge and resources for sales and export activities

Table 2. Number of used models

Direct export	Subsidiary company/own sales office	Reseller/ dealer	Agent	Partner	Joint venture	Export house/ring
10	13	6	5	7	1	4

provider, one automation provider, a mall shipyard and an engineering design company. Ten companies are involved in project-type deliveries and the rest in assembly to order and standardise product-type deliveries. With the exception of the software provider and the engineering company, all the other companies deliver tangible products together with a varying degree of intangible services and software.

Table 2 shows the number of used models in the interviewed companies. As a summary of the interviews, we can state that;

- All model types are used. The used model depends on production volumes, maturity and time, geographical area and local legislation.
- Subsidiary company/own sales office is the most used distribution network model within this group of companies.
- Direct export is the second most used model.
- Collaboration with international partners is an important channel to the international customers.
- When companies have production abroad, then it is mostly organised in own subsidiary companies.
- A majority of project-type deliveries go by direct export.
- Joint venture is an unpopular model.
- A development pattern can be distinguished as the time passes and local presence matures. Dealer and agent → Direct export → Subsidiary.

The following list is a collection of success factors for sales and distribution. The comments have been collected from the interviews and grouped together.

Suitability of products and service. In direct sales, it is important to have the right customer contacts, knowledge about the customers, asking the customers about their problems and needs, and finding the decision-maker; plus, understanding customer needs and sales professionalism are also needed.

Small services to customers can grow to large sales. Start small and grow.

Joint R&D projects and technology development with customer groups are important to strengthen customer relationships. Collaboration, communication and common understanding in service/delivery network are vital.

Partnership with another global company with a connected product may be a successful path for entering a new market.

Sales and marketing partnerships with joint products have been useful.

The Rule No. 1 in international business is to *respect local habits and customers*. The internal company processes and rules must guarantee this.

Trustworthy of sales channels. Communication and interaction are needed; and, personal relationships support the trustworthy cooperation between partners.

4. Research on CNO

Much knowledge and understanding of CNO has been acquired through European Union-supported research during the last decades. There have been several research initiatives in Europe in the field of CNOs (Camarinha-Matos, Afsarmanesh, & Ollus, 2008) and also in the context of global operation (Putnik & Cruz-Cunha, 2008). The research objects span from traditional supply chains, collaborative networks and business ecosystems to virtual organisations. Several international initiatives, e.g. VOSTER; ECOLEAD and COIN have developed reference models, collaboration platforms and software tools to support CNOs.

The collaboration within CNOs fits into the concept of a business ecosystem introduced already in the 1990s by Moore (1993): "An economic community supported by a foundation of interacting organisations and individuals—the organisms of the business world. This economic community produces goods and services of value to customers, who are themselves members of the eco-system".

In the European CNO research cluster VOSTER (2006), two main concepts for inter-enterprise collaboration were identified according to the objective and duration of the collaboration (Kurumluoglu, Nostdal, & Karvonen, 2005):

- Network/breeding environment, which is a more stable, though not static, group of organisations which have developed a preparedness to cooperate.
- Virtual organisation (VO)/virtual enterprise, which is a temporary consortium of partners from different organisations established to fulfil a value-adding task, for example, a product or service to a customer.

Accordingly, a VO is usually created within a network composed of organisations committed to collaborate. To achieve efficient collaboration, some degree of preparedness and sufficient preparations are needed. This preparation takes place within the Breeding Environment. The concept of the Breeding environment is used to characterise the network behind a VO. The Virtual organisation's Breeding Environment (VBE) represents a long-term "strategic" alliance, cluster, association or pool of organisations that provide the needed conditions for collaboration. (Camarinha-Matos et al., 2008), Figure 3.

The ARCON Reference model for Collaborative Networks, created in the ECOLEAD Project (2008), has enhanced previous frameworks for understanding the relationships between entities in a CNO (Camarinha-Matos & Afsarmanesh, 2008). An earlier reference model is VERAM—Virtual Enterprise Reference Architecture and Methodology (Toelle, 2004) Zwegers, Toelle, & Vesterager, 2003). The VERAM reference model is based on the Generalised Enterprise Reference Architecture and Methodology (GERAM), which was developed for modelling of single enterprises and has been accepted as an ISO standard (GERAM, 2000). In VERAM, this model was extended for modelling virtual and networked enterprises. According to GERAM, the processes of an entity can be distinguished into two different types:

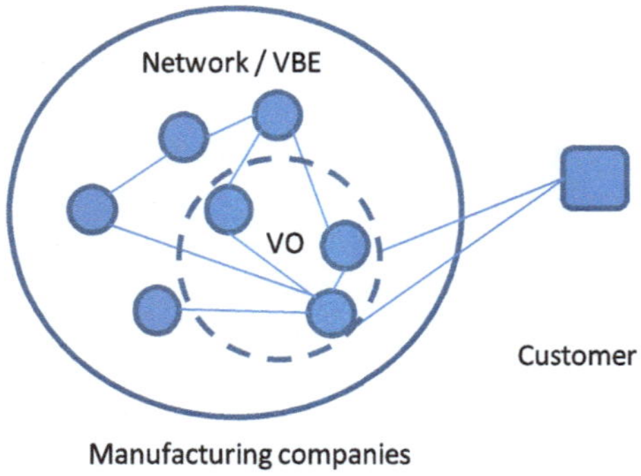

Figure 3. VBE and VO.

- Service to the customer (creating the value, product or service).
- Management and control process (coordinating and managing the value creation).

The processes are interdependent. The more complex and time or cost-critical the operational processes are, the more challenging is the management process. The outcome of a project can be improved either by investing in the project management or by developing the project operations. Naturally, both efficient management and processes are preferable.

Also several scholars and research projects have, during the last decade, studied the economic and organisational transition of industrial enterprises towards a service economy (DuTertre, 2007). This transition, also called servicisation (Balin, 2007), to the step-by-step integration of service and value-adding activities in industrial companies together with their more traditional manufacturing-oriented activities induces (Baines & Lightfoot, 2011) a makeover of enterprise models. Servicisation deals with the development of integrated product/service offers, also named product service systems (PSS). Currently, such new enterprise models are increasingly being used within the international economy.

The CNO concept has also been used for special purposes and application niches, for example, offering opportunities to provide ageing care and elderly population assistance (Camarinha-Matos, Rosas, Oliveira, & Ferrada, 2012) and to promote the sharing and recycling of resources, such as information, materials, water, energy and/or infrastructure, with the intention of achieving sustainable development in a collaborative way (Romero & Molina, 2012).

As a conclusion, we can state that the field of knowledge in the area of CNO is extensive but highly fragmented. There is still no consensus on the definitions. As an example, in literature, the term VO is found in a variety of more or less synonyms, e.g. virtual enterprise, extended enterprise, agile enterprise, smart organisation, value net, concurrent enterprise, supply chain and one product-integrated manufacturing. However,

- All these models use information and communication technology (ICT) systems as a prerequisite and enabler.
- All these models belong to the CNO concept, in a more or less broader sense (Putnik & Cruz-Cunha, 2008).

Most of the previous research has been focused on manufacturing networks and supply chains. Few examples have been studying networks in the distribution, sales, marketing or market opportunity identification. Only few exceptions can be found, e.g. Anastasiou and Tsagkas (2003) developed tools and approaches to support the management of information in collaborative sales. In the ECOLEAD and COIN (2010), projects tools for collaboration opportunity identification and characterisation were developed.

5. Application of CNO approaches in outbound logistics

Much of the mentioned CNO research work has centred around the question on how to organise work in the best way to deliver products and services to a customer in the most efficient and dynamic way. In this paper, the aim is to review the outbound logistics for SME type of organisations, using the CNO concepts established in previous and existing research results. In this context, the outbound logistics is used as a large concept. The meaning of outbound logistics is broad, as mentioned in the introduction. In this context, we call it a customer interface network (CIN), Figure 4, covering also the activities, such as final product configuration and assembly, inventories, customer pick-up points, spare parts, repair and maintenance service, and training.

Figure 1 introduces a taxonomy of organisation models in international distribution networks and Table 1 summarises the features of these models. The emphasis is on creating and using the best

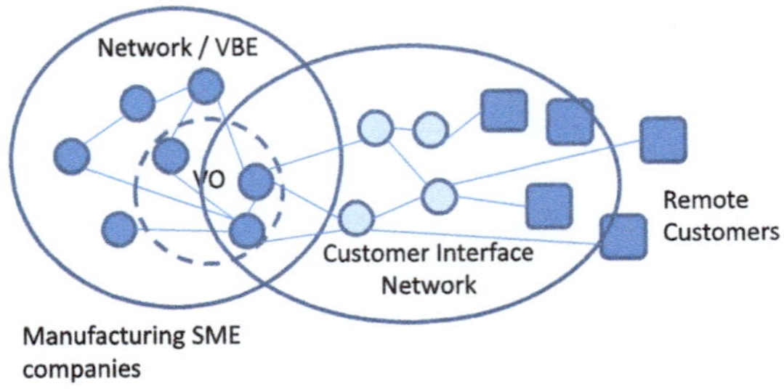

Figure 4. Customer interface network.

suitable and cost/resource-efficient organisation model that increases sales and new order intake. The concept of VO is best suited for project-type and one-of-a-kind deliveries. From an order fulfilment point of view the delivery of a one-of-a-kind and project-type delivery is quite different from a mass product or continuous supply chain. However, when we take the broader view involving lifecycle services as well, then the difference is not so dramatic. All types of products need services, spare parts, maintenance, repair and potential (software and hardware) updates. Support to the long usage phase of a one-of-a-kind product throughout its full lifecycle is always needed. The CIN concept can respond to all lifecycle support needs and create the preparedness for delivering the product and service as well as ongoing support. The CIN, thus, forms the correspondence to the VBE as mentioned above. The CIN can be a long-term strategic alliance.

In Figure 4, the left-hand side VBE and VO represents a network with manufacturing-oriented business process and the right-hand side, a network with service-oriented business processes.

The business approaches of services delivery as an integrated part of product is a vast research topic. Manufacturers are moving more towards the business approach of PSS in order to achieve new customers and increased sales. PSS have many advantages such as achieving closer customer connection and generating increased profit from manufactured products. The importance of services is also recognised in the marketing field. According to Vargo and Lusch (2004), there is an emerging trend in the marketing and business thinking of firms where firms are moving from traditional goods-centred dominant logic to service-centred dominant logic. In the G-D logic, customers are seen as targets, and they are acted on to create transactions with resources. In the S-D logic, customers are active participants in relational exchanges and co-production. Paasi, Lappalainen, Rantala, and Pikkarainen (2014) have, from broad empirical interview data, studied challenges faced by product and service providers, when innovating openly with customers in business-to-business markets. Four types of open innovation with customers are presented to categorise product and service providers through their main business and innovation logics, and relating customer involvement of companies. The types are customer-oriented providers, customer-solution integrators, developer partners and facilitators of co-innovation.

According to Sakao and Lindahl (2009), the following grouping of PSS is frequently used:

- Product-oriented Services: the product is owned by the user/consumer.
- Use-oriented Services: the product is owned by the service provider, who sells functions instead of products by means of modified distribution and payment systems, for example, sharing, pooling and leasing.
- Result-oriented Services: the products are substituted by new services, often driven by new technologies. In these cases, the supplier provides incentives for the customer to consume services by using results-orientated payment systems.

However, when we view the challenges in outbound logistics as presented in the introduction section and the proposed CIN concept, only the first group in the list above is relevant.

A CIN is a long-term strategic collaboration that should be based on agreed business models and operational rules. The main issues to be agreed on are how to create value for all CIN stakeholders including customers, how to measure the value, how to manage and support the interest of the CIN partners, and operation of the network. Thus, setting up a CIN requires the following main elements;

- Business model.
- Operational model.

The selected distribution network models, as presented in Figure 1, naturally form the starting point for defining the business model. The Osterwalder Business Model Canvas (Osterwalder, 2004) is an excellent tool for defining the business models' elements e.g. offerings, customers, infrastructure and finances (what, who, how and how much).

Setting up the operation model of a CIN helps the stakeholders to know who can make decisions, who is accountable for which effect and how each actor must work to conform to the defined business model. The operational model, thus, contains at least the agreement on CIN principles and a set of rules. The definition of the CIN principles includes agreement on not only values that guide the operation within CIN such as honesty, trust, openness, responsibility, accountability, mutual respect, but also cultural behaviours, values and ethics. The CIN operational rules are related to used managerial processes such as accounting rules, ICT technology, information sharing, risk management, rights, duties, profit sharing and conflict resolution. The overall objective of defining the business model and operational model is to create the needed preparedness for efficient and profitable operation of a CIN.

For a European SME, it is indeed very difficult to provide alone all the functions that are expected for an efficient outbound logistics system on overseas markets. Often, this requires proximity to the customer, speaking the customer's language, technical knowledge, spare part inventory, etc. The purpose of a CIN is thus, to establish the necessary prerequisites to serve a remote customer in a cross company and collaborative way.

For the CIN to be operable, there are certain task and duties that need to be fulfilled. The stakeholders in the various organisation models, Figure 1, can take responsibility through different roles in the CIN. The following is a (minimum) list of roles needed:

- CIN Manager (localisation of operations).
- Marketing and sales (contact to overseas customers).
- Transportation (overseas and local).
- Warehouse (final products, spares and consumables).
- Pick-up point (customer interface point).
- Final Assembly (customisation, localisation).
- Training (to overseas customers).
- Service and maintenance.
- IT systems provider.
- Invoicing.

The following Table 3 contains a set of required roles in the CIN and how the organisation models from taxonomy in Table 1 could fulfil the different roles. Note, the *local logistics operator* is not a model in the taxonomy, but usage of a local logistics operator is a form of partnership. The local logistics operator is a local or international operator that can be included in the network but not sufficiently alone as an international distribution network.

Table 3. Roles in a CIN and organisation models

Roles	Possible organisation models
CIN Manager	Agent, dealer, own sales office
Marketing and sales	Agent, dealer, own sales office
Transportation	Local logistics operator
Warehouse	Local logistics operator
Pick-up point	Dealer, local logistics operator
Assembly	Own subsidiary company, partner
Training	Agent, dealer, own sales office
Service and maintenance	Dealer
IT systems	
Invoicing	Dealer, own sales office

From the organisation model viewpoint, it can be stated that the selection of the organisation model of the distribution network is always dependent on the context, environment and the development phase; and, the model may evolve along time. Thus, there is no one recommended solution good for all. The different options have different challenges regarding needed resources, identification and commitment of partners, contracting, pricing and controlling the market and contacts to the customers. Understanding the customer needs is essential. The understanding may be achieved through direct contacts, common development projects and collaboration and communication in the distribution network.

6. Conclusions

It is very well known that the outbound logistics is a challenge for SME type of organisations seeking to expand from local to international markets. The conducted interviews have collected information on how this problem has been successfully tackled. The paper presents a taxonomy of used organisation models in international distribution networks. The paper proposes approaches on how the currently used organisation models in international distribution networks could be further developed based on the CNO concept to fulfil the requirements of efficient outbound logistics. Further research is needed for the different models to assess in more detail, the roles and how to establish the appropriate preparedness to act in the roles. For practical feedback, more real case studies are needed.

Authorship Note

§Aino Vaittinen is no longer affiliated with VTT Technical Research Centre of Finland and could not be reached to review the final version of this article. However, the authors are indebted to her for her contribution to Section 2 of this article, "International distribution network models".

Acknowledgements

We wish to give our gratitude and appreciation to all companies for their contribution to the interviews during the development of concepts presented in this paper.

Funding

This work has been partly funded by the Tekes—the Finnish Funding Agency for Technology and Innovation Finland, through the FIMECC Ltd, I&N Programme [grant number 40210/09] and through the Tarvo project [grant number 40100/12]. The authors wish to thank Tekes for their support to the both projects.

Author details

Kim Jansson[1]

E-mail: Kim.Jansson@vtt.fi

Iris Karvonen[1]

E-mail: Iris.Karvonen@vtt.fi

[1] Business Ecosystems, Value Chains and Foresight, VTT Technical Research Centre of Finland, Tekniikantie 2, P.O. Box 1000, Espoo FI-02044 VTT, Finland.

References

Anastasiou, M., & Tsagkas, E. (2003). Management of information in collaborative sales. In I. Karvonen, R. van den Berg, P. Bernus, Y. Fukuda, M. Hannus, I. Hartel, & J. Vesterager (Eds.), *Global engineering and manufacturing in enterprise networks (GLOBEMEN), VTT Symposium 224* (pp. 207–217). Espoo, Finland. ISBN 951-38-6275-5.

Baines, T. S., & Lightfoot, H. W. (2011). Towards an operations strategy for the infusion of product-centric services into manufacturing. In H. Demirkan, J. C. Spohrer, & V. Krishna (Eds.), *Service systems implementation* (pp. 89–112). New York, NY: Springer.

Balin, S. (2007). *Amélioration de processus de production de services par la simulation* (PhD thesis). University Paris Dauphine.

Camarinha-Matos, L. M., & Afsarmanesh, H. (Eds.). (2008). *Collaborative networks: Reference modeling*. New York, NY: Springer. ISBN-13 978-0-387-79425-9.

Camarinha-Matos, L. M., Afsarmanesh, H., & Ollus, M. (Eds.). (2008). *Methods and tools for collaborative networked organizations*. New York, NY: Springer. ISBN 978-0-387-79423-5.

Camarinha-Matos, L. M., Rosas, J., Oliveira, A. I., & Ferrada, F. (2012, October 1–3). A collaborative services ecosystem for ambient assisted living in collaborative networks in the internet of services. In L. M. Camarinha-Matos, L. Xu , & H. Afsarmanesh (Eds.), *Proceedings PRO-VE 2012*. Bournemouth, UK. http://dx.doi.org/10.1007/978-3-642-32775-9

Caraiani, G., & Potecea, V. (2010). Agent and distribution agreements in international business. *Metalurgia International, 15*, 129–133.

COIN Project. (2010). Retrieved May 12, 2014, from http://www.coin-ip.eu/

DuTertre, C. (2007, September 13–15). Modèle industriel et "modèle serviciel" de performance [Industrial models and "service model" performance]. In *XVIIth International Conference of RESER*. Tampere, Finland.

ECOLEAD Project. (2008). Retrieved May 12, 2014, from http://virtual.vtt.fi/virtual/ecolead

Finnish Metals and Engineering Competence Cluster Ltd. (2009). Retrieved March 26, 2013, from http://www.fimecc.com/

Generalised Enterprise Reference Architecture and Methodology. (2000). *IFAC/IFIP task force on architectures for enterprise integration*. Author. Retrieved October 8, 2010, from http://www.cit.griffith.edu.au/~bernus. ISO15704:2000.

Kurumluoglu, M., Nostdal, R., & Karvonen, I. (2005). Base concepts. In L. Camarinha-Matos, H. Afsarmanesh, & M. Ollus (Eds.), *Virtual organizations. Systems and practices* (pp. 11–28). New York: Springer-Verlag.

Moore, J. (1993, May/June). Predators and prey: A new ecology of competition. *Harvard Business Review*.

Osterwalder, A. (2004). *The business model ontology—A proposition in a design science approach* (PhD thesis). University of Lausanne, Lausanne.

Paasi, J., Lappalainen, I., Rantala, T., & Pikkarainen, M. (2014, April). Challenges for product and service providers in open innovation with customers in business-to-business markets. *International Journal of Innovation Management, 18*(2), p. 1450012. Imperial College Press.

Putnik, G., & Cruz-Cunha, M. (Eds.). (2008). Services and logistics. In *Encyclopedia of networked and virtual organizations*. (Vols. I–III, Article number 1450012). IGI Global. http://dx.doi.org/10.4018/978-1-59904-885-7

Romero, D., & Molina, A. (2012, October 1–3). Green virtual enterprise breeding environments: A sustainable industrial development model for a circular economy. In L. M. Camarinha-Matos, X. Lai, & H. Afsarmanesh (Eds.), *Collaborative networks in the internet of services. Proceedings PRO-VE 2012*. Bournemouth, UK.

Sakao, T., & Lindahl, M. (Eds.). (2009). *Introduction to product/ service-system design*. Springer. Retrieved from http://www.springer.com/978-1-84882-908-4

Tarvo Project. (2013). Website in Finnish. Retrieved September 18, 2013, from http://interali.vtt.fi/tarvo/

Tekes. (2013). *The finnish funding agency for technology and innovation Finland*. Retrieved March 26, 2013, from http://www.tekes.fi/en/community/Home/351/ Home/473

Toelle, M. (2004). *Management and engineering of virtual enterprises* (PhD dissertation). Technical University of Denmark, Department of Manufacturing Engineering and Management.

Vargo, S. L., & Lusch, R. F. (2004, January). Evolving to a new dominant logic for marketing. *Journal of Marketing, 68*(1), 1–17. ISSN 0022-2429. http://dx.doi.org/10.1509/jmkg.68.1.1.24036

VOSTER. (2006). Retrieved May 12, 2014, from http://cic.vtt.fi/ projects/voster/public.html

Zwegers, A., Toelle, M., & Vesterager, J. (2003). VERAM: Virtual enterprise reference architecture and methodology. In *Global engineering and manufacturing in enterprise networks. VTT SYMPOSIUM 224*. Otamedia. ISBN 951-38-5739-5.

13

Solving a multi-objective chance-constrained hub covering location problem by discrete invasive weed optimization

Seyed Hossein Nikokalam-Mozafar[1], Behzad Ashjari[1], Reza Tavakkoli-Moghaddam[2]* and Aida Omidvar[3]

*Corresponding author: Reza Tavakkoli-Moghaddam, School of Industrial Engineering, College of Engineering, University of Tehran, Tehran, Iran

E-mail: tavakoli@ut.ac.ir

Reviewing editor: Zude Zhou, Wuhan University of Technology, China

Abstract: This paper presents a stochastic bi-objective model for a single-allocation hub covering problem (HCP) with the variable capacity and uncertainty parameters. Locating hubs can influence the performance of hub and spoke networks, as a strategic decision. The presented model optimizes two objectives minimizing the total transportation cost and the maximum transportation time from an origin to a destination simultaneously. Then, due to the NP-hardness of the multi-objective chance-constrained HCP, the presented model is solved by a well-known meta-heuristic algorithm, namely multi-objective invasive weed optimization. Additionally, the associated results are compared with a well-known multi-objective evolutionary algorithm, namely non-dominated sorting genetic algorithm. Furthermore, the computational results of the foregoing algorithms are reported in terms four well-known metrics, namely quality, spacing, diversification, and mean ideal distance. Finally, the conclusion is reported.

Subjects: Evolutionary Computing; Logistics; Operations Research; Optimization; Stochastic Models & Processes; Supply Chain Management

ABOUT THE AUTHORS

Seyed Hossein Nikokalam-Mozafar obtained his BSc and MSc degrees in Industrial Engineering from Islamic Azad University and Tafresh University in Iran, respectively. His interest researches are in facility location, optimization and management.

Behzad Ashjari is an assistant professor of Industrial Engineering at Tafresh University. He obtained his PhD degree from AmirKabir University of Technology in Iran. His research interests are in operations research, maintenance and supply chain management.

Reza Tavakkoli-Moghaddam is a professor of Industrial Engineering at University of Tehran. He obtained his PhD in Industrial Engineering from Swinburne University of Technology in Australia. His research interests include facility layouts and location design, scheduling, and metaheuristics.

Aida Omidvar received her BSc and MSc degrees in Applied Mathematics and Industrial Engineering from University of Isfahan and Islamic Azad University in Iran, respectively. Her interest researches are in facility location, optimization, metaheuristics and strategic planning.

PUBLIC INTEREST STATEMENT

Nowadays, transportation and communication networks are of a vital role in modern societies. Thus, optimization of the network design for managing better the traffics and decreasing transport time and cost in acceptable service quality is more important. In this paper, we optimize a network structure for achieving two objectives, namely minimizing the total transport cost and maximum transport time in each allocation when a demand is uncertain. Because of the complexity of the problem, we propose a meta-heuristic algorithm that hss inspired the behavior of invasive weed colonization.

Keywords: stochastic hub covering problem; multi-objective optimization; chance-constrained programming; invasive weed optimization

1. Introduction

A hub location problem is one of the significant classes of facility location problems that after more than two decades from appearance nevertheless many papers of various journals are titled to hub networks design and applications. The background of these problems is to manage traffic better in the network that causes decrease in the transportation cost or time in an acceptable quality level of servicing. In hub and spoke networks, supply and demand points are usually connected together via special nodes, called hubs, which are equipped with some facilities. Therefore, the initiation flow from origin nodes routes destinations through hubs. Hub covering problems (HCPs) are one of the subproblems in a hub location area, which are applied in transportation systems and telecommunication networks. Campbell (1994) introduced three types of HCPs by defining three covering criteria depending on the part of flow route that is focused by the decision-maker for optimizing. Hence, according to the definition an origin–destination pair (i, j) is covered by hubs k and l if: (1) the cost from i to j via k and l $(i \rightarrow k \rightarrow l \rightarrow j)$ does not exceed a specified value; (2) the cost of each link of the path $(i \rightarrow k, k \rightarrow l, l \rightarrow j)$ does not exceed a specified value; and (3) the cost of origin-hub or hub-destination does not exceed a specified value. Also, he presented the first mixed-integer programming model for both single- and multiple-hub set covering problems, whose objective was to locate hub nodes with minimum fixed costs in the network satisfying all demand points. Kara and Tansel (2003) studied the single-allocation HCP and proposed various linear formulations. Wagner (2004) presented a model for single- and multiple-allocation HCPs. Afterward, Ernst, Jiang, and Krishnamoorthy (2005) proposed a different model for single HCPs using a new concept, radius.

Recently, most of the investigations in the optimization area are interested in studying the problems under uncertainty, and hub location problems are no exception in this matter. One of the approaches in decision-making under uncertainty is stochastic programming. Drezner and Scott (2013) presented a location-inventory model with a perishable product for a distribution center. They considered the transportation cost and time. Marianov and Serra (2003) proposed a stochastic model for hub location in an airline network by the queuing theory and chance-constraint approach. Sim, Lowe, and Thomas (2009) introduced a p-hub center problem and used a chance-constrained formulation for minimizing the service-level requirement. Lium, Crainic, and Wallace (2009) addressed the stochastic demand in the service network design. Contrerase, Cordeau, and Laporate (2011) studied stochastic uncapacitated hub location problems with uncertainty in demand, dependent, and independent transportation costs and described a Mont-Carlo simulation to solve the problem. Alumur, Nickel, and Saldanha-da-Gama (2012) addressed several aspects concerning hub location problems under uncertainty, such as setup cost and demand. Most of the optimization problems involve optimizing more than one objective function simultaneously. Hence, the decision-maker usually deals with trade-offs between some conflictive objectives to make an optimal decision. It has been another challenge to researchers on study of the problem with realistic assumptions. There are some researches about multi-objective hub location problems (Costa, Captivo, & Clímaco, 2008).

Mohammadi, Tavakkoli-Moghaddam, and Rostami (2011) proposed a new weight-based multi-objective imperialist competitive algorithm (MOICA) for the HCP. The total transportation cost and the time service in the hub nodes as two objective functions were minimized. Furthermore, Mohammadi, Jolai, and Tavakkoli-Moghaddam (2013) presented a new stochastic multi-mode transportation model for a hub covering location problem under uncertainty. They minimized two objectives, the total current investment costs and the maximum transportation time between each origin–destination pair in the network simultaneously. In addition, they proposed a MOICA that compared two well-known multi-objective evolutionary algorithms. Mohammadi, Torabi, and Tavakkoli-Moghaddam (2014) considered a new hub location problem (HLP) considering environmental aspects for air and noise pollution of vehicles. They proposed a mixed possibilistic–stochastic programming method along with two meta-heuristics, namely simulated annealing and imperialist competitive algorithm. Ghodratnama, Tavakkoli-Moghaddam, and Azaron (2013) presented a new mathematical model for a hub location with heterogeneous fleet vehicles and proposed three meta-heuristics; namely, genetic algorithm, particle swarm optimization, and simulated annealing to solve the model. Ghodratnama et al. (2013) proposed a novel fuzzy bi-objective HCP for minimizing the total cost of transportation, consisting of transportation, covering, opening and reopening the facilities in hubs, activating the facilities of hubs and transporter purchasing, and the sum of times of shipping commodities by transporters from origin to destination. Table 1 is a short extract of other literature studies in HCPs (Calik, Alumar, Kara, & Karasan, 2009; Hamacher & Meyer, 2006; Karimi & Bashiri, 2011; Lowe & Sim, 2012; Mohammadi et al., 2011; Qu & Weng, 2009; Tan & Kara, 2007; Weng & Wang, 2008.

Nowadays, transportation and communication networks are a vital role in modern societies. Thus, optimization of the network design for managing better the traffics and decreasing transport time and cost in acceptable service quality is more important. In this paper, we optimize a network structure for achieving two objectives, namely minimizing the total transport cost and maximum transport time in each allocation when a demand is uncertain. Because of the complexity of the problem, we propose a meta-heuristic algorithm that has inspired the behavior of invasive weed colonization.

This paper surveys multi-objective single-allocation HCPs with the uncertain demand and transportation time using chance-constrained programming (CCP). CCP is one of the powerful approaches in modeling stochastic systems by replacing the deterministic constraints involving the effects of uncertainty with probabilistic ones. These new constraints hold at least the reliability level. The authors assume that the demand and transportation time between the nodes are the only uncertain parameters. The target of the problem is to find the location of hubs and allocate the non-hub nodes to hubs such that the transportation cost between each origin-hub or hub-destination links is within a considered economical bound. Moreover, the maximum transportation time between origin–destination is minimized simultaneously. The variable capacity of hub nodes is one of the different aspects of this problem with the other hub location models in the literature. The authors propose a new formulation for the problem and solve the multi-objective stochastic HCP using a novel artificial intelligent technique called as discrete invasive weed optimization (IWO) that is more efficient in comparison to the older famous

Table 1. Hub set covering literature

Authors	Notes
Hamacher and Meyer (2006)	Comparing various formulations for the HCP
Weng et al. (2006)	Solving a new formulation for hub maximal covering problems and by genetic algorithm and tabu search
Tan and Kara (2007)	Applying the HCP for cargo delivery system
Weng and Wang (2008)	Using scatter search and genetic algorithm for solving the HCP
Qu and Weng (2009)	Proposing a path relink method for solving multiple allocation hub maximal covering problems
Calik et al. (2009)	Solving the HCP over incomplete hub networks by a heuristic based on tabu-search
Karimi and Bashiri (2011)	Proposing a new formulation and two heuristic solving methods for the HCP with different coverage types
Mohammadi et al. (2010)	Solving the HCP under capacity constraints by genetic algorithm and simulated annealing
Low and Sim (2012)	Study hub covering flow problems

evolutionary algorithms. To the best of authors' knowledge, there are no similar studies in the hub location literature. The proposed algorithm is tested with various networks that the results show its efficiency.

The rest of this paper is organized as follows. In the next section, we first describe the deterministic model and then employ a CCP approach to model a stochastic environment. Some definitions and concepts of multi-objective optimization are presented in Section 3. Section 4 consists of an overview on the proposed multi-objective invasive weed optimization (MOIWO) algorithm and its structure. Subsequently, we present our proposed solving method. All the computational results and conclusions are brought in Sections 5 and 6, respectively.

2. Problem definition

This section intends to define the studied hub location problem and its assumptions. At first, the notations are introduced as follow.

N	Set of nodes ($N = \{1, 2, \ldots, n\}$)
f_k	Fixed setup cost for installing a hub at node k
g_k	The cost for each unit of capacity development of hub k
W_{ij}	Flow between origin–destination pair (i, j)
O_i	Total flow originated at node i
D_i	Total flow destined to node i
w_{ij}	Mean flow with a normal distribution between origin–destination pair (i, j)
o_i	Mean of the total flow with a normal distribution originated at node i
d_i	Mean of the total flow with a normal distribution destined to node i
σ_{ij}	Standard deviation of the flow with a normal distribution between o–d pair (i, j)
σ_{o_i}	Standard deviation of the total flow with a normal distribution originated at node i
σ_{d_i}	Standard deviation of the total flow with a normal distribution destined to node i
T_{ij}	Transportation time of o–d pair (i, j)
t_{ij}	Mean of transportation time of o–d pair (i, j)
$\sigma_{t_{ij}}$	Standard deviation of time with normal distribution between o–d pair (i, j)
C_{ij}	Transportation cost of o–d pair (i, j)
r_k	Coverage radius for the hub node k
χ	Collection cost unit flow per unit of distance
α	Discount factor between the hubs
δ	Distribution cost unit flow per unit of distance
x_{ik}	1, if node i is allocated to a hub at node k; 0, otherwise (decision variable)
y^i_{kl}	Amount of flow originated at node i is routed via hub nodes (decision variable)
z_k	Variable capacity of hub node k (decision variable)

Let a complete network with n nodes, some of them, have potential to being a hub. All nodes are connected together via hub nodes. Every node is only assigned one hub; usually this type of problem is known as a single-allocation hub location problem in the literature. In this model, the transportation cost consists of collection (i.e. from origin node to hub, $i \rightarrow k$), transfer (i.e. between the hubs, $k \rightarrow l$), and distribution (i.e. from hub to destination, $l \rightarrow j$), which are represented by the coefficients χ, α, and δ, respectively. Our deterministic model is based on one of the effective three-index flow variables for hub location problems proposed by Ernst and Krishnamoorthy (1999) and improved by Correia, Nickel, and da-Gama (2010). They interpreted the traffic flow originating from a particular node i as a commodity by defining the variables y^i_{kl}. Often in transportation networks, opening costs of hub nodes are dependent on the amount of traffic flow in them so that the cost of equipping hub nodes with facilities increases by rise in the amount of flows. On the other hand, the hub capacity affects the import flow in hub nodes directly and, like most of researches in the literature of capacitated hub location problems, only adding a capacity constraint to the model can be restricted the minimum and maximum of input/ output flow in a hub. Hence, in this paper, we consider variable capacities for hubs and moreover define a new parameter, g_k, that represents the cost for each unit of capacity development of hub k. By this assumption, the proposed model is different with the other ones in hub location literature.

2.1. Mathematical formulation

Following this, the authors firstly elaborate the formulation of the studied problem in deterministic conditions. According to the past explanations, the deterministic model is as follows.

$$\text{Min} \sum_{k\in N} f_k X_{kk} + \sum_{k\in N} g_k z_k + \sum_{i\in N}\sum_{k\in N} \chi C_{ik} O_i X_{ik} + \sum_{i\in N}\sum_{k\in N}\sum_{l\in N} \alpha C_{kl} y^i_{kl} + \sum_{i\in N}\sum_{k\in N} \delta C_{ki} D_i X_{ik} \tag{1}$$

$$\text{Min}\ \beta \tag{2}$$

s.t.

$$\sum_{k\in N} X_{ik} = 1 \qquad \forall i \in N \tag{3}$$

$$X_{ik} \le X_{kk} \qquad \forall i,k \in N \tag{4}$$

$$y^i_{kl} \ge \sum_{j\in N} W_{ij} X_{jl} X_{ik} \qquad \forall i,k,l \in N, k \ne l \tag{5}$$

$$C_{ik} X_{ik} \le r_k \qquad \forall i,k \in N \tag{6}$$

$$\sum_{i\in N} (O_i + D_i) X_{ik} \le Z_k \qquad \forall k \in N \tag{7}$$

$$(T_{ik} + T_{kl} + T_{lj}) X_{kk} X_{ll} \le \beta \qquad \forall i,j,k,l \in N \tag{8}$$

$$x_{ik} \in \{0,1\} \qquad \forall i,k \in N \tag{9}$$

$$y^i_{kl}, z_k \ge 0 \qquad \forall i,k,l \in N \tag{10}$$

It is a quadratic formulation with three decision variables for finding location of hub nodes, allocations, and the hub capacities represented by x_{ik}, y^i_{kl} and z_k, respectively. The objective function (1) calculates the total transportation costs of the hub and spoke network. The objective function (2) and Constraint (8) ensure to minimize the maximum transportation time between the nodes for every origin–destination pair (*i, j*). Constraint (3) ensures each node is a hub or allocated to a hub node. Constraint (4) represents that non-hub nodes are only assigned to one hub. Constraint (5) indicates for every origin–destination pair (*i, j*), if node *j* is allocated to hub l then the maximum amount of flow between *i* and *j* is equal y^i_{kl}. Constraint (6) ensures that node *i* can only be allocated to *k*, if cost c_{ik} is, at most, the radius r_k. Constraint (7) is the capacity constraint for each hub node that represents the amount of hub traffics. Constraints (9) and (10) show the domain of decision variables. The main difference in this formulation and the proposed one (Ernst & Krishnamoorthy, 1999) are in Constraint (5). Indeed, the base of both formulations is equivalent. For evaluating this constraint, we test the above formulation with different examples specially one that Correia et al. (2010) proposed for showing the Ernst and Krishnamoorthy's model that had defect. The obtained results witness the two being similar.

2.2. Stochastic multi-objective single-allocation HCP

In the real world, some parameters of the models are uncertain. Therefore, stochastic modeling approaches may be more efficient for studying the natural environments. CCP is one of the powerful approaches in modeling stochastic systems, in which the variation of the uncertain parameter is based on the known distribution. Charnes and Cooper (1959) introduced CCP, in which the main idea of their proposed method is to relax the constraints in deterministic programming and replace them with probabilistic ones. CCP holds the constraints with at least reliability level ($\alpha < 1$). Indeed, an uncertain parameter has variations in an interval and the expected value of it is considered for deterministic programming. In the other words, by changing the parameter value, the solution space is different from deterministic programming. When $\alpha = 1$, then the probabilistic constraints are equal deterministic ones. W_{ij} is assumed to be normally distributed with a mean (w_{ij}) and a standard deviation (σ_{ij}). With this assumption, the total flow originated at node i (O_i) and destined to node i (D_i) are also uncertainty parameters with normal probability distributions. Pairs (o_i, σ_{o_i}), and (d_i , σ_{D_i}) show the mean and standard deviation of it, respectively, that can be calculated simply as follows:

$$O_i \sim N\left(\sum_{j\in N} w_{ij}, \sqrt{\sum_{j\in N} \sigma_{ij}^2}\right) = (o_i, \sigma_{o_i}) \tag{11}$$

$$D_i \sim N\left(\sum_{j\in N} w_{ji}, \sqrt{\sum_{j\in N} \sigma_{ji}^2}\right) = (d_i, \sigma_{D_i}) \tag{12}$$

In addition, T_{ij} is assumed an uncertain parameter with a normal distribution.

$$T_{ij} \sim N(t_{ij}, \sigma_{t_{ij}}) \tag{13}$$

According to the explanations, the stochastic model obtained by replacing Constraints (5), (7), and (8) in the deterministic model with three following probabilistic constraints are given below:

$$p\left(y_{kl}^i \geq \sum_{j\in N} W_{ij} X_{jl} X_{ik}\right) \geq \alpha_1 \quad \forall i,k,l \in N,\ k \neq l \tag{14}$$

$$p\left(\sum_{iN} (O_i + D_i) X_{ik} \leq Z_k\right) \geq \alpha_2 \quad \forall k \in N \tag{15}$$

$$p\left((T_{ik} + T_{kl} + T_{lj}) X_{kk} X_{ll} \leq \beta)\right) \geq \alpha_3 \quad \forall i,j,k,l \in N \tag{16}$$

Clearly, according to the statistics theories, Constraints (14–16) can be simplified to Constraints (17–19), respectively.

$$y_{kl}^i - \sum_j w_{ij} X_{jl} X_{ik} - \varphi^{-1}(\alpha_1) \sqrt{\sum_i \sigma_{ij} X_{jl} X_{ik}} \geq 0 \quad \forall i,k,l \in N\ k \neq l \tag{17}$$

$$Z_k - \sum_{i\in N} (O_i + D_i) X_{ik} - \varphi^{-1}(\alpha_2) \sqrt{\sum_i X_{ik}^2 \left(\sigma_{Oi}^2 + \sigma_{Di}^2\right)} \geq 0 \quad \forall k \in N \tag{18}$$

$$p - (T_{ik} + T_{kl} + T_{lj}) u_{kl} - \varphi^{-1}(\alpha_3) u_{kl} \sqrt{\left(\sigma_{tij}^2 + \sigma_{tkl}^2 + \sigma_{tlj}^2\right)} \geq 0 \quad i,j,k,l \in N \tag{19}$$

So, the stochastic model can be written as follows:

$$\text{Min} \sum_{k\in N} (f_k X_{kk} + z_k g_k) + E\left(\sum_{i\in N}\sum_{k\in N} \chi C_{ik} O_i X_{ik} + \sum_{i\in N}\sum_{k\in N}\sum_{l\in N} \alpha C_{kl} y_{kl}^i + \sum_{i\in N}\sum_{k\in N} \delta C_{ki} D_i X_{ik}\right) \tag{20}$$

$$\text{Min}\, E(\beta) \tag{21}$$

s.t.

Constraints (3), (4), (6), (9), (10), (17–19).

3. Multi-objective optimization

Considering more than one objective function to optimize a problem leads to conceptual changes in meaning of the term "optimize" to "Pareto-optimal set." Without loss of generality, let an optimization problem with n objective functions as follows (Coello Coello, Lamont, & van Veldhuizan, 2007):

Minimize $y = F(x) = (f_1(x), \dots, f_n(x))$

where $x = (x_1, \dots, x_m)$ is the vector of decision variable and $y \in \mathfrak{R}^n$.

Definition 1 (dominance): A solution s_1 dominates another solution s_2 iff:

$$f_i(s_1) \leq f_i(s_2) \quad \forall i = 1, \dots, n$$

$$f_i(s_1) < f_i(s_2) \quad \exists i \in \{1, \dots, n\}$$

Definition 2 (Pareto-optimality set): If a solution of the feasible space is not dominated by other solutions, then it is known as a Pareto solution and the set of these solutions is nominated as Pareto-optimal set.

Convergence to the Pareto-optimality set and maintenance of diversity in its solutions has two general goals, in which multi-objective optimization will be achieved.

All non-dominated solutions are classified in a front, which is called the first front (front1). Afterward, the remaining solutions are categorized again based on non-domination as a new front (front2) and so on. Figure 1 depicts three non-dominated sorting fronts. To estimate the density of solutions' another attribute, the crowding distance is calculated by the following formula (22) for solutions in the same front (Deb, Pratap, Agarwal, & Meyarivan, 2000). The minimum and maximum objective function values are assigned an infinite distance value.

$$\text{Croding distance } (s_t) = \frac{\sum_{i=1}^{n} |f_i(s_{t+1}) - f_i(s_{t-1})|}{f_i^{max} - f_i^{min}} \tag{22}$$

4. Discrete IWO

Recently, many researchers in optimization areas tend to solve problems under uncertainty and dynamic information that cause high complexity and difficulty. However, often classical approaches cannot be more efficient in a dynamic environment. Here, alternative techniques (e.g. meta-heuristics) solve these problems effectively in acceptable time and quality by patterning out the behavior of nature to conform events. The IWO algorithm is introduced by Mehrabian and Lucas (2006) as an efficient evolutionary algorithm, especially in nonlinearity and non-convexity. They evaluated the

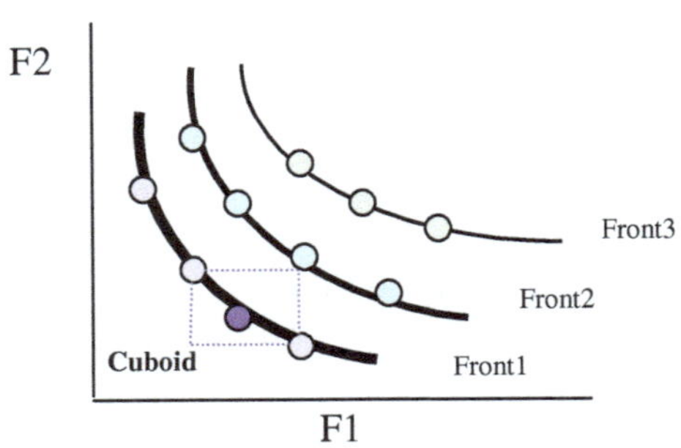

Figure 1. Illustration of non-dominated sorting fronts for a minimization problem.

capability of the algorithm in finding global extermum of optimization test functions and compared with genetic, memetic, shuffled frog leaping, particle swarm optimization, simplex, and direct search-simulated annealing algorithms. Until now, various studies used this novel algorithm for solving their problems, such as array antenna synthesis problems (Karimkashi & Kishk, 2010), time-cost trade-off problems (Ramezani-Ghalenoei & Hajimirsadeghi, 2009), no wait-two-stage multiprocessor flow shop scheduling problems (Moradi-Nasab, Shafaei, Rabiee, & Mazinani, 2012) and multi-objective problems (Kundu, Suresh, Ghosh, Das, & Panigrahi, 2011). All the results show that IWO outperforms most of the oldest powerful meta-heuristics. It is a population-based algorithm inspired by natural behavior of colonizing weeds. The solving process of an optimization problem by IWO can be summarized in four steps fundamentally below:

Initialization: A finite number of feasible solutions as initial population of weeds is generated and being dispread uniformly randomly in the search space.

Reproduction: In each iteration, seeds grow to flowering plants then according to the considered criterion, all plants are evaluated and ranked on the base of fitness value. The maximum and minimum numbers of seeds (S_{max} and S_{min}) belong to the best and worst fitness values, respectively. The number of seeds can be calculated by:

$$\text{seed}_i = \text{floor}(s_{min} + (s_{max} - s_{min}) \times (\text{popsize}_{iter} - \text{rank}_i)/(\text{popsize}_{iter} - 1)) \tag{23}$$

where popsize_{iter} and rank_i show the number of population in an iteration and the rank of the *i*th weed, respectively.

Spatial dispersal: This step provides a selection mechanism for recognizing appropriate plants. Produced seeds must be distributed normally in every iteration with different decreasingly standard deviations (sd). This ensures generated seeds be around their plants. Equation 24 calculates the sd for each iteration (sd_{iter}).

$$\text{sd}_{iter} = \left(\frac{\text{iter}_{max} - \text{iter}}{\text{iter}_{max}}\right)^{pow} (\text{sd}_{initial} - \text{sd}_{final}) + \text{sd}_{final} \tag{24}$$

where pow, $\text{sd}_{initial}$, sd_{final}, and iter_{max} show the predefined nonlinear modulation index, initial and final standard deviations, and the maximum number of iterations, respectively ($\text{sd}_{final} < \text{sd}_{initial}$).

Competitive exclusion: Newly generated seeds in the previous step are added to the colony. When the maximum number of populations is obtained in the colony, the algorithm ranks all the weeds for screening better ones. Therefore, according to the defined number of colony population, the surplus is eliminated.

Here, we illustrate the proposed multi-objective stochastic algorithm based on discrete IWO for solving multi-objective single-allocation HCPs with CCP. The generality of this algorithm is similar to the main structure since we focus on differences of implementation. The algorithm is as follows:

Step 1. *Initialization:* The initial population of weeds is generated and is dispread uniformly randomly in the search space. Each solution or weed is represented by two separated strings of lengths n and p, respectively, for coding a hub and spoke network with n nodes and p unknown hubs. The first string of length n is formed of different n random numbers into the set N, which denotes the allocations indirectly. The second string of length p consisting of p random natural numbers concerns the locations of hub nodes in the first string. Always, two assumptions must be observed for the second string: (1) the value of last array is equal to n and (2) the random generated natural numbers sorted in to ascending order. Hub nodes are marked in the new string according to the proposed locations of hubs in the hub location string. To achieve a non-hub nodes assignment to hubs, the unmark nodes are allocated to the right-hand side distinct array or hubs. Figure 2 shows the structure of a

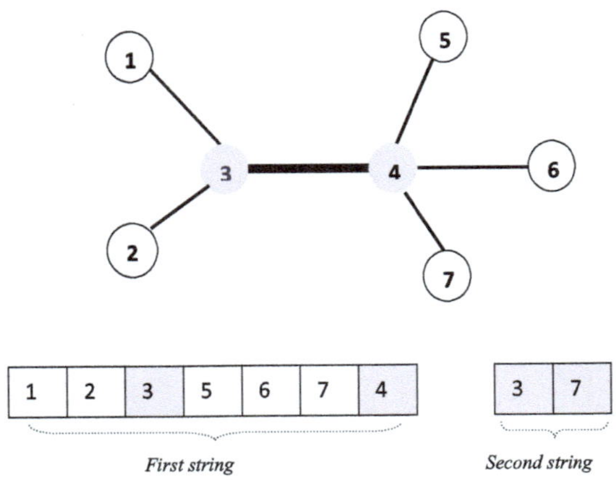

Figure 2. Representation of each weed in the colony.

weed for coding a hub and spoke network with seven nodes and two hubs. The advantage of this representation is to decrease the probability of generating infeasible solutions. So, it is not necessary to check or repair weeds after each operation and finally influence computational time-saving.

Step 2. *Reproduction and spatial dispersal:* As described in steps b and c, in the previous section, all plants are evaluated and ranked according to fitness values. Therefore, the population must be ranked using non-dominated sorting as described in the previous sections. We note all constraints instead of covering, and the probabilistic ones are satisfied by representation mechanism. Covering restriction imposes on the solution space by employing penalty function. The difference in probabilistic constraints in comparison with deterministic ones is only a constant because of considering a reliability level and simulating an uncertain condition. Thus, it is clear that this subject leads to translation of the feasible space without any fundamental changes on it. So, the proposed formulation causes to simplify chance-constrained implementation. Flowering plants reproduce seeds and distribute them normally in the search space by Equations (23) and (24), respectively. The algorithm generates a random number (num) in the interval $[sd_{final}, sd_{iter}]$. Then, the num random positions of allocation string are selected and swapped. Figure 3 illustrates how the reproduction mechanism of an invasive weed works when the num is equal to 3. The cells containing nodes 1, 3, and 6 are randomly selected by the algorithm. The contents of the cells are exchanged one by one, specifically that of No. 1 is exchanged for that of No. 3 and then that of No. 3 for that of No. 5.

Step 3. *Mutation:* The idea of using the mutation operator for the IWO algorithm was proposed by Moradi-Nasab et al. (2012) that inspired herbicide-resistant weeds. In this step, the mutation is done on the percent of the previous step population, (p_m). The algorithm generates a random number. If this value is less than or equal to the predefined probability of mutation, the invasive weed is mutated. Generally, five types of mutation operators can occur in the algorithm.

Type one: The number of hubs (*p*) increases or decreases randomly. Figure 4(a) shows a mutated weed. The number of hubs is increased.

Type two: The μ percent of non-hub nodes reallocated where μ is a predefined parameter. The algorithm generates *n* random numbers in unit interval as a new string. The maximum of the generated numbers is found and assigned to the array, in which the node 1 is located in the main string. This rule is observed for the other assignments. μ percent of nodes are changed by new random numbers. Hub nodes locate according to the main string and non-hub nodes are reallocated by the

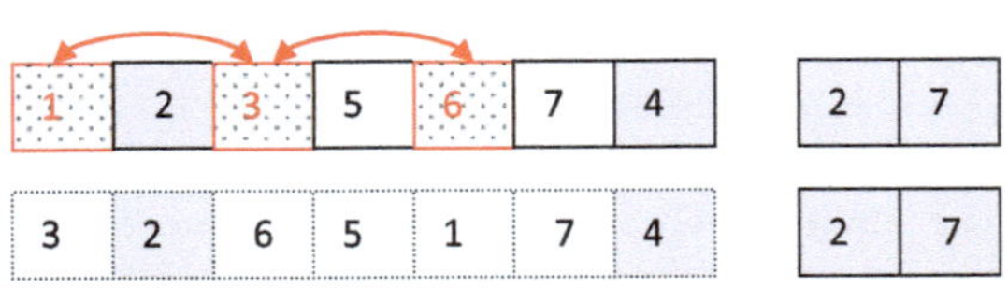

Figure 3. Representation of weed reproduction.

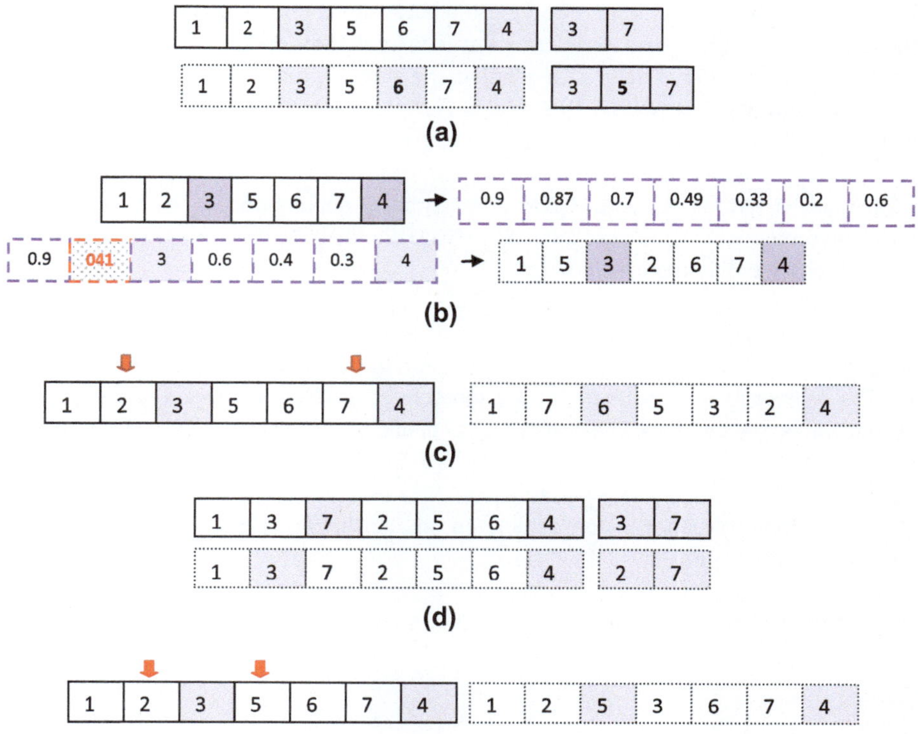

Figure 4. Five types of probabilistic mutations.

above rule. The algorithm generates a new string of random numbers in initial interval [0, 1]; for example [0.9, 0.2, 0.6, 0.87, 0.49, 0.7, 0.33]. In this new coding, the maximum of generated numbers (0.9) refers to node 1 and the maximum remaining numbers refers to node 2. Consequently, the invasive weed is coded as follows: [1, 2, 3, 5, 6, 7, 4] → [0.9, 0.87, 0.7, 0.49, 0.33, 0.2, 0.6]. The algorithm randomly selected 20% of the cells assigned to the non-hub nodes and replaces them with the new random numbers. In Figure 4(b), the cell No. 2 is selected and replaced with 0.41. The objective of this mutation is to change the structure of the spokes, so there is no change in the hub nodes. The mutated weed is decoded by algorithm in the above-mentioned manner.

Type three: Selecting two non-hub nodes in allocation string and reversing all the nodes between these positions (see Figure 4(c)).

Type four: Select a hub node randomly and change with a practicable non-hub node. What restrict this substitution are two assumptions (see Figure 4(d)).

Type five: Two non-hub nodes are selected randomly. The second node is located adjacent to the first non-hub node position. Therefore, the other nodes shifted right side (see Figure 4(e)).

Step 4. *Competitive exclusion:* In this step, the initial weeds, produced seed, and mutated weeds are merged together. Only pop size of this huge population with the best fitness can win this competition as new population for the next iteration. The algorithm employs this evaluation using non-dominated sorting.

Step 5. Go to Step 2 and repeat the algorithm until the stopping criteria are satisfied.

5. Computational experiments

In this section, the performance of the proposed MOIWO algorithm is tested for different instances. The authors use the Civil Aeronautics Board data-set and a random generation data-set for a network up to 100 nodes. For the random data-set, x and y coordinates are uniformly generated from U [0, 100000]. We assume that w_{ij} and t_{ij} take the values given in the data-sets. All the standard variations are considered as a coefficient (v) of means. In addition, the studied problem is solved by the non-dominated sorting genetic algorithm (NSGA-II). Deb et al. (2002) proposed an improved version of the NSGA-II. It ranks the populations due to the non-domination level.

Each problem is solved 10 times for both MOEAs. The number of population for the problems with up to 50 nodes is assumed less than larger ones. The NSGA-II and MOIWO parameter values for a given run are used for the problem provided in Table 2.

Good distribution of the proposed solutions, a wide-range convergence of non-dominated solutions and less distance of a non-dominated solution set to Pareto front are measured in multi-objective problems. The computational results of the two foregoing algorithms are reported in terms of the following four well-known metrics. The quality, spacing, and diversification metrics are used for the validation of the multi-objective evolutionary algorithm (MOEA) reliability.

Quality metric (*QM*): The ratio between non-dominated solutions found by algorithms is employed to measure the quality (see Figure 5).

Spacing metric (*SM*): It measures the uniformity of Pareto solutions' distribution as follows:

$$SM = \frac{\sum_{i=1}^{n-1} \left| \bar{d} - d_i \right|}{(n-1)\bar{d}} \tag{25}$$

where d_i denotes Euclidean distance between consecutive solutions, and the mean of d_i is shown by $\bar{d}$ (see Figure 6).

Table 2. Parameter values of the NSGA-II and the proposed MOIWO

	Quantity	Value
MOIWO	Initial numbers of populations	80,120
	Maximum number of seeds	3
	Minimum number of seeds	1
	Initial value of standard deviation	$0.4 \times n$
	Final value of standard deviation	2
	Maximum numbers of populations	80,120
	Nonlinear modulation index	4
	Mutation rate	0.1
NSGA-II	Numbers of populations	100,150
	Crossover rate	0.7
	Mutation rate	0.1

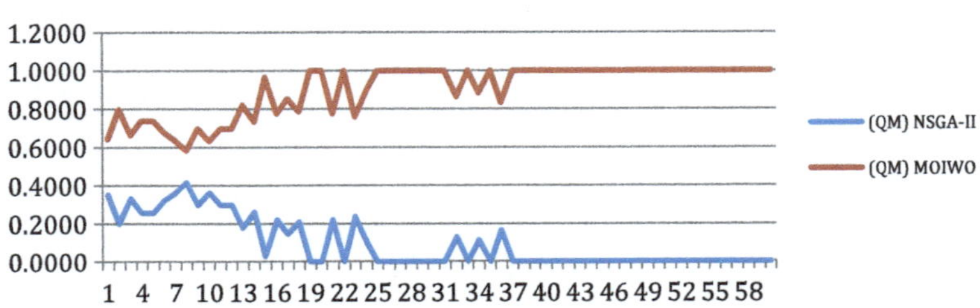

Figure 5. Comparison of the MOIWO and NSGA-II in terms of the QM.

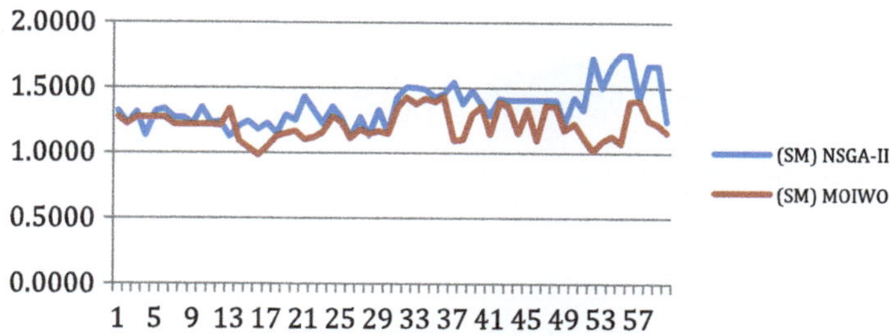

Figure 6. Comparison of the MOIWO and NSGA-II in terms of the SM.

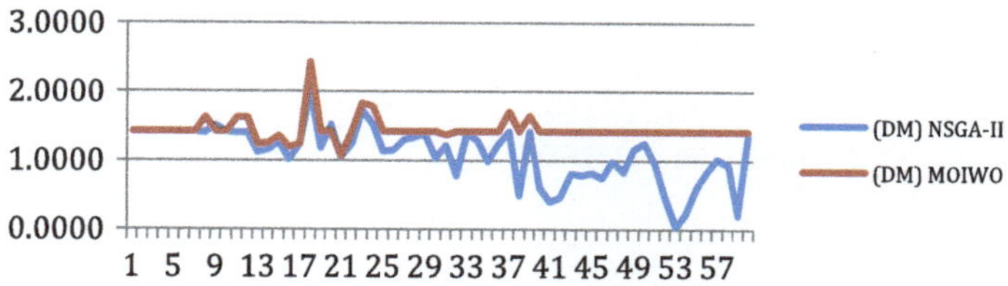

Figure 7. Comparison of the MOIWO and NSGA-II in terms of the DM.

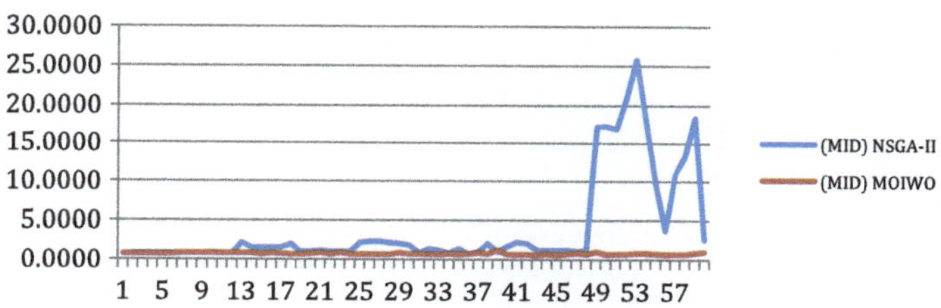

Figure 8. Comparison of the MOIWO and NSGA-II in terms of the mean ideal distance.

Diversification metric (*DM*): A high degree of Pareto-solution diversification is another sign of the MOEA efficiency (see Figure 7). Let ($\max f_{1,i}, \min f_{1,i}$) and ($f_{1,total}^{max}, f_{1,total}^{min}$) represent the maximum and minimum of the first objective values for each Pareto solution *i* and all of them, respectively. It is similar for another objective function. Therefore, the spread of proposed Pareto solutions can be measured by:

$$\mathrm{DM} = \sqrt{\left(\frac{\max f_{1,i} - \min f_{1,i}}{f_{1,total}^{max} - f_{1,total}^{min}}\right)^2 + \left(\frac{\max f_{2,i} - \min f_{2,i}}{f_{2,total}^{max} - f_{2,total}^{min}}\right)^2} \tag{26}$$

Mean of distances between ideal solution and Pareto solution (*MID*): Figure 8 shows the mean of distances between ideal solution and Pareto solutions obtained from MOIWO and NSGA-II. The mean ideal distance is calculated by the following formula (27). Less MID is another MOIWO pre-eminence (see Figure 7).

$$\mathrm{MID} = \frac{1}{n}\sum_{i=1}^{n}\sqrt{\left(\frac{f_{1,i} - f_1^{best}}{f_{1,total}^{max} - f_{1,total}^{min}}\right)^2 + \left(\frac{f_{2,i} - f_2^{best}}{f_{2,total}^{max} - f_{2,total}^{min}}\right)^2} \tag{27}$$

Table 3 illustrates the results obtained by the NSGA-II and proposed MOIWO in terms of the foregoing four metrics.

The concluding results are as follows:

(a) Higher quality of the proposed MOIWO algorithm for small, median, and especially large networks.

Table 3. Results of the NSGA-II and the proposed MOIWO in terms of four metrics

Node	α	Reliability level	v	QM		SM		DM		MID	
				NSGA-II	MOIWO	NSGA-II	MOIWO	NSGA-II	MOIWO	NSGA-II	MOIWO
10	0.2	0.9	0	0.3548	0.6452	1.3215	1.2679	1.4142	1.4142	0.7330	0.6850
10	0.2	0.99	0	0.2000	0.8000	1.2211	1.2179	1.4142	1.4142	0.7908	0.6850
10	0.2	0.9	1	0.3333	0.6667	1.3114	1.2679	1.4142	1.4142	0.7653	0.6859
10	0.2	0.99	1	0.2593	0.7407	1.1307	1.2679	1.4142	1.4142	0.7558	0.6866
10	0.2	0.9	4	0.2593	0.7407	1.3159	1.2679	1.4142	1.4142	0.7640	0.6867
10	0.2	0.99	4	0.3220	0.6780	1.3394	1.2678	1.4142	1.4142	0.7500	0.6880
10	0.8	0.9	0	0.3636	0.6364	1.2701	1.2154	1.4142	1.4142	0.8212	0.7815
10	0.8	0.99	0	0.4167	0.5833	1.2679	1.2154	1.4070	1.6142	0.7569	0.7815
10	0.8	0.9	1	0.3000	0.7000	1.2253	1.2147	1.5111	1.4142	0.7995	0.7817
10	0.8	0.99	1	0.3636	0.6364	1.3478	1.2141	1.4142	1.4142	0.8404	0.7819
10	0.8	0.9	4	0.3000	0.7000	1.2331	1.2139	1.4065	1.6142	0.7979	0.7819
10	0.8	0.99	4	0.3000	0.7000	1.2372	1.2128	1.4062	1.6142	0.7995	0.7822
25	0.2	0.9	0	0.1800	0.8200	1.1241	1.3315	1.1073	1.2283	2.1038	0.7812
25	0.2	0.99	0	0.2632	0.7368	1.2049	1.0851	1.1535	1.2487	1.5452	0.7864
25	0.2	0.9	1	0.0313	0.9688	1.2425	1.0372	1.2632	1.3457	1.4777	0.6408
25	0.2	0.99	1	0.2222	0.7778	1.1806	0.9763	1.0055	1.1849	1.5137	0.7950
25	0.2	0.9	4	0.1463	0.8537	1.2258	1.0466	1.2284	1.2286	1.5167	0.6954
25	0.2	0.99	4	0.2128	0.7872	1.1498	1.1243	1.9998	2.4196	1.9517	0.6548
25	0.8	0.9	0	0.0000	1.0000	1.2893	1.1491	1.1739	1.4142	0.9709	0.6632
25	0.8	0.99	0	0.0000	1.0000	1.2494	1.1620	1.5177	1.4142	1.0438	0.7273
25	0.8	0.9	1	0.2222	0.7778	1.4283	1.1010	1.0448	1.0526	1.1160	0.8846
25	0.8	0.99	1	0.0000	1.0000	1.3266	1.1138	1.2385	1.4142	1.0422	0.6449
25	0.8	0.9	4	0.2381	0.7619	1.2197	1.1527	1.7216	1.8224	1.0765	0.8893
25	0.8	0.99	4	0.1053	0.8947	1.3566	1.2762	1.5158	1.7747	0.9526	0.6274
50	0.2	0.9	0	0.0000	1.0000	1.2626	1.2328	1.1369	1.4142	2.1961	0.6226
50	0.2	0.99	0	0.0000	1.0000	1.1110	1.1089	1.1378	1.4142	2.2809	0.5988
50	0.2	0.9	1	0.0000	1.0000	1.2725	1.1763	1.3023	1.4142	2.2708	0.6112
50	0.2	0.99	1	0.0000	1.0000	1.1292	1.1432	1.3367	1.4142	2.1608	0.6004
50	0.2	0.9	4	0.0000	1.0000	1.3266	1.1638	1.3824	1.4142	2.0641	0.8096
50	0.2	0.99	4	0.0000	1.0000	1.1523	1.1395	1.0332	1.4142	1.7741	0.6912
50	0.8	0.9	0	0.0000	1.0000	1.4301	1.3442	1.2114	1.3597	0.8228	0.6862
50	0.8	0.99	0	0.1333	0.8667	1.5008	1.4245	0.7672	1.4142	1.3988	0.6633
50	0.8	0.9	1	0.0000	1.0000	1.5000	1.3722	1.4040	1.4142	1.2385	0.6551
50	0.8	0.99	1	0.1154	0.8846	1.4884	1.4119	1.2808	1.4142	0.8137	0.7094
50	0.8	0.9	4	0.0000	1.0000	1.4253	1.3876	0.9784	1.4142	1.3605	0.6629
50	0.8	0.99	4	0.1667	0.8333	1.4496	1.4283	1.2390	1.4142	0.7491	0.6740
81	0.2	0.9	0	0.0000	1.0000	1.5385	1.0917	1.4142	1.7004	0.9486	0.9048
81	0.2	0.99	0	0.0000	1.0000	1.3750	1.0984	0.4899	1.4142	2.0069	0.6667
81	0.2	0.9	1	0.0000	1.0000	1.4788	1.2974	1.4142	1.6364	0.9808	1.1076
81	0.2	0.99	1	0.0000	1.0000	1.3750	1.3528	0.6033	1.4142	1.5974	0.6252
81	0.2	0.9	4	0.0000	1.0000	1.2857	1.1430	0.3967	1.4142	2.2021	0.6193
81	0.2	0.99	4	0.0000	1.0000	1.4137	1.3869	0.4521	1.4142	2.0241	0.6172
81	0.8	0.9	0	0.0000	1.0000	1.4000	1.3513	0.8022	1.4142	1.1680	0.5759
81	0.8	0.99	0	0.0000	1.0000	1.4000	1.1496	0.7807	1.4142	1.1907	0.7050

Table 3. Continued

Node	α	Reliability level	v	QM		SM		DM		MID	
				NSGA-II	MOIWO	NSGA-II	MOIWO	NSGA-II	MOIWO	NSGA-II	MOIWO
	0.8	0.9	1	0.0000	1.0000	1.4000	1.3309	0.8120	1.4142	1.1547	0.5149
	0.8	0.99	1	0.0000	1.0000	1.4000	1.0947	0.7375	1.4142	1.2441	0.7291
	0.8	0.9	4	0.0000	1.0000	1.4000	1.3600	0.9863	1.4142	1.0294	0.7644
	0.8	0.99	4	0.0000	1.0000	1.4000	1.3554	0.8278	1.4142	1.1393	0.7196
100	0.2	0.9	0	0.0000	1.0000	1.2222	1.1666	1.1592	1.4142	17.2488	1.0000
100	0.2	0.99	0	0.0000	1.0000	1.4192	1.2199	1.2443	1.4142	17.2645	0.6210
100	0.2	0.9	1	0.0000	1.0000	1.3263	1.1141	0.9403	1.4142	16.9136	0.7296
100	0.2	0.99	1	0.0000	1.0000	1.7265	1.0101	0.4829	1.4142	20.7587	0.7325
100	0.2	0.9	4	0.0000	1.0000	1.5000	1.0825	0.0346	1.4142	25.7222	0.7866
100	0.2	0.99	4	0.0000	1.0000	1.6667	1.1276	0.2203	1.4142	18.0727	0.8272
100	0.8	0.9	0	0.0000	1.0000	1.7500	1.0703	0.5889	1.4142	9.5704	0.7191
100	0.8	0.99	0	0.0000	1.0000	1.7500	1.3918	0.8384	1.4142	3.5995	0.6286
100	0.8	0.9	1	0.0000	1.0000	1.4000	1.3968	1.0230	1.4142	11.0077	0.7126
100	0.8	0.99	1	0.0000	1.0000	1.6666	1.2488	0.9327	1.4142	13.2720	0.7347
100	0.8	0.9	4	0.0000	1.0000	1.6667	1.2109	0.2009	1.4142	18.3454	0.8536
100	0.8	0.99	4	0.0000	1.0000	1.2415	1.1535	1.3497	1.4142	2.5618	0.9970

(b) Lower uniformity of the Pareto solutions of the proposed MOIWO algorithm for different networks.

(c) Higher diversification of the Pareto solutions of the proposed MOIWO algorithm for different networks.

(d) Lower mean of distances between the ideal solution and Pareto solutions of the proposed MOIWO algorithm for different networks.

The Pareto solutions distributed more uniformly and the distance between ideal solutions and Pareto ones are lower with fewer variations.

6. Conclusion

In this paper, the authors have considered a hub location problem by considering realistic assumptions as uncertain parameters and two conflictive goals for optimization. A quadratic formulation has been proposed for multi-objective single-allocation HCP with the variable capacity in uncertainty demand and transportation time using CCP. Additionally, the authors have developed the MOIWO algorithm to solve the given problem. The reliability of the proposed MOIWO algorithm was evaluated by four well-known metrics; namely QM, SM, DM, and MID. The associated results are in evidence, in which the proposed MOIWO performance is appreciable, compared with the NSGA-II. These algorithms rank the individuals by non-dominated sorting; however, the proposed MOIWO algorithm can propose better Pareto solution sets regarding quality and diversification.

Acknowledgements
The authors would like to thank the Editor-in-Chief of the Cogent Engineering ("Production & Manufacturing" section) and anonymous referees for their helpful comments and suggestions, which greatly improved the presentation of this paper.

Funding
The third author would like to acknowledge the partially financial support of the University of Tehran for this research [grant number 8106043/1/23].

Author details
Seyed Hossein Nikokalam-Mozafar[1]
E-mail: nikokalam110@gmail.com
Behzad Ashjari[1]
E-mail: ashied@aut.ac.ir
Reza Tavakkoli-Moghaddam[2]
E-mail: tavakoli@ut.ac.ir
Aida Omidvar[3]
E-mail: omidvar.aida@gmail.com

[1] Department of Industrial Engineering, Tafresh University, Tafresh, Iran.
[2] School of Industrial Engineering, College of Engineering, University of Tehran, Tehran, Iran.
[3] School of Industrial Engineering, South Tehran Branch, Islamic Azad University, Tehran, Iran.

References

Alumur, S. A., Nickel, S., & Saldanha-da-Gama, F. (2012). Hub location under uncertainty. *Transportation Research - Part B: Methodological, 46*, 529–543. doi:10.1016/j.trb.2011.11.006

Calik, H., Alumar, S., Kara, B., & Karasan, O. (2009). A tabu-search based heuristic for the hub covering problem over incomplete hub. *Computers & Operations Research, 36*, 3088–3096. doi:10.1016/j.cor.2008.11.023

Campbell, J. F. (1994). Integer programming formulations of discrete hub location problems. *European Journal of Operational Research, 72*, 387–405. doi:10.1016/0377-2217(94)90318-2

Charnes, A., & Cooper, W. W. (1959). Chance constrained programming. *Management Science, 6*, 73–79. doi:10.1287/mnsc.6.1.73

Coello Coello, C. A., Lamont, G. B., & van Veldhuizan, D. A. (2007). *Evolutionary algorithms for solving multi-objective problems* (2nd ed.). New York, NY: Springer.

Contrerase, I., Cordeau, J. F., & Laporate, G. (2011). Stochastic uncapacitated hub location. *European Journal of Operational Research, 212*, 518–528. doi:10.1016/j.ejor.2011.02.018

Correia, I., Nickel, S., & da-Gama, F. S. (2010). The capacitated single-allocation hub location problem revisited: A note on a classical formulation. *European Journal of Operational Research, 207*, 92–96. doi:10.1016/j.ejor.2010.04.015

Costa, M., Captivo, M., & Clímaco, J. (2008). Capacitated single allocation hub location problem a bicriteria approach. *Computers & Operations Research, 35*, 3671–3695. doi:10.1016/j.cor.2007.04.005

Deb, K., Pratap, A., Agarwal, S., & Meyarivan, T. (2000). A fast elitist non-dominated sorting genetic algorithm multi-objective: NSGA-II. *Lecture Notes in Computer Science, 1917*, 849–858. doi:10.1007/3-540-45356_83

Deb, K., Pratap, A., Agarwal, S., & Meyarivan, T. (2002). A fast and elitist multi-objective genetic algorithm: NSGA-II. *IEEE Transactions on Evolutionary Computation, 6*, 182–197. doi:10.1109/4235.996017

Drezner, Z., & Scott, C. H. (2013). Location of a distribution center for a perishable product. *Mathematical Methods of Operations Research, 78*, 301–314. doi:10.1007/s00186-013-0445-6

Ernst, A. T., Jiang, H., & Krishnamoorthy, M. (2005). *Reformulations and computational results for uncapacitated single and multiple allocation hub covering problems*. Unpublished Report. CSIRO Mathematical and Information Sciences, Australia.

Ernst, A. T., & Krishnamoorthy, M. (1999). Solution algorithms for the capacitated single allocation hub location problem. *Annals of Operations Research, 86*, 141–159. doi:10.1023/A:1018994432663

Ghodratnama, A., Tavakkoli-Moghaddam, R., & Azaron, A. (2013). A fuzzy possibilistic bi-objective hub covering problem considering production facilities, time horizons and transporter vehicles. *International Journal of Advanced Manufacturing Technology, 66*, 187–206. doi:10.1007/s00186-013-0445-6

Ghodratnama, A., Tavakkoli-Moghaddam, R., & Baboli, A. (2013). Comparing three proposed meta-heuristics to solve a new *p*-hub location-allocation problem. *International Journal of Engineering, Transactions C: Aspects, 26*, 787–797.

Hamacher, H. W., & Meyer, T. (2006). *Hub cover and hub center problems* (Working Paper). Kaiserslautern: Department of Mathematics, University of Kaiserslautern.

Kara, B. Y., & Tansel, B. C. (2003). The single-assignment hub covering problem: Models and linearization. *Journal of Operational Research Society, 54*, 59–64. doi:10.1057/palgrave.jors.2601473

Karimi, H., & Bashiri, M. (2011). Hub covering location problems with different coverage types. *Scientia Iranica – E, 18*, 1571–1578. doi:10.1016/j.scient.2011.09.018

Karimkashi, A., & Kishk, A. (2010). Invasive weed optimization and its features in electromagnetic. *IEEE Transactions on Antennas and Propagation, 58*, 1269–1278. doi:10.1109/TAP.2010.2041163

Kundu, D., Suresh, K., Ghosh, S., Das, S., & Panigrahi, B. K. (2011). Multi-objective optimization with artificial weed colonies. *Information Sciences, 181*, 2441–2454. doi:10.1016/j.ins.2010.09.026

Lium, A. G., Crainic, T. G., & Wallace, S. (2009). A study of demand stochasticity in service network design. *Transportation Science, 43*, 144–157. doi:10.1287/trsc.1090.0265

Lowe, T. J., & Sim, T. (2012). The hub covering flow problem. *Journal of the Operational Research Society, 64*, 973–981. doi:10.1057/jors.2012.122

Marianov, V., & Serra, D. (2003). Location models for airline hubs behaving as M/D/c queues. *Computers and Operations Research, 30*, 983–1003. doi:10.1016/S0305-0548(02)00052-7

Mehrabian, A. R., & Lucas, C. (2006). A novel numerical optimization algorithm inspired from weed colonization. *Ecological Informatics, 1*, 355–366. doi:10.1016/j.ecoinf.2006.07.003

Mohammadi, M., Jolai, F., & Tavakkoli-Moghaddam, R. (2013). Solving a new stochastic multi-mode *p*-hub covering location problem considering risk by a novel multi-objective algorithm. *Applied Mathematical Modeling, 37*, 10053–10073. doi:10.1016/j.apm.2013.05.063

Mohammadi, M., Tavakkoli-Moghaddam, R., & Rostami, H. (2011). A multi objective imperialist competitive algorithm for a capacitated hub covering location problem. *International Journal of Industrial Engineering Computations, 2*, 671–688. doi:10.5267/j.ijiec.2010.08.003

Mohammadi, M., Tavakkoli-Moghaddam, R., Tolouei, H., & Yousefi, M. (2010). Solving a hub covering location problem under capacity constraints by a hybrid algorithm. *Journal of Applied Operational Research, 2*, 109–116.

Mohammadi, M., Torabi, S. A., & Tavakkoli-Moghaddam, R. (2014). Sustainable hub location under mixed uncertainty. *Transportation Research - Part E, 62*, 89–115. doi:10.1016/j.tre.2013.12.005

Moradi-Nasab, N., Shafaei, R., Rabiee, M., & Mazinani, M. (2012). Minimization of maximum tardiness in a no-wait two stage flexible flow shop. *International Journal of Artificial Intelligence, 8*, 166–181.

Qu, B., & Weng, K. (2009). Path relinking approach for multiple allocation hub maximal covering problem. *Computers and Mathematics with Applications, 57*, 1890–1894. doi:10.1016/j.camwa.2008.10.004

Ramezani-Ghalenoei, M., & Hajimirsadeghi, H. (2009, December 16–18). Discrete invasive weed optimization algorithm, Application to Cooperative Multiple Task Assignment of UAVs. In *48th of IEEE Conference on Decision and Control and 28th Chinese Control Conference*, Shanghai, P. R. China. doi: 10.1109/CDC.2009.5400938

Sim, T., Lowe, T. J., & Thomas, B. W. (2009). The stochastic *p*-hub center problem with service-level constraints. *Computers and Operations Research, 36*, 3166–3177. doi: 10.1016/j.cor.2008.11.020

Tan, P. Z., & Kara, B. Y. (2007). A hub covering model for cargo delivery systems. *Networks, 49*, 28–39. doi:10.1002/net.20139

Wagner, B. (2004). *Model formulations for hub covering problems* (Working Paper). Darmstadt: Institute of Operations Research, Darmstadt University of Technology.

Weng, K., & Wang, Y. (2008, April 6–8). Evolutionary algorithms for multiple allocation hub set covering problem. In *IEEE Conference on Networking, Sensing and Control* (pp. 408–411). Sanya. doi:10.1109/ICNSC.2008.4525250

14

Small-angle neutron scattering investigations of nanocrystalline alloy chips obtained by machining

Elwyn Rebello[1], Frazer Vaz[1], Avino Colaco[1], Wellon Rodrigues[1], Geethalakshmi K[1*], Debasis Sen[2], S. Mazumder[2] and A.O. Surendranathan[3]

*Corresponding author: Geethalakshmi K, Department of Mechanical Engineering, P. C. College of Engineering, Verna, Goa, India
E-mail: geethashyam@gmail.com
Reviewing editor: Zude Zhou, Wuhan University of Technology, China

Abstract: Ultrafine-grained (UFG) materials exhibit significantly enhanced mechanical properties. This has brought renewed attention on the use of large strain or severe plastic deformation as a means for achieving microstructural refinement in metals and alloys. Large plastic strains imposed in a machine chip result in significant microstructural refinement, including the creation of UFG and nanocrystalline materials. It looks to be an economical route for realizing nanocrystalline materials. In the present study, small-angle neutron scattering (SANS) was employed to investigate the modifications in the microstructure of the chips produced via machining. Double crystal-based medium resolution SANS instrument has been used for this purpose. Significant scattering intensity at small enough angles reveals the presence of mesoscopic density fluctuations produced because of the machining. Atomic force microscopy images also corroborate the existence of such small length scale density fluctuations.

Subjects: Engineering & Technology, Manufacturing Technology, Technology

Keywords: nanocrystalline alloys, machining, severe plastic deformation, small-angle neutron scattering

1. Introduction

It has been proved by time and again by researchers that nanocrystalline materials have enhanced mechanical properties in terms of hardness and strength over the bulk materials. However, the production cost of nanocrystalline materials is excessively high due to the equipment used and the probable low output. Recently, researchers have proved that nanocrystalline materials can be produced through machining (Ni & Alpas, 2003; Shankar, Chandrasekar, Compton, et al., 2005; Shankar, Chandrasekar, King, et al., 2005; Swaminathan et al., 2005). They have demonstrated that under certain combinations of machining parameters, large shear strains can be imposed in the machining

ABOUT THE AUTHOR

Geethalakshmi K is working as an associate professor in the Department of Mechanical Engineering at P. C. College of Engineering, Verna, Goa, India. She has more than 15 years of teaching experience and her areas of research interest include open extrusion, severe plastic deformation, and polymer nanocomposites.

PUBLIC INTEREST STATEMENT

Nanostructured materials derived from machining chips, in combination with some consolidation technique such as powder metallurgy, spraying, and composite processing methods, offer the potential for creating advanced materials with new and interesting combinations of properties. The possibility that the chips produced in machining operations, usually disposed as scrap, may be having nanocrystalline structure makes this process very exciting, as it can be used as an abundant source for producing nanomaterial at low cost.

process, resulting in nanoscale microstructures. The availability of large volumes of metal chips, as bi-products either from existing manufacturing operations or from specific chip-making processes, offers opportunities to create nanocrystalline materials with mechanical properties that are influenced by the ultrafine-grained microstructure of the chips. Furthermore, the production costs for nanocrystalline materials created using the machining-based deformation processes are much less than those of conventional methods (Iglesias et al., 2008). However, this extremely fine microstructure is difficult to image without the use of Transmission electron microscopy (TEM) or Field emission-Scanning electron microscopy (FE-SEM).

Small-angle neutron scattering (SANS) can be used to determine the shape and the organization, averaged in time, of particles or aggregates dispersed in a continuous medium (Borsali & Pecora, 2008). The term particle is applied to a wide range of objects from the fraction of a nanometer to the micrometer level. The use of large samples without any special preparation and of a non-destructive character gives it an edge over other techniques, like TEM, FE-SEM, Field ion microscopy, etc.

In the present work, machined chips were produced from two commercially available and frequently used alloys through orthogonal machining. Investigations were carried out on the chips through SANS along with Scanning electron and Atomic force microscopy (AFM), followed by mechanical characterization through shear strain and microhardness testing.

2. Experimental

2.1. Materials

Two commonly used alloys, viz. commercially available steel and brass with respective near equivalent specification IS 45C8 and IS 319/74, were used for the present investigation, keeping in view of the fact that these alloys are most often shaped through regular machining operations. The nominal chemical compositions of these two alloys are given in Table 1.

2.2. Machining

Machining chips were produced by using with high-speed steel and carbide tools under controlled plain strain conditions with a range of rake angles and machining speeds. Feed rate was kept at .2 mm/rev and rake angle was varied from +8° to −15° with a depth of cut of 40 μm. The machining velocity was kept at 15–20 mm/s to avoid the influence of temperature. The initial grain sizes of the bulk material in both cases were greater than 10 μm. The bulk hardness of steel was ~140 HV and

Table 1. Chemical composition

Material	Composition (wt %)										
Steel	Element	C	Si	Mn	P	S	Cr	Ni	Mo	Al	Cu	Fe
	%	.463	.277	.81	.016	.02	.177	.016	.02	.03	.01	Balance
Brass	Element	Zn	Pb	Sn	Fe	Ni	As	Sb	Bi	Ag	Co	Cu
	%	37.82	2.8	.38	.259	.157	.17	.048	.018	.023	.024	Balance

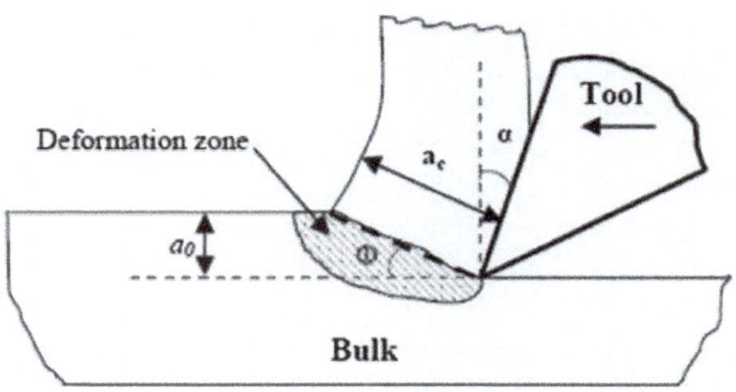

Figure 1. Schematic of large strain machining.

that of brass was around ~110 HV. Typical chip samples were 50–80 μm in width, 40–60 μm in thickness, and 100–200 μm in length.

During the cutting process, the tool and work piece are forced against each other and a compressive force is set up, which causes the metal to deform in front of the tool point. The deformation takes place in a zone along the shear plane and the metal is forced to slide over the tool surface. In doing this, the shear stress causes the material to separate as a chip, based on the rake angle. With positive rake angle, the area under shear decreases leaving less deformation on the surface of the work piece (Geels, 2007). A schematic view of large strain machining is shown in Figure 1.

From the measured deformed chip thickness (*ac*) and undeformed chip thickness (*ao*) and the shear plane angle (φ), the shear strain values (γ) imposed in the chip were estimated using Equations 1 and 2 (Merchant, 1945; Shaw, 1984).

$$\tan\varphi = \frac{\frac{a_o}{a_c}\cos\propto}{1-\frac{a_o}{a_c}\sin\propto} \tag{1}$$

$$\gamma = \frac{\cos\alpha}{\sin\varphi\cdot\cos(\varphi-\propto)} \tag{2}$$

2.3. Characterization

The chips were characterized by SANS for estimation of crystallite/particulate size followed by the determination of Vickers microhardnesss and microstructural characteristics. Microhardness measurements were done on a Matsuzawa Micro Vickers Hardness Tester Model MMT-X7. SANS measurements were performed using a double crystal-based medium resolution SANS instrument (Mazumder, Sen, Sarvanan, & Vijayraghavan, 2001a, 2001b) at the guide tube laboratory of DHRUVA reactor, India. The scattered intensities [$I(q)$] have been recorded as a function of wave vector transfer q [$=4\pi\sin(\theta)/\lambda$, where 2θ is the scattering angle and λ (=.312 nm.) is the incident neutron wavelength for the present experiment]. The specimens were placed on a sample holder with a circular slit of 1.5 cm diameter. The SANS data were corrected for background, transmission, and instrument resolution effects (Lake, 1967; Schmidt, 1988). Microstructural characterizations were done using an optical microscope, JEOL Scanning Electron Microscope, operating at 20 kV and NTEGRAPRIMA/NTMDT Atomic Force Microscope. For metallographic observation, the as-machined chips were collected and mounted in a resin and mechanically polished with wet SiC paper of 2000 grit and 3000 grit and then finely polished with .05 and .02 μm Al_2O_3 powder, followed by chemical etching.

3. Results and discussion

3.1. Morphology and microstructure

The optical micrographs of the bulk specimen are shown in Figure 2, with (a) and (b) representing the respective microstructures of steel and brass in the bulk state. The average grain size of steel specimen is ~15 μm, while that of brass is ~25 μm.

The representative of scanning electron micrographs of the as-machined steel chip at different magnification is shown in Figure 3. Figure 3(a) represents the SEM micrograph of typical as cut chip specimen exhibiting a "flow-line" type microstructure, characteristic of large strain deformation. Figure 3(b) and (c) represent the cross section of chip segments that are attached loosely to each other, exhibiting intense shear localization and formation of micro shear bands. There exists a distinct direction of shearing throughout the chip cross section and a thin secondary shear zone along the back of the chip. These shear bands are formed due to high strain rate and large plastic deformation.

AFM could be a faster alternative to TEM techniques for trying to image the microstructure within shear bands. Figure 4 represents the atomic force micrographs of metallographically prepared specimens. Figure 4(a) is a typical AFM 2D image of the machined chip prepared by embedding the chips

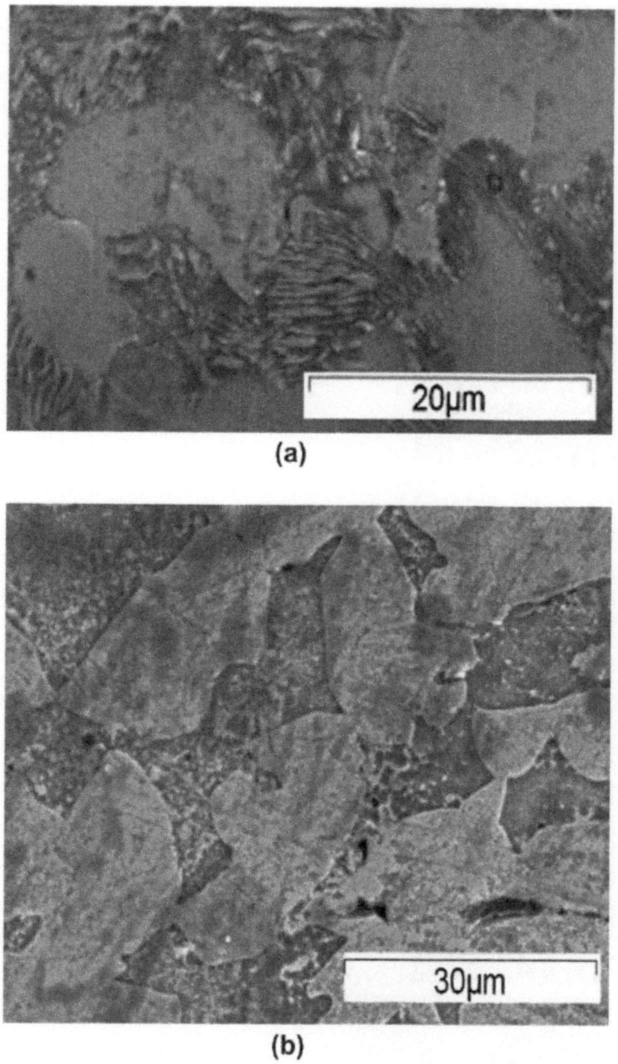

Figure 2. Optical micrograph of bulk specimen (a) Steel and (b) Brass.

in a polymer matrix. The image reveals the distinctive shear bands that are formed during machining as mesoscopic traces, aligned with the shear plane. Figure 4(b) shows the microstructure of brass chip having an average grain size of ~40 ± 10 nm, while Figure 4(c) shows that of the steel chip having an average grain size of ~60 ± 10 nm. The 3D image (Figure 4(d)) shows the average height and roughness of the chip surface.

3.2. Shear strain and hardness

The shear strain values in the chip as estimated from Equations 1 to 2 were found to vary from 1 to 3.9 with a change of rake angle from +5° to −15°. The dislocations that are generated to accommodate these substantial strains assemble themselves into complex and elaborate cellular structures in the chip (Shekhar, Cai, Lee, Wang, & Shankar, 2009) leading to the formation of deformation shear bands. The Vickers hardness values of the chips were determined with a load of 300 g on steel and 200 g on brass samples. The Vickers hardness values of the chips machined at −5° and −15° rake angle were found to be 235 ± 5 and ~256 ± 5 HV, respectively, and that of brass were found to be 175 ± 5 and 198 ± 5 HV, respectively. So, the machined chips exhibit a remarkable increase in the hardness as compared with the measured bulk hardness values of the two materials. This increase in hardness values is undoubtedly a consequence of the reduced grain size and dislocation substructures in the chips.

(a)

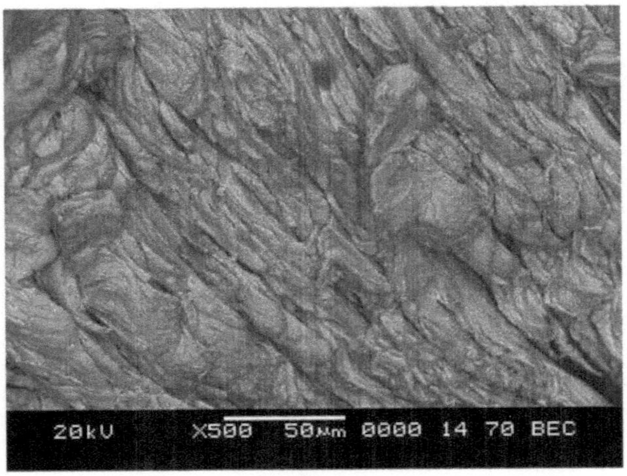

(b)

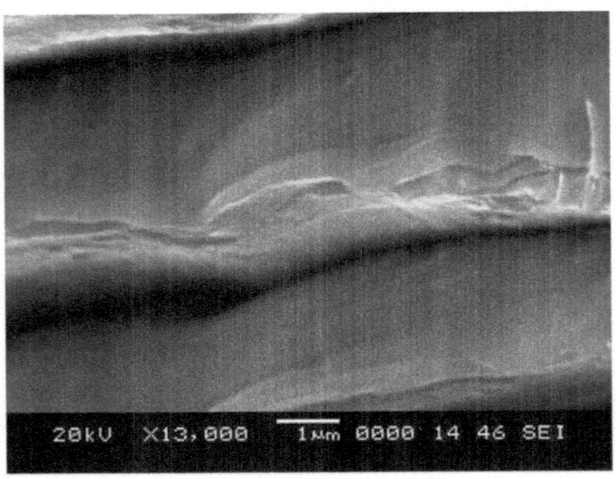

(c)

Figure 3. SEM micrographs of the as-machined chips at different magnification (a) Steel and (b) Brass.

3.3. Small-angle neutron scattering

Small-angle scattering signal originates from the density fluctuations in the mesoscopic length scale. In the present case, it is assumed that the scattering intensity at small angle, as observed experimentally, originates from the density fluctuation because of the presence of the sub-micrometric

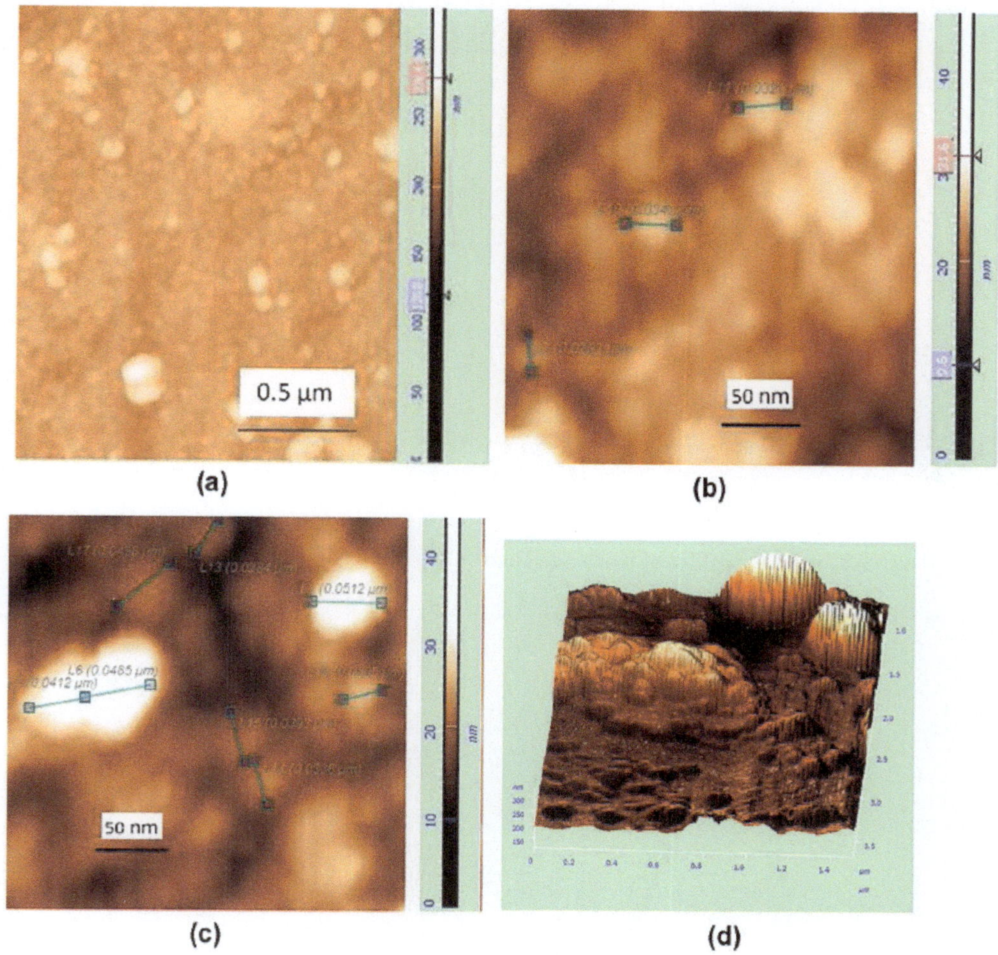

Figure 4. Typical AFM images and a representative surface profile of the machined chips (a) Image of a typical chip prepared by embedding in a polymer matrix, (b) Microstructure of a brass chip, (c) Microstructure of a steel chip, and (d) Typical 3D image of a chip surface.

structure after machining. It is observed from the scattering profiles that the scattering profile becomes significantly broader due to small-angle scattering after the machining. The scattering intensity $I(q)$ at small-angle region corresponding to polydisperse pores may be approximated as:

$$I(q) = C\left(\int_0^\infty D(R)\,R^6 P(q,R)\,S(q,R)\,dR\right) \tag{3}$$

where $D(R)$ represents the size distribution of the basic constituents of the fractal, i.e. $D(R)dR$ represents the probability of finding a particle with radius R to $R+dR$. $P(q, R)$ is the form factor of a particle of radius r. In present case, a spherical form factor is assumed for the data analysis,

$$P(q,R) = 9\left(\frac{\sin(qR) - qR\cos(qR)}{(qR)^3}\right)^2 \tag{4}$$

C is the scale factor and is independent of q. $S(q, r)$ is the interparticle structure factor. A fractal like $S(q, r)$ was considered in the present case to take into account of the particle cluster. The neutron scattering profiles of the machined chips are presented in Figure 5.

The above model was fitted to the experimental profile and the particle size distribution was estimated using nonlinear least-square method.

The estimated particle size distribution for the machined samples is plotted in Figure 6. The average radius of the grains in machined samples is nearly 45 nm. It was found that a consideration

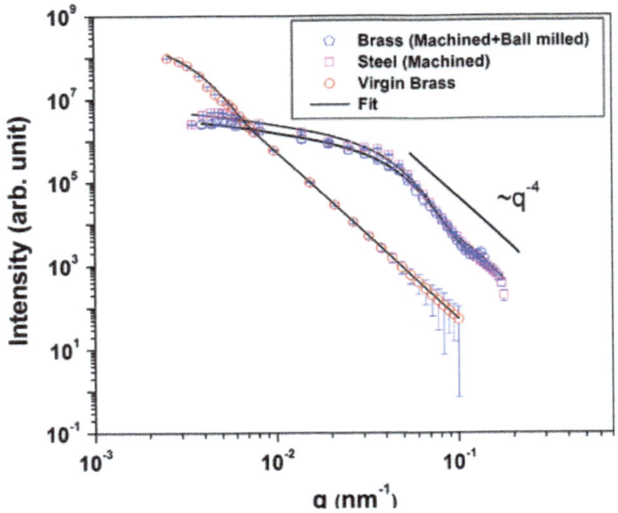

Figure 5. Neutron scattering profiles of machined chips.

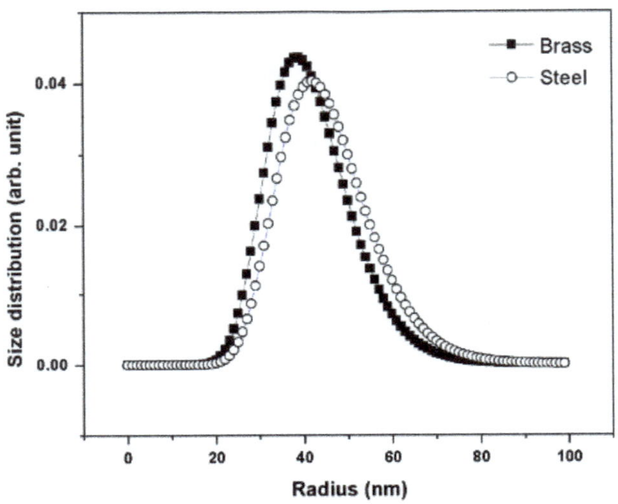

Figure 6. Estimated particle size distribution for the machined chip samples.

of a fractal-like structure factor, due to the clustered particles, fits the data fairly well. The upper cutoff of the fractal was found to be above 300 nm, although no distinct value could be obtained due to low q limit. From the analysis of scattering data, it was found that for the un-machined sample, the grain sizes are large enough (>1 micron) compared with the grain size in the machined samples, but no conclusive decision about the grain size of the un-machined sample could be drawn as the larger grain size demands an access to further low q values.

The average particle size obtained from USANS experiment is found to be in fair corroboration with the grain size as obtained from AFM. It is to be mentioned that the grain morphology obtained from AFM represents nearly coherent arrangement within the local zones, whereas the particle size obtained from SANS represents the average size of the primary particles of the aggregates in the machined chip. It should be mentioned that the presence of mesoscopic traces of shear bands in the machined chips, formed due to large strain deformation, may also lead to the density fluctuations as far as scattering data are concerned.

4. Conclusion

Fine microstructure, composed of grains even smaller than 100 nm, is found to result from chip formation in commercial steel and brass due to the induced shear strain leading to large strain deformation. The machined chips exhibit a remarkable increase in the hardness as compared with the measured bulk hardness values in both materials. This increase in hardness values is attributed to the reduced grain size and dislocation substructures in the chips. The average particle size obtained from neutron scattering experiment is found to be in corroboration with the particle size obtained from AFM. Furthermore, the presence of mesoscopic traces of shear bands in the machined chips formed due to large strain deformation leads to the density fluctuations during neutron scattering.

Funding
The authors received no direct funding for this research.

Author details
Elwyn Rebello[1]
E-mail: ellwynreb66@gmail.com
Frazer Vaz[1]
E-mail: frazervaz@gmail.com
Avino Colaco[1]
E-mail: avino.colaco@yahoo.com
Wellon Rodrigues[1]
E-mail: wellor@yahoo.com
Geethalakshmi K[1]
E-mail: geethashyam@gmail.com
Debasis Sen[2]
E-mail: Debases.sen@gmail.com
S. Mazumder[2]
E-mail: smazu@barc.gov.in
A.O. Surendranathan[3]
E-mail: aos_nathan@yahoo.com
[1] Department of Mechanical Engineering, P. C. College of Engineering, Verna, Goa, India.
[2] Solid State Physics Division, Bhabha Atomic Research Centre, Mumbai, India.
[3] Department of Metallurgical & Materials Engineering, National Institute of Technology Karnataka, Surathkal, India.

References

Borsali, R., & Pecora, R. (Eds.). (2008). *Soft-matter characterization: Small-angle neutron scattering and applications in soft condensed matter* (Vol. 2). New York, NY: Springer Science + Business Media LLC.

Geels, K. (2007). *Metallographic and materialographic specimen preparation, light microscopy, image analysis and hardness testing*. West Conshohocken, PA: ASTM International. http://dx.doi.org/10.1520/MNL46-EB

Iglesias, P., Moscoso, W., Mann, J. B., Saldana, C., Shankar, M. R., Chandrasekar, S., ... Compton, W. D. (2008). Production analysis of new machining-based deformation processes for nanostructured materials. *International Journal of Material Forming, 1*, 459–462. doi:10.1007/s12289-008-0094-0

Lake, J. A. (1967). An iterative method of slit—Correcting small angle X-ray data. *Acta Crystallographica, 23*, 191–194. doi:10.1107/S0365110X670002440

Mazumder, S., Sen, D., Sarvanan, T., & Vijayraghavan, P. R. (2001a). A medium resolution double crystal based small-angle neutron scattering instrument at Trombay. *Current Science, 81*, 257–262.

Mazumder, S., Sen, D., Sarvanan, T., & Vijayraghavan, P. R. (2001b). Performance and calibration of the newly installed medium resolution double crystal based small angle neutron scattering instrument at Trombay. *Journal of Neutron Research, 9*, 39–57. doi:10.1080/10238160108200241

Merchant, M. E. (1945). Mechanics of metal cutting process 1: Orthogonal cutting and type 2 chip. *Journal of Applied Physics, 16*, 267–275. http://dx.doi.org/10.1063/1.1707586

Ni, H., & Alpas, A. T. (2003). Sub-micrometer structures generated during dry machining of copper. *Materials Science and Engineering A, 361*, 338–349. doi:10.1016/S0921-5093(03)00530-6

Schmidt, P. W. (1988). Collimation effects in small-angle X-ray and neutron scattering. *Journal of Applied Crystallography., 21*, 602–612. doi:10.1107/S0021889888006375

Shankar, M. R., Chandrasekar, S., Compton, W. D., & King, A. H. (2005). Characteristics of aluminum 6061-T6 deformed to large plastic strains by machining. *Materials Science and Engineering A, 410–411*, 364–368. doi:10.1016/j.msea.2005.08.137

Shankar, M. R., Chandrasekar, S., King, A. H., & Compton, W. D. (2005). Microstructure and stability of nanocrystalline aluminum 6061 created by large strain machining. *Acta Materialia, 53*, 4781–4793. doi:10.1016/j.actamat.2005.07.006

Shaw, M. C. (1984). *Metal cutting principles*. Clarendon: Oxford University Press.

Shekhar, S., Cai, J., Lee, S., Wang, J., & Shankar, M. R. (2009). How strains and strain-rates are accommodated by dislocations and twins during chip formation by machining. *Transactions of North American Manufacturing Research Institute/Society of Manufacturing Engineers (NAMRI/SME), 37*, 637–644.

Swaminathan, S., Shankar, M. R., Lee, S., Hwang, J., King, A. H., Kezar, R. F., ... Rao, B. C. (2005). Large strain deformation and ultra fine grained materials by machining. *Materials Science and Engineering A, 410–411*, 358–363. doi:10.1016/j.msea.2005.08.139

15

A novel approach toward coordinated inventory management of an agile multi-echelon multi-product supply chain

Maryam Alimardani[1], Hamed Rafiei[1]* and Masoud Rabbani[1]

*Corresponding author: Hamed Rafiei, School of Industrial Engineering, College of Engineering, University of Tehran, Tehran, Iran
E-mail: hrafiei@ut.ac.ir
Reviewing editor: Zude Zhou, Wuhan University of Technology, China

Abstract: Inventory control systems development is one of the most appealing issues of the supply chain management which have recently been studied by academicians and practitioners. In this paper, an inventory management framework and deterministic cost models within the context of this framework are structured for effective supply chain (SC) in an agile environment. To do so, we propose a three-echelon multi-product SC including multiple suppliers, multiple manufacturing firms, and multiple retailers. Moreover, a fuzzy-ANP framework is employed to evaluate suppliers' agility level and to elicit suitable suppliers aiming at achieving an agile network. To develop operational measures of SC flexibility, a cost structure is defined which captures some different operational cost factors including inventory holding, backordering, lost sell, and ordering costs. Furthermore, a hybrid metaheuristic algorithm is developed to tackle the computational complexity of a production-inventory planning task. Finally, numerical experiments validate the proposed solution methodology.

Subjects: Industrial Engineering & Manufacturing; Operations Research; Production Systems

ABOUT THE AUTHORS

Maryam Alimardani graduated from the School of Industrial Engineering, University of Tehran. She has published several papers in international scholarly journals and conference proceedings in the field of production planning, inventory control, and enterprise resource planning.

Hamed Rafiei obtained his PhD from School of Industrial Engineering, University of Tehran. He received his BSc and MSc degrees of Industrial Engineering from University of Tehran as an honored student. His research interests include applications of operations research in operations management on which he has published several papers in international journals and conference proceedings.

Masoud Rabbani is a professor of Industrial Engineering at School of Industrial Engineering, University of Tehran. He has published more than 50 papers in international journals, e.g. *European Journal of Operational Research, International Journal of Production Research, International Journal of Production Economics*. His current research interests comprise production planning, applied graph theory, productivity management, and EFQM.

PUBLIC INTEREST STATEMENT

This paper attempts to solve the inventory management problem of a supply chain. The considered supply chain includes two layers of the manufacturers and retailers. It is challenging to manage inventory orders and cycles in both layers of the supply chain simultaneously and in accordance with each other. In this regard, an integrated mathematical program is developed. The developed model consists of three sub-models each of which attempts to cope with the inventory system of the retailer, warehouse of the manufacturer, and production of the manufacturer. Since the proposed model is not computationally solvable, a metaheuristic algorithm is developed to solve the proposed integrated model.

Keywords: inventory management; multi-echelon multi-product inventory models; agile supply chain; fuzzy analytic network process; differential evolution algorithm; particle swarm optimization

1. Introduction

Supply chain management (SCM) as an emerging field has recently attracted industrial community. SCM is implemented in order to manage multiple relationships across the supply chain (SC) with respect to the entire flow of information, materials, and services to fulfill customer demand in an efficient manner. This philosophy led to fundamental changes in the field of business management (Li & Wang, 2007). In this regard, commodities should be produced and distributed at the right quantities, to the right locations, and at the right time resulting in minimization of the entire costs of system wide (Routroy & Kodali, 2005).

Beside diverse decisions involved in the field of SCM, SC inventory management is one of the major challenges of SCM. While, researchers and industrial practitioners have recently devoted major attention to develop inventory problems in SCM by considering a whole SC as a network of procurement, transformation, and delivering firms (Routroy & Kodali, 2005). Some of the important reasons of efficient inventory management in SC operations could be declared as follows. First, 20–60% of the total assets of manufacturing firms belong to inventory (Gümüs & Güneri, 2007). Second, there are various forms of inventories throughout the SC including raw materials, semi-finished or finished goods, as well as in-process between locations. Third, holding of inventories can cost anywhere between 20 and 40% of their value (Ganeshan & Harrison, 1995). More importantly, lack of comprehensive inventory management throughout the SC can lead to the bullwhip effect as well as, excessive inventory investments, lost revenues, misguided capacity plans, ineffective transportation, missed production schedules, and poor customer service (Giannoccaro, Pontrandolfo, & Scozzi, 2003). On the other hand, traditional SC was a sequence of weakly connected activities both within and outside of the organization, whose most important results are capacity-infeasible plans with unbalanced inventories and poor overall performance (Torabi & Hassini, 2009). Inventory replenishment decisions are made taking into account only those cost parameters that could be directly influenced by the planning company. An optimal solution is concluded for only a single company. Hence, based on traditional inventory management, optimality is not guaranteed when multiple companies or a whole SC are taken into consideration (Glock, 2012). Tackling with this lack of coordination is a spread era for SC managers (Torabi & Hassini, 2009). So, having multi-echelon inventory management applied to continue competitive and reduce inventory, operation costs are decreased and customer service is improved (Gümüs & Güneri, 2007).

In this regards, the classic square root economic-order quantity (EOQ) model developed by Harris is one of the earliest models to cover two basic questions that any inventory control system has to answer when and how much to order in every basic textbook on operations research/management science, operations management, and SCM (Pentico & Drake, 2009). Next, the basic EOQ model has been extended to as the economic production quantity (EPQ) or economic manufacturing quantity (EMQ) model. In the extension, the replenishment order is assumed to be received at a constant finite rate over time against EOQ which is instantaneous replenishment (Pentico, Drake, & Toews, 2009).

In this study, we attempt to cope with the shortcomings of the traditional inventory management. Hence, this paper deals with the inventory management of SCs. In this regard, an inventory management framework is proposed to satisfy customers' needs in a three-echelon SC in an agile manner. The supposed SC consists of multi-supplier, multi-retailer, and multi-manufacturing plant which offers multiple product types to the customers. Upon the proposed inventory management framework, different levels of the supposed network are integrated to decrease total production-inventory costs of the SC. In order to achieve a more flexible and responsible network, manufacturing firms purchase raw materials from variant suppliers, and assemble them into final products, then the final products are split and delivered to the retailers. Moreover, a fuzzy-ANP framework is proposed to evaluate suppliers' agility levels in responding to the downward firms' needs.

Rest of the paper is structured as follows. Section 2 briefly reviews the previously conducted researches. Section 3 presents the proposed three-echelon multi-item SC structure. The models of the proposed inventory policies as well. And the proposed solution methodology is described thoroughly in Section 4. This paper follows in Section 5 on computational experiments to validate the proposed model. Finally, some conclusions and future research directions are provided in the last section.

2. Literature review

Since numerous researchers addressed different aspects of the multi-echelon inventory system and the integrated SC, we provide only the studies which have been recently published in order to conclude contributions of this paper in comparison with the body of literature. Additionally, literature body of agile SCs is briefly reviewed in this section.

2.1. The multi-echelon inventory system

A *multi-echelon* inventory system is a system in which an item reaches to the final customer after moving more than one stage (Rau, Wu, & Wee, 2003). The problems related to the multi-echelon inventory control have been perused since 1950s. The main aim of these problems is to control the inventory levels by determining the size of the orders for each echelon during each period so that objective function is optimized (Zhou, Chen, & Ge, 2012). The first research toward the two-echelon inventory model dates back to the one by Clark and Scarf (1960). In their study, the optimality of a base stock policy for the pure serial inventory system was demonstrated and an efficient decomposing method was introduced in order to compute the optimal base stock ordering policy. In another seminal work, Sherbrooke (1968 cited by Rau et al., 2003) introduced an ordering policy of two-echelon inventory model for warehouse and retailer while pure shortage was allowed at the retailers. Moreover, Mitra (2009) and Hill, Seifbarghy, and Smith (2007) presented single-item, two-echelon inventory models applying continuous review. In addition to the two-level systems, abundant researches have been presented on multi-echelon inventory control. Routroy and Kodali (2005) developed a single product, three-echelon inventory planning model. They applied the differential evolution algorithm to minimize the total system-wide cost by determining ordering/production quantity and inventory/service level. Two other instances of multi-item, multi-echelon systems are modeled by Caggiano, Jackson, Muckstadt, and Rappold (2009) and Yang and Lin (2010).

2.2. The integrated SC

Another interesting issue related to SCM is to optimize SC integrated systems. In particular, the integrated inventory replenishment decisions between buyer and vendor are called the joint economic lot sizing problem (JELP) upon which order, production, and shipment quantities are determined from the perspective of the SC with the objective of minimizing total system costs (Ben-Daya, As'ad, & Seliaman, 2010). Goyal (1976) was the first who introduced the idea of joint total cost for the buyer and the supplier. Thereafter, numerous researches have been appeared to coordinate inventory replenishment decisions among individual companies to benefit the entire supply network. Sajadieh and Akbari Jokar (2009) developed a model in which demand at the buyer followed a linear demand function. In Sawik's (2009) study, supplier operates multiple production lines in parallel and supplies multiple products to a manufacturer. And, the manufacturer runs a production facility with multiple stages in series. In the case of more than two actors in an integrated inventory model, Joglekar and Tharthare (1990) were the first who modeled a system with a single-vendor and multiple identical buyers (SVMB). Chan and Kingsman (2007) developed Banerjee and Burton's model (1994), in which a basic cycle time was considered for the vendor and the order interval of each buyer was assumed to be an integer multiple of the basic cycle. Hoque (2008) developed a single-manufacturer multi-buyer integrated inventory SC applying close relationship between the actors of the SC for a costless way of benefit sharing. The case of multi-vendor in the integrated inventory models was analyzed by Chen and Sarker (2010), who studied a buyer sourcing a product from multiple vendors supposing a single truck to collect the products from the vendors in a milk run. The authors modeled a freight cost function which considers both product weights and travel distances.

With respect to the multi-echelon inventory models, Banerjee, Kim, and Burton (2007) offered a common delivery cycle in order to reduce the complexity of the problem. The first research toward three stages SC with one or more firms at each stage was introduced by Khouja (2003) whose considered system comprised of a single raw material supplier, multiple manufacturers, and multiple buyers, which was further extended by Ben-Daya and Al-Nassar (2008). Finally, some papers took into account more cost factors in multi-echelon inventory models, such as deteriorating inventory models by Yang and Wee (2000) and stochastic demand arrivals by Zhou et al. (2012).

2.3. Agile SC

The concept of the agile SC has been taken into account as means of achieving a competitive edge in rapidly changing business environments. In this regard, one of the most important issues is supplier selection. According to the related body of literature, supplier selection process comprises of three main steps: criteria formulation, qualification, and choice. The last stage involves selecting suitable partners through qualified partners for specific purchases with respect to the criteria which are related to the suppliers' level of agility (Wu & Barnes, 2011).In the context of agility, as the capability of a process to respond to unanticipated changes, the first research toward agility measurement dates back to the one by Dove (1994 cited by Sherehiy, Karwowski, & Layer, 2007). An integrated agility index which was defined as combination of agile capabilities intensity levels has been applied by some authors such as Tsourveloudis & Valavanis, 2002. Moreover, the linguistic expressions and fuzzy logic were augmented to overcome the uncertainty and ambiguity of the agility indicators and definitions (Lin, Chiu, & Tseng, 2006; Tsourveloudis & Valavanis, 2002).

In order to evaluate agility level of suppliers, numerous research papers have been published so far, among which Tam and Tummala (2001) presented an AHP-based model to select a vendor for a telecommunications system applying some criteria with respect to cost, technical, and operational factors. Moreover, the developed models have been equipped with fuzzy set theory in order to cope with the involved fuzziness in the problem, as Mikhailov (2002) which developed a fuzzy-AHP model to tackle supplier selection in an ambiguous environment. In addition to AHP, ANP has been another method used in this field, since it compensate a great drawback of AHP. ANP is capable to model interrelations among decision criteria and alternatives. Similar to many applications of ANP, this method has been adopted in the field of agile supplier selection as in Büyüközkan and Çifçi (2011) and Sarkis, Talluri, and Gunasekaran (2007).

With respect to the literature, no attention has been dedicated to multi-item, multi-vendor, multi-buyer SC in an agile environment. Hence, this paper is devoted to inventory management by developing EOQ and EPQ models subject to stockouts in a three-echelon SC for the first time. The proposed SC embraces multi-supplier, multi-manufacturer, and multi-retailer, while SC's agility is enhanced by assessing suppliers with respect to the agility criteria.

3. Proposed multi-echelon SC

The present study attempts to develop an inventory management framework in which decisions on inventory control system are made in a multi-item, multi-buyer, multi-vendor, multi-echelon SC. This framework is developed to seek when and how much product should be placed as an order/production with respect to the related inventory policy. To do so, a three-echelon SC including various manufacturing plants with capacitated storage, multiple retailers at the bottom level, and multiple suppliers at the top level is proposed. In this regard, it is supposed several manufacturing firms operate based upon multi-product production system.

As illustrated in Figure 1, each stage in the SC is fed from upper echelon and feed lower ones. The entire customers' demand is firstly entered into the retailers so that is responded through retailers' on-hand inventory. And, retailers' needs are forwarded to manufacturing plants to be produced in an economic manner. To do so, multiple manufacturing firms are considered to supply retailers' demands leading to a more flexible and responsible network. In addition, manufacturing firms' requirements are fulfilled

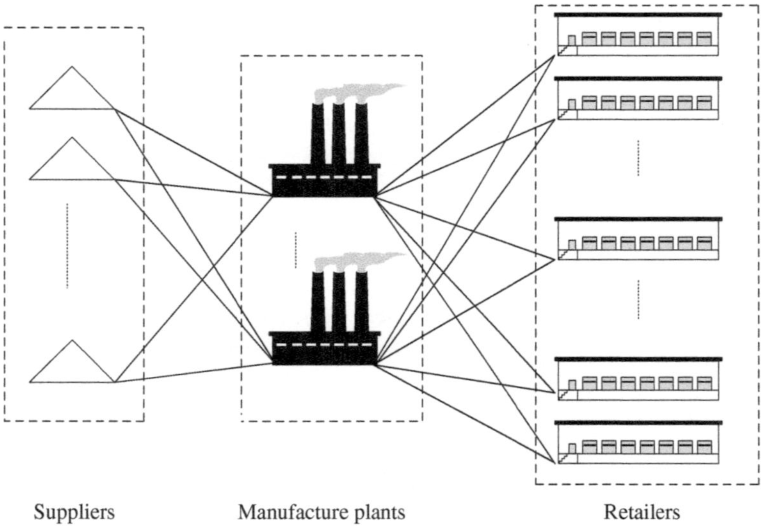

Figure 1. A three-echelon supply chain.

through various suppliers, and are delivered in economic batch size assuring agility in terms of flexibility and performance. In other words, the supposed network suggests multiple product types which are produced at different costs at various manufacturing plants to the customers arrived to the retailer nodes. Furthermore, the connections between two partners have been suggested in subsequent levels allowing physical transportation and information relationships between entities for the sake of the product flow efficiency and effectiveness. This section elaborates the proposed model including components depicted in Figure 2.

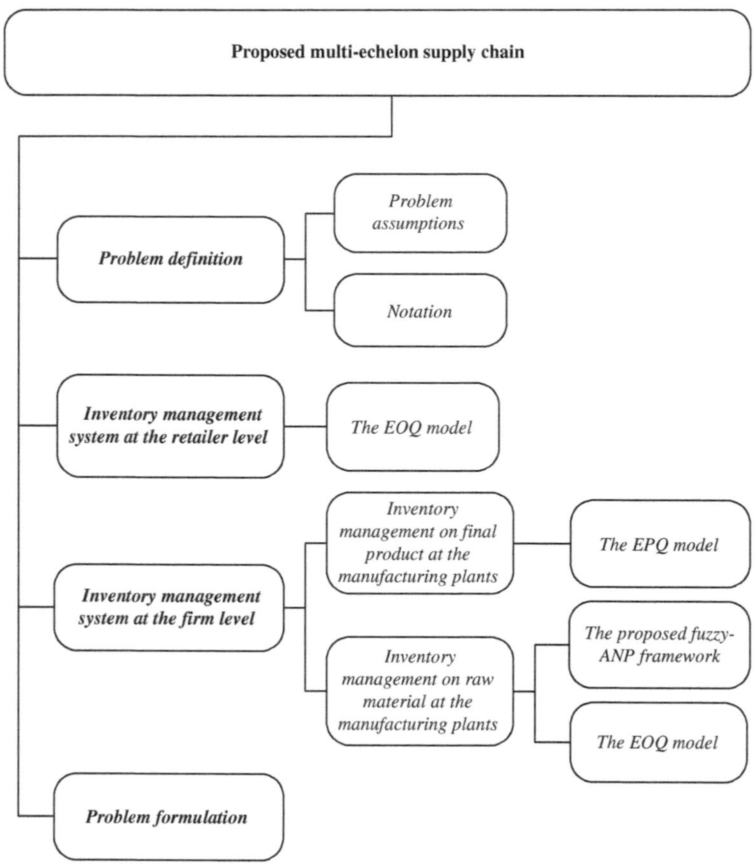

Figure 2. General schema of Section 3.

3.1. Problem definition

In order to manage inventory system corresponding to retailer level, a deterministic EOQ policy with partial backordering is supposed. Also, relevant inventory control system of manufacturing firms is analyzed applying two inventory policies including EOQ and EPQ from which an EOQ policy with pure backordering is implemented to investigate the inventory system of raw material as well as an EPQ policy with pure backordering for the final products. In this regard, the problem is defined so as to minimize total cost of the corresponding SC.

3.1.1. Problem assumptions

Assumptions of the problem as followings:

- Customer demand rates for all products are deterministic and constant during the planning period;
- Production rates of all products are deterministic and constant;
- Production rate is greater than demand rate for every product in the SC;
- Replenishment lead times for retailers and firms are zero;
- Replenishment batch sizes for retailers and manufacturing plants are equal;
- Raw material warehouses of manufacturing plants have finite capacity;
- Shortage is allowed; and
- Shortage, ordering, and holding costs at each level are known.

3.1.2. Notations

The proposed EOQ and EPQ models are dedicated to determine how much and when product should be placed as an order and a production. In this regard, the used notations are as followings:

Indices

i	Index for final products ($i = 1, ..., N$);
i'	Index for raw materials ($i' = 1, ..., n$);
j	Index for retailer nodes ($j = 1, ..., nr$);
k	Index for manufacturing plant nodes ($k = 1, ..., nm$);
z	Index for supplier nodes ($z = 1, ..., ns$);

Parameters

D_{ij}^{r}	Demand per year for final product i at retailer j;
P_{ik}^{m}	Production rate per year for final product i at manufacturing plant k;
sr_{ij}	The unit selling price of final product i at retailer j;
Co_{ij}^{r}	The fixed cost of placing an order for final product i at retailer j;
Co_{ik}^{m}	The fixed cost of receiving an order for final product i at manufacturing plant k;
$Co_{i'k}^{m}$	The fixed cost of placing an order for raw material i' at manufacturing plant k;
Cpu_{ij}^{r}	The variable cost of purchasing a unit of final product i at retailer j;
Ch_{ij}^{r}	The cost to hold a unit of final product i at retailer j in inventory for a cycle;
Ch_{ik}^{m}	The cost to hold a unit of final product i at manufacturing plant k in inventory for a cycle;
$Ch_{i'k}^{m}$	The cost to hold a unit of raw material i' at manufacturing plant k in inventory for a cycle;
Cb_{ij}^{r}	The cost to keep a unit backordered of final product i at retailer j for a cycle;
Cb_{ik}^{m}	The cost to keep a unit backordered of final product i at manufacturing plant k for a cycle;
$Cb_{i'k}^{m}$	The cost to keep a unit backordered of raw material i' at manufacturing plant k for a cycle;

Cg_{ij}^{r} The goodwill loss on a unit of unfilled product *i* demand at retailer *j*;

Cl_{ij}^{r} The cost for a lost sale of final product *i* at retailer *j*, including the lost profit on that unit and any goodwill loss ($sr_{ij} - Cpu_{ij}^{r} + Cg_{ij}^{r}$);

$f_{i'k}$ Space occupied by each unit of raw material *i'* at manufacturing plant *k*;

f_{k} Available warehouse space for all raw materials at manufacturing plant *k*;

$q_{i'kz}$ The percentage of order quantity of raw material *i'* from manufacturing plant *k* toward supplier z (i.e. $0 \leq q_{i'kz} \leq 1$);

$cc_{ii'k}$ The consumption coefficient of final product *i* from raw material *i'* at manufacturing plant *k*;

β_{ijk}^{r} The fraction of stockouts of final product *i* at retailer *j* from manufacturing plant *k* that will be backordered

Decision variables

D_{ik}^{m} Demand per year for final product *i* at manufacturing plant *k*;

$D_{i'k}^{m}$ Demand per year for raw material *i'* at manufacturing plant *k*;

Q_{ijk}^{r} The order quantity for final product *i* at retailer *j* toward manufacturing plant *k*;

Q_{ik}^{m} The production quantity for final product *i* at manufacturing plant *k*;

$Q_{i'kz}^{m}$ The order quantity for raw material *i'* at manufacturing plant *k* toward supplier *z*;

q_{ijk} The percentage of order quantity of final product *i* from retailer *j* toward manufacturing plant *k* (i.e. $0 \leq q_{ijk} \leq 1$);

$x_{ii'k}$ 1 if $cc_{ii'z} \geq 0$, otherwise 0;

T_{ij}^{r} The time between orders or the length of an order cycle of final product *i* at retailer *j*;

T_{ik}^{m} The time between productions or the length of a planning cycle of final product *i* at manufacturing plant *k*;

$T_{i'k}^{m}$ The time between orders or the length of an order cycle of raw material *i'* at manufacturing plant *k*;

K_{ijk}^{r} The integer multiple of the length of order cycle of final product *i* at retailer *j* from manufacturing plant *k*;

K_{ik}^{m} The integer multiple of the length of production cycle of final product *i* at manufacturing plant *k*;

F_{ijk}^{r} The fill rate or the percentage of demands for final product *i* at retailer *j* from manufacturing plant *k* that will be filled from stock;

F_{ik}^{m} The fill rate or the percentage of demands for final product *i* at manufacturing plant *k* that will be filled from stock;

$F_{i'kz}^{m}$ The fill rate or the percentage of demands for raw material *i'* at manufacturing plant *k* from supplier *z* that will be filled from stock;

TC_{EOQ}^{r} Total cost per year for the retailers;

TC_{EOQ}^{m} Total related cost to *EOQ* per year for the manufacturing plants;

TC_{EPQ}^{m} Total related cost to *EPQ* per year for the manufacturing plants;

TC^{m} Total cost per year for the manufacturing plants (i.e. $TC^{m} = TC_{EOQ}^{m} + TC_{EPQ}^{m}$);

TC Total cost per year for the whole of the **SC**

3.2. Inventory management system at the retailer level

In this study, retailer level is proposed to offer *N* types of product to the customers. Customer demands are provided through on-hand inventory of the retailers and their orders are partially backordered in the case of shortage. There are several manufacturing firms in the SC to supply retailers in order to enhance flexibility and agility of the network because each retailer places an order for the final products toward variant manufacturing plants. While, the length of order cycle of each final product at each retailer is multiple of each other. And, the order quantity of any product *i* placed in every cycle of time is equal. However, the order quantity placed by a retailer may vary from one retailer to the others. To do so, the retailer-level proposed model is described in this section.

3.2.1. The EOQ model

According to the previous researches on inventory management frameworks, there are six basic decision variables to develop EOQ model with partial backordering: *Q* (the order quantity), *T* (the length of an order cycle), *I* (the maximum inventory level), *S* (the maximum stockout level), *B* (the maximum backordered position), and *F* (the fill rate or the percentage of demand). However, it is adequate to find two variables among the six above, hence, the remainder might be calculated. Therefore, T^r_{ij}, the length of an order cycle of product *i* at retailer *j*, and F^r_{ijk}, the fraction of demand for final product *i* at retailer *j* from manufacturing plant *k* to be filled from stock, and integer value of the multiplier of the order cycles (K^r_{ijk}) are supposed as decision variables. In order to develop operational measures of retailer level's flexibility, an objective function which captures some different operational cost factors such as inventory holding, backordering, ordering, and goodwill loss is defined.

Partial backordering is supposed for customers who face unavailable production inventory at the retailer level. It must be noted, the backordering rate of each product at each retailer from each firm is assumed a constant rate, which $0 \leq \beta^r_{ijk} \leq 1$). Figure 3 illustrates the inventory level of EOQ problem with partial backordering for final product *i* at retailer *j* from firm *k*.

As depicted in Figure 3, order cycle for final product *i* at retailer *j* from manufacturing plant *k* is divided into two sub-interval including t^1_{ijk} which is demand time with $t^1_{ijk} = F^r_{ijk}\, q_{ijk}\, T^r_{ij}$, and t^2_{ijk} which is shortage time with $t^2_{ijk} = \left(1 - F^r_{ijk}\right) q_{ijk}\, T^r_{ij}$. Therefore, based upon these values, I^r_{ijk} is the maximum inventory level of final product *i* at retailer *j* from manufacturing plant *k* with $I^r_{ijk} = D^r_{ij}\, t^1_{ijk} = D^r_{ij}\, F^r_{ijk}\, q_{ijk}\, T^r_{ij}$, S^r_{ijk} is the maximum stockout quantity of product *i* at retailer *j* from manufacturing plant *k* with $S^r_{ijk} = D^r_{ij}\, t^2_{ijk} = D^r_{ij}\left(1 - F^r_{ijk}\right) q_{ijk}\, T^r_{ij}$, B^r_{ijk} is the maximum backorder position of product *i* at retailer *j* from manufacturing plant *k* with $B^r_{ijk} = \beta^r_{ijk}\, S^r_{ijk} = \beta^r_{ijk}\, D^r_{ij}\left(1 - F^r_{ijk}\right) q_{ijk}\, T^r_{ij}$. With respect to these expressions, order quantity of each product at each retailer node from each manufacturing plant is calculable using

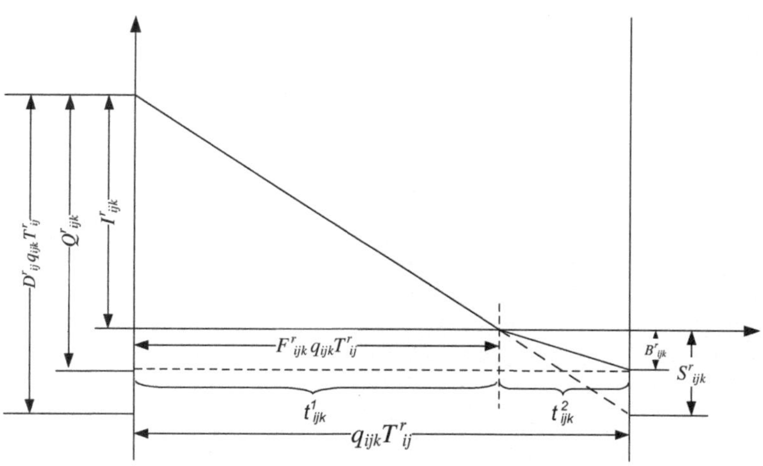

Figure 3. Inventory level of EOQ problem with partial backordering.

Equation 1. Hence, the relevant average profit of retailer level per year (i.e. Average profit per year = income − [(ordering + purchasing + inventory holding + backordering + goodwill loss) cost])) is maximized through Equation 2 and the constraints are taken into account (Equation 3).

$$Q^r_{ijk} = I^r_{ijk} + B^r_{ijk} = D^r_{ij} F^r_{ijk} q_{ijk} T^r_{ij} + \beta^r_{ijk} D^r_{ij} (1 - F^r_{ijk}) q_{ijk} T^r_{ij} \quad (1)$$

$$\begin{aligned} \max \quad \Pi\left(T^r_{ij}, F^r_{ijk}, q_{ijk}, K^r_{ijk}\right) = &\left[\textstyle\sum_i \sum_j \sum_k sr_{ij} D^r_{ij} \left[F^r_{ijk} + \beta^r_{ijk}\left(1 - F^r_{ijk}\right)\right]\right] \\ - &\left[\begin{array}{l} \sum\limits_i \sum\limits_j \sum\limits_k \frac{Co^r_{ij}}{q_{ijk} T^r_{ij}} + \sum\limits_i \sum\limits_j \sum\limits_k Cpu^r_{ij} q_{ijk} D^r_{ij} \left[F^r_{ijk} + \beta^r_{ijk}\left(1 - F^r_{ijk}\right)\right] \\ + \sum\limits_i \sum\limits_j \sum\limits_k \frac{Ch^r_{ij} D^r_{ij} q_{ijk} T^r_{ij} F^{r^2}_{ijk}}{2} + \\ \sum\limits_i \sum\limits_j \sum\limits_k \frac{\beta^r_{ijk} Cb^r_{ij} D^r_{ij} q_{ijk} T^r_{ij} \left(1 - F^r_{ijk}\right)^2}{2} + \sum\limits_i \sum\limits_j \sum\limits_k Cg^r_{ij} D^r_{ij} \left(1 - \beta^r_{ijk}\right)\left(1 - F^r_{ijk}\right) \end{array}\right] \end{aligned} \quad (2)$$

$$K^r_{ijk} q_{ijk} T^r_{ij} = K^r_{i,j+l,k} q_{i,j+l,k} T^r_{i,j+l} \qquad \forall \quad i, k, j, l = 1, \ldots, nr - j \quad (3)$$

The value of the order quantity, the maximum inventory level, and the maximum backorder position of product i at retailer j are, respectively, yield as follows:

$$Q^r_{ij} = \sum_k Q^r_{ijk};\ I^r_{ij} = \sum_k I^r_{ijk};\ B^r_{ij} = \sum_k B^r_{ijk}$$

As shown in Appendix A, $\Pi\left(T^r_{ij}, F^r_{ijk}, q_{ijk}, K^r_{ijk}\right)$ is maximized by the $T^r_{ij}, F^r_{ijk}, q_{ijk}, K^r_{ijk}$ that minimizes the average cost per year, as illustrated in Equation 4.

$$\begin{aligned} Min \quad & TC^r_{EOQ}(T^r_{ij}, F^r_{ijk}, q_{ijk}, K^r_{ijk}) = \sum_i \sum_j \Big(\sum_k \frac{Co^r_{ij}}{q_{ijk} T^r_{ij}} + \\ & \frac{Ch^r_{ij} D^r_{ij} q_{ijk} T^r_{ij} F^{r^2}_{ijk}}{2} + \frac{\beta^r_{ijk} Cb^r_{ij} D^r_{ij} q_{ijk} T^r_{ij} (1 - F^r_{ijk})^2}{2} + \\ & Cl^r_{ij} D^r_{ij} (1 - \beta^r_{ijk})(1 - F^r_{ijk})\Big) \end{aligned} \quad (4)$$

3.3. Inventory management system at the firm level

As mentioned earlier, the proposed SC offers N types of products to the customers based upon a coordinated mechanism among suppliers, manufacturing firms, and retailers. In the considered SC of the proposed model, a manufacturing firm is supplied from its suppliers (upper level) and processes the raw materials to the final products. Therefore, two inventory control policies including EOQ and EPQ are simultaneously modeled to analyze the related inventory system to the manufacturing plants. A comprehensive description of the supposed inventory control policies is presented in the next sub-sections.

3.3.1. Inventory management on final products at the manufacturing plants

In order to model the inventory system of the final products at the manufacturing plants, an EPQ model with complete backordering is adopted in which the plants have constant production rate P^m_{ik} and demand rate D^m_{ik} ($P^m_{ik} > D^m_{ik}$). Having the manufacturing plant inventory system analyzed, T^m_{ik}, the length of an planning cycle of product i at manufacturing plant k, and F^m_{ik}, the fraction of demand for final product i at manufacturing plant k to be filled from stock, and K^m_{ik}, the integer multiple of the length of production cycle of final product i at manufacturing plant k are obtained in such away to minimize total inventory system costs. So, holding costs, ordering, and backordering costs of the finished products are modeled as objective function. As depicted in Figure 4, manufacturing plant k produces Q^m_{ik} of final product i during production cycle T^m_{ik} upon EPQ model with full backordering. In this model, the economic quantity of each finished product may vary from a firm to another one.

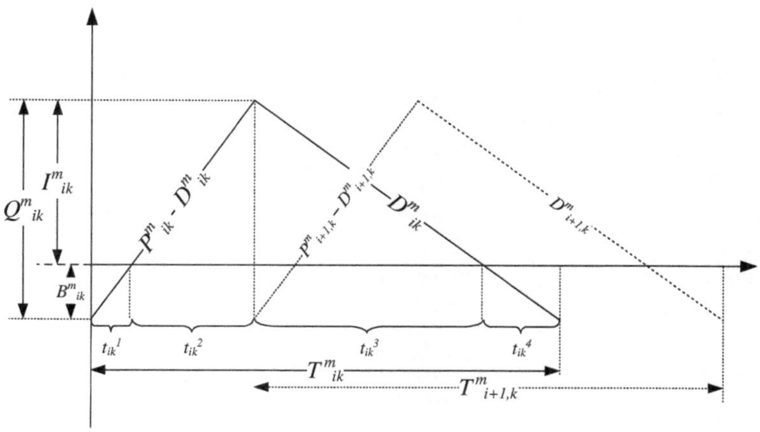

Figure 4. Inventory level of EPQ model with full backordering.

The length of planning cycle of each final product at each firm is divided into four intervals including production/demand time, demand time, and shortage time as elaborated in Figure 4 for which $t_{ik}^{1} = \left(1 - F_{ik}^{m}\right) T_{ik}^{m} D_{ik}^{m}/P_{ik}^{m}$; $t_{ik}^{2} = F_{ik}^{m} T_{ik}^{m} D_{ik}^{m}/P_{ik}^{m}$; $t_{ik}^{3} = F_{ik}^{m} T_{ik}^{m}\left(1 - D_{ik}^{m}/P_{ik}^{m}\right)$; $t_{ik}^{4} = \left(1 - F_{ik}^{m}\right) T_{ik}^{m}\left(1 - D_{ik}^{m}/P_{ik}^{m}\right)$.

With respect to the above values, the value of the maximum inventory level, the maximum backorder position, the production quantity of final product *i* at manufacturing plant *k*, and total costs of this model are found applying Equations 5–8, respectively. In addition, Equation 9 is extended to compute the relationship between the production cycles of final products at each firm which are multiple of each other.

$$I_{ik}^{m} = \left(P_{ik}^{m} - D_{ik}^{m}\right) t_{ik}^{2} = F_{ik}^{m} T_{ik}^{m} D_{ik}^{m}\left(1 - D_{ik}^{m}/P_{ik}^{m}\right) \tag{5}$$

$$B_{ik}^{m} = D_{ik}^{m} t_{ik}^{4} = \left(1 - F_{ik}^{m}\right) T_{ik}^{m} D_{ik}^{m}\left(1 - D_{ik}^{m}/P_{ik}^{m}\right) \tag{6}$$

$$Q_{ik}^{m} = I_{ik}^{m} + B_{ik}^{m} = F_{ik}^{m} T_{ik}^{m} D_{ik}^{m}\left(1 - D_{ik}^{m}/P_{ik}^{m}\right) + \left(1 - F_{ik}^{m}\right) T_{ik}^{m} D_{ik}^{m}\left(1 - D_{ik}^{m}/P_{ik}^{m}\right) \tag{7}$$

$$Min \quad TC_{EPQ}^{m}\left(T_{ik}^{m}, F_{ik}^{m}\right) = \sum_{i}\sum_{k} \frac{Co_{ik}^{m}}{T_{ik}^{m}} + \sum_{i}\sum_{k}\left(\frac{Ch_{ik}^{m} D_{ik}^{m} T_{ik}^{m} F_{ik}^{m^2}}{2}\right)\left(1 - D_{ik}^{m}/P_{ik}^{m}\right) + \sum_{i}\sum_{k}\left(\frac{Cb_{ik}^{m} D_{ik}^{m} T_{ik}^{m}\left(1 - F_{ik}^{m}\right)^2}{2}\right)\left(1 - D_{ik}^{m}/P_{ik}^{m}\right) \tag{8}$$

$$K_{ik}^{m} x_{ii'k} T_{ik}^{m} D_{ik}^{m}/P_{ik}^{m} = K_{i+l,k}^{m} x_{i+l,i',k} T_{i+l,k}^{m} D_{i+l,k}^{m}/P_{i+l,k}^{m} \quad \forall \; i', k, i, l = 1, \dots, N - i\,; \quad x_{ii'k} \neq 0 \tag{9}$$

3.3.2. Inventory management on raw materials at the manufacturing plants

As mentioned earlier, various suppliers provide manufacturing plants demands in order to enhance flexibility and responsibility in the turbulent and rapidly changing market. In this regard, suppliers' agility levels are assessed to elicit suitable suppliers applying a fuzzy-ANP framework which is developed according to the relevant criteria of agile SCs.

3.3.2.1. The proposed fuzzy-ANP framework. In the supposed network, raw materials which are consumed in producing the final products at firms are provided by variant suppliers so that the network flexibility is obtained. In order to achieve a more flexible and responsible SC, supplier's agility level is investigated applying analytic network process (ANP), leading to elicit agile suppliers. In this regard, an ANP framework is developed to assess suppliers' agility level. Since most of the decisions made in diverse contexts are driven upon expert preferences, fuzzy sets theory is utilized to overcome the uncertainty and ambiguity of the preference-based comparisons of decision-makers as well.

For the sake of brevity, description of the developed the proposed fuzzy-ANP model is skipped and the readers are referred to Rafiei and Rabbani (2011). Additionally, decision criteria are adopted from the agility context with respect to the agile supplier's characteristics (Christian et al., 2001; Lin et al., 2006; Sharifi & Zhang, 1999; Tsourveloudis &Valavanis, 2002). In this regard, decision criteria and their corresponding clusters are listed in Table 1, where the network structure of the proposed ANP model is shown in Figure 5.

Having the proposed fuzzy-ANP model implemented to assess suppliers' agility level, the related value to the alternatives' (suppliers) agility is concluded. The obtained values are normalized to the weights, $q_{i'kz}$. In this regard, $q_{i'kz}$ is introduced as the percentage of order quantity of each raw material at each manufacturing plant which is provided through each supplier. Since $q_{i'kz}$ = the value of supplier's agility level/∑supplier the value of supplier's agility level, then $0 \leq q_{i'kz} \leq 1$, and $\sum_k q_{i'kz} = 1$; hence,

Table 1. Agility measurement factors

Cluster	Criteria	Code
Technology/process	Manufacturing speed	T1
	Products design and manufacture flexibility	T2
	Operations efficiency and effectiveness (leanness)	T3
	Appropriate/new technology (hard and soft)	T4
	Skill and knowledge enhancing technologies	T5
	IT skills and knowledge	T6
Product	Product flexibility	P1
	Product/services quality	P2
	Product volume flexibility	P3
Market	Products and services delivery quickness and timelines	M1
	Customer-driven innovations	M2
	Production support, production, supply & distribution, and new product introduction responsiveness	M3
	Inventory turnover	M4
	Customer satisfaction	M5
Employee	Multi-skilled and flexible people	E1
	Employee skills utilization	E2
	Continuous training and development	E3
	Learning organization	E4
	Employee satisfaction	E5
Information/organization management	Networking	I1
	Perfect degree of enterprise information system	I2
	Information network extensiveness	I3
	Information accessible to Beneficiary	I4
	Change management	I5
	The degree of cooperation with other enterprises	I6
	Concurrent execution of activities	I7

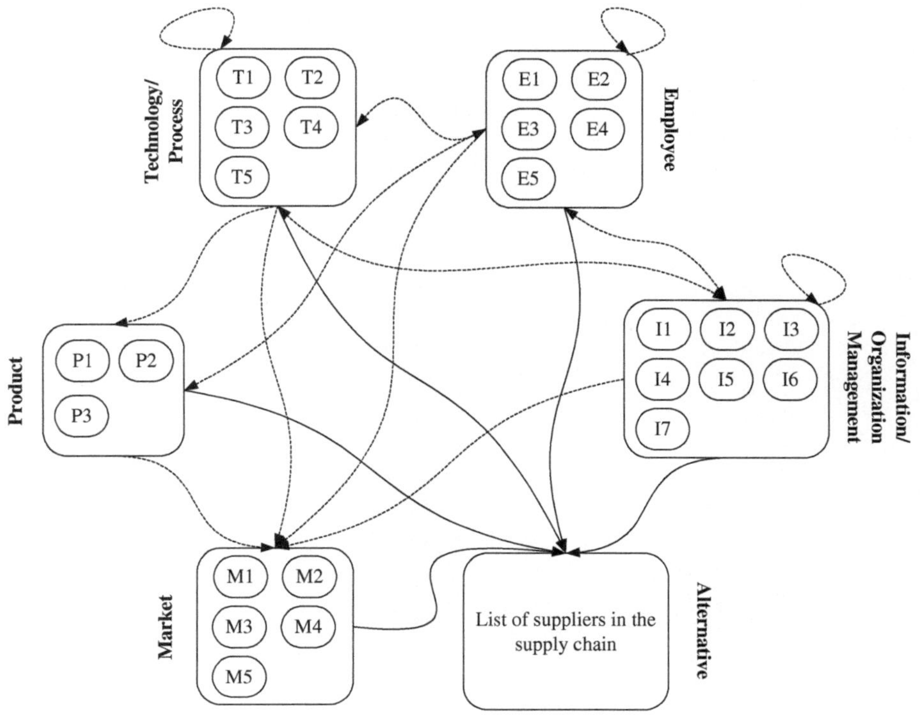

Figure 5. ANP network to rank suppliers.

the order quantity for raw materials at the firms is split and forwarded to different suppliers upon the calculated weights ($Q^m_{i'kz} = q_{i'kz} Q^m_{i'k}$). Based upon this method, firms' demand could be satisfied rapidly from different suppliers.

3.3.2.2. The EOQ model. The EOQ model with complete backordering considers each manufacturing plant with a constant demand rate $D^m_{i'k}$ ($i' = 1, \ldots, n$; $k = 1, \ldots, nm$). The proposed manufacturing plants have finite capacity to store raw materials. Based on the supposed EOQ policy, manufacturing plant k places an order for raw material i' to supplier z. Although, the order quantities placed by firms might vary from one firm to another, the order quantity of any raw material i' at any manufacturer k is assumed equal. In this regard, this section is developed to seek the length of an order cycle of raw material i' at manufacturing plant k; $T^m_{i'k}$ and the fill rate or of demands for raw material i' at manufacturing plant k from supplier z that will be filled from stock; $F^m_{i'kz}$ due to minimize the corresponding inventory system costs. To do so, the summation of holding costs, ordering and backordering costs of raw materials is assumed as objective function. Additionally, it is assumed the length of order cycle of each raw material at each firm is an integer multiple of the length of production cycle of the final products which consume that raw material.

Figure 6 illustrates the inventory level of EOQ problem with complete backordering for raw material i' at manufacturing plant k from supplier z. As shown in Figure 6, the summation of the demand time and the shortage time corresponds to the length of an order cycle of raw material i' at manufacturing plant k from supplier z (i.e. $q_{i'kz} T^m_{i'k} = t^1_{i'kz} + t^2_{i'kz}$), where $t^1_{i'kz} = F^m_{i'kz} q_{i'kz} T^m_{i'k}$; $t^2_{i'kz} = \left(1 - F^m_{i'kz}\right) q_{i'kz} T^m_{i'k}$.

As depicted in Figure 6, $I^m_{i'kz}$ is the maximum inventory level of raw material i' at manufacturing plant k from supplier z with $I^m_{i'kz} = D^m_{i'k} F^m_{i'kz} q_{i'kz} T^m_{i'k}$, $S^m_{i'kz}$ is the maximum stockout quantity of raw material i' at manufacturer k from supplier z with $S^m_{i'kz} = D^m_{i'k} \left(1 - F^m_{i'kz}\right) q_{i'kz} T^m_{i'k}$, $B^m_{i'kz}$ is the maximum backorder position of raw material i' at manufacturer k from supplier z with $B^m_{i'kz} = S^m_{i'kz} = D^m_{i'k} \left(1 - F^m_{i'kz}\right) q_{i'kz} T^m_{i'k}$. The EOQ model with complete backordering is a particular case

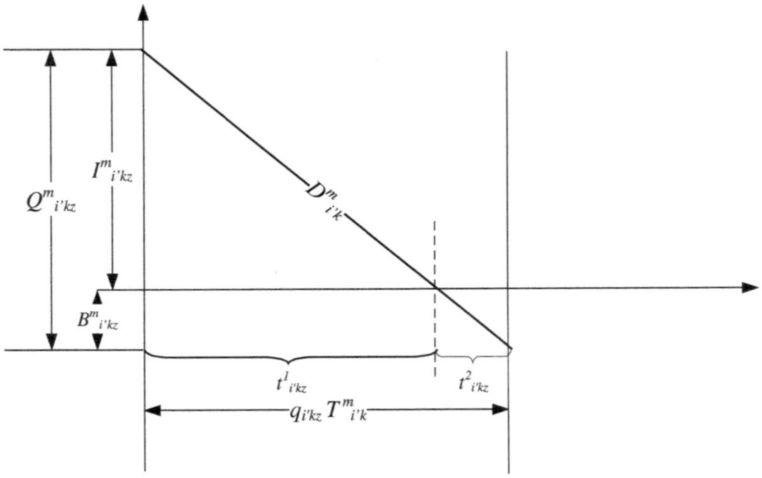

Figure 6. The inventory level of EOQ problem with full backordering for at the firm level.

of EOQ model with partial backordering in which the backordering rate of product that will be backordered equals one. Further, the order quantity of each product at each firm from each supplier is yield applying Equation 10 with respect to the values of $I^m_{i'kz}$ and $B^m_{i'kz}$.

$$Q^m_{i'kz} = D^m_{i'k}\,F^m_{i'kz}\,q_{i'kz}\,T^m_{i'k} + D^m_{i'k}\left(1 - F^m_{i'kz}\right)q_{i'kz}\,T^m_{i'k} \tag{10}$$

The order quantity, the maximum inventory level, and the maximum backorder position of raw material i' at manufacturer k could be found using $Q^m_{i'k} = \sum_z Q^m_{i'kz}$; $I^m_{i'k} = \sum_z I^m_{i'kz}$; $B^m_{i'k} = \sum_z B^m_{i'kz}$.

Based on Equation 11, the average cost of manufacturing plants' inventory under EOQ policy per year is obtained.

$$\begin{aligned} Min \quad TC^m_{EOQ}\left(T^m_{i'k}, F^m_{i'kz}\right) &= \sum_{i'}\sum_{k}\sum_{z}\frac{Co^m_{i'k}}{q_{i'kz}\,T^m_{i'k}} \\ &+ \sum_{i'}\sum_{k}\sum_{z}\frac{Ch^m_{i'k}\,D^m_{i'k}\,q_{i'kz}\,T^m_{i'k}\,F^{m^2}_{i'kz}}{2} + \sum_{i'}\sum_{k}\sum_{z}\frac{Cb^m_{i'k}\,D^m_{i'k}\,q_{i'kz}\,T^m_{i'k}\left(1-F^m_{i'kz}\right)^2}{2} \end{aligned} \tag{11}$$

3.4. Problem formation

As mentioned earlier, it was supposed the order interval of each final product at each retailer is an integer multiple of each other. Also, length of the production cycle of final products at each manufacturing plant is assumed to be an integer multiple of each other. Having the various inventory policies integrated for the sake of total SC cost reduction, some relationships between different levels of the network are taken into consideration. The characteristics at the upper levels have a relationship with the lower level ones. These associations comprise of relation between the value of order and planning cycle, demand, and shortage of final products as well as raw materials. Figures 7 and 8 illustrate the relationships among the order/planning cycle at retailer and firm levels, and order/production cycle at firm level, respectively.

In this regard, mathematical model in Equations 13–20 is proposed.

$$min \quad TC = TC^r_{EOQ} + TC^m_{EOQ} + TC^m_{EPQ} \tag{12}$$

s.t:

$$\sum_{i'=1}^{N} f_{i'k}\,D^m_{i'k}\,T^m_{i'k} \le f_k \qquad \forall \quad k \tag{13}$$

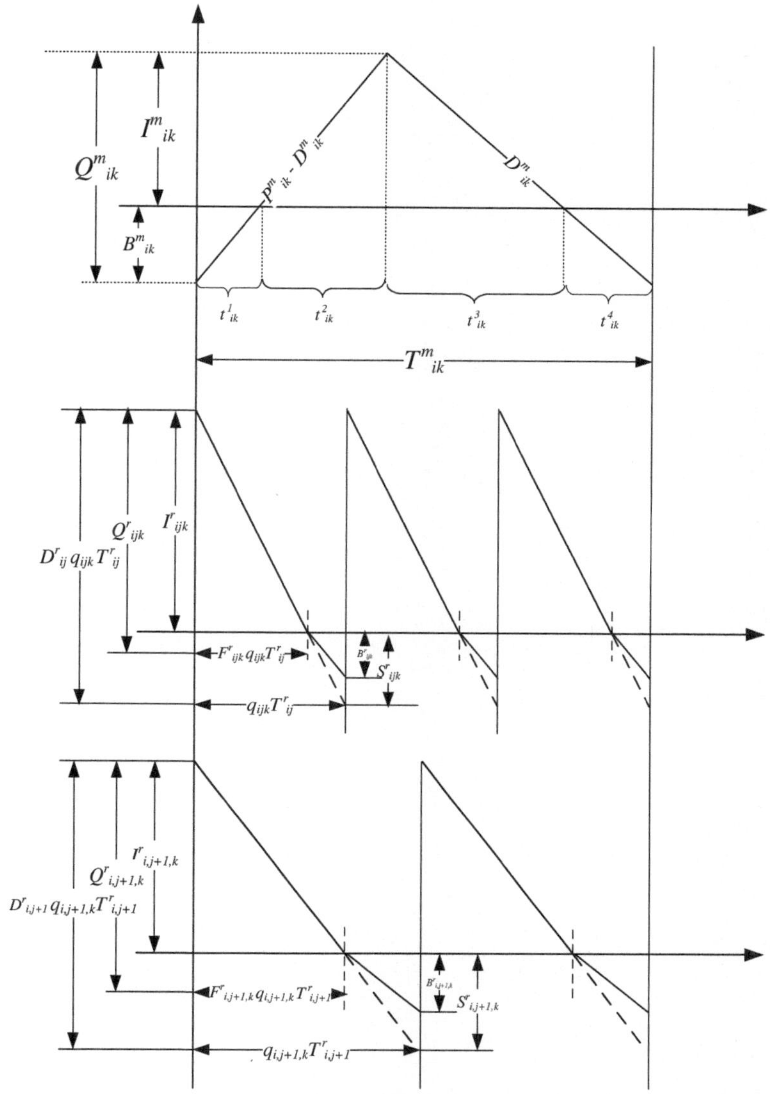

Figure 7. The combined inventory profile for the level of the retailer and manufacturing plant.

$$D^m_{ik} = \sum_{j=1}^{nr} q_{ijk} D^r_{ij}; \quad \forall i, k \tag{14}$$

$$D^m_{i'k} = \sum_{i=1}^{N} cc_{ii'k} D^m_{ik}; \quad \forall i', k \tag{15}$$

$$T^m_{ik} = K^r_{ijk} q_{ijk} T^r_{ij} = K^r_{i,j+l,k} q_{i,j+l,k} T^r_{i,j+l} \qquad \forall \quad i,k,j,l = 1,\ldots, nr - j \tag{16}$$

$$T^m_{i'k} = K^m_{ik} x_{ii'k} T^m_{ik} D^m_{ik} / P^m_{ik} = K^m_{i+l,k} x_{i+l,i',k} T^m_{i+l,k} D^m_{i+l,k} / P^m_{i+l,k} \quad \forall \quad i', k, i, l = 1, \ldots, N - i; \quad x_{ii'k} \neq 0 \tag{17}$$

$$B^m_{ik} = \max\left\{ \left[B^m_{i'k} \times cc_{ii'k} / \sum_i cc_{ii'k} \right]; \quad \forall i' \right\}; \quad \forall i, k \tag{18a}$$

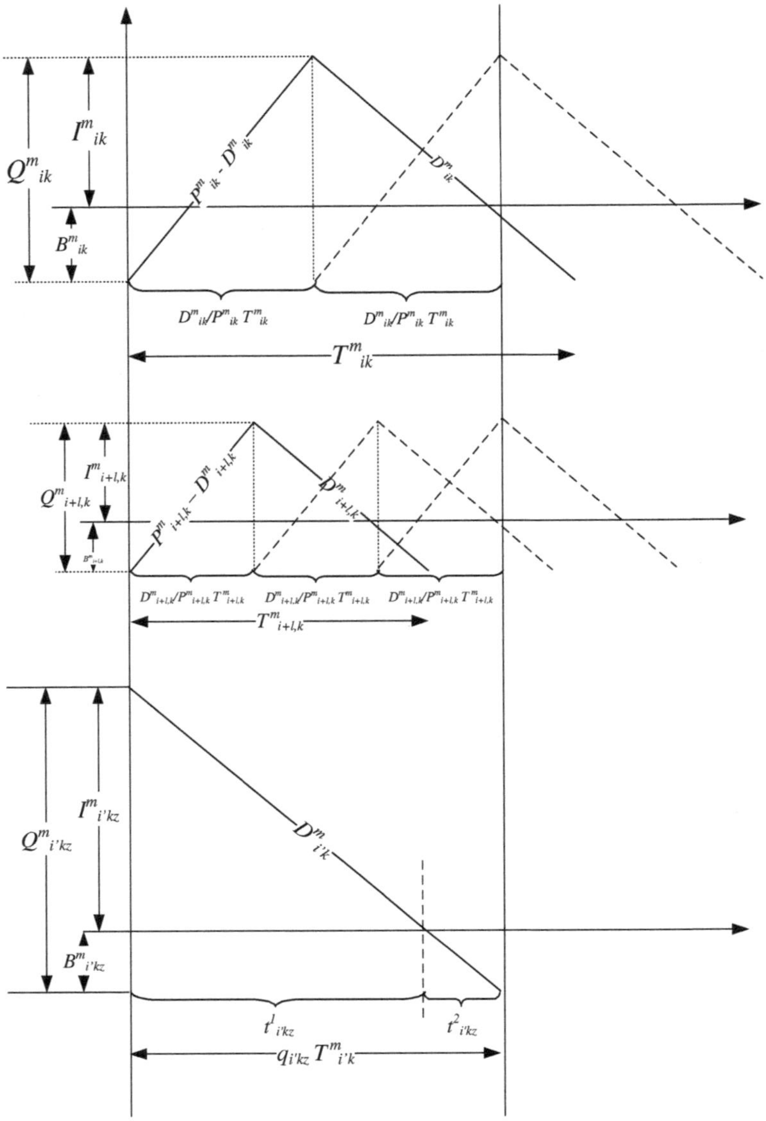

Figure 8. The combined inventory profile for the level of manufacturing plant.

$$\left(1-F^m_{ik}\right) T^m_{ik}\, D^m_{ik} \left(1-{D^m_{ik}}/{P^m_{ik}}\right) = \max\left\{ \left[\left(\sum_z D^m_{i'k} \left(1-F^m_{i'kz}\right) q_{i'kz}\, T^m_{i'k} \right) \times {cc_{ii'k}}/{\sum_i cc_{ii'k}} \right];\quad \forall i' \right\};\quad \forall i,k \tag{18b}$$

$$B^r_{ij} = \sum_{k=1}^{nm} q_{ijk}\, B^m_{ik};\quad \forall i,j \tag{19a}$$

$$\sum_{k=1}^{nm} \beta^r_{ijk}\, D^r_{ij} \left(1-F^r_{ijk}\right) q_{ijk}\, T^r_{ij} = \sum_{k=1}^{nm} q_{ijk} \left(1-F^m_{ik}\right) T^m_{ik}\, D^m_{ik} \left(1-{D^m_{ik}}/{P^m_{ik}}\right) \quad \forall i,j \tag{19b}$$

$$T^r_{ij}, T^m_{ik}, T^m_{i'k}, q_{ijk}, q_{i'kz}, F^r_{ijk}, F^m_{i'kz}, F^m_{ik} \in [0,1];\quad D^m_{i'k} D^m_{ik}, TC > 0;$$
$$K^r_{ijk}, K^m_{ik} > 0 \quad \text{and} \quad \text{integer};\qquad x_{ii'k} = 0,1;\quad \forall i,i',j,k,z \tag{20}$$

In the proposed model, the objective function seeks to minimize total average cost of SC per year given in Equation 12. As mentioned earlier, Constraints Equation 13 is devoted to meet the warehouse capacity limitation at the firms. As shown in Equations 14 and 15, demand rate of final products at the retailer stage influences on the demand rate of final products as well as raw material at the firm stage. Equation 16 expresses the length of planning cycle of each final product at each firm is assumed to be an integer multiple of the order interval of final product at each retailer from the firm. Equation 17 shows the associations between the relevant order and production cycles of the manufacturing firms. Based upon Equation17, the length of order cycle of each raw material at each firm is an integer multiple of the length of production cycle of the final products which consume that raw material. Since, all raw materials are used with different consumption coefficients in producing the finished product. Hence, deficiency on final products at manufacturing plant level and retailer level depends on the shortage of raw materials as shown in Equations 18 and 19, respectively. Finally, Equation 20 demonstrates the real and integer variables.

4. Solution methodology

This section is developed in order to provide a proper methodology to tackle the computational complexity of the problem. To do so, a hybrid metaheuristic algorithm including a differential evolution (DE) algorithm and a particle swarm optimization (PSO) algorithm is applied. The algorithm starts through DE with random initial population up to the selection step where the current population is compared with its corresponding trial vector based on their objective function values. If the trial vector satisfies the conditions given by Equation 25, then it is copied to the next generation; otherwise, it creates a new solution by conducting the PSO phase. Having the PSO phase implemented, a perturbation is created in the population, leading to maintain the diversity of the population as well as to reach a good optimal solution. In advanced, the pseudo-code of the hybrid optimization algorithm is briefly presented in Figure 9.

4.1. Differential evolution algorithm

DE was firstly introduced by Storn and Price in 1995 as a powerful heuristic method for solving nonlinear, non-differentiable, and multi-modal optimization problems. This technique has been structured based upon a combination of simple arithmetic operators, the classical crossover, mutation, and selection operators. In this method, the purpose of mutation and crossover is to generate new vectors, and the vectors will survive to the next generation are determined upon selection operator (Damak, Jarboui, Siarry, & Loukil, 2009).

Broad scope and applicability of DE have been challenged in diverse disciplines, such as digital filter design, batch fermentation process, dynamic optimization of continuous polymer reactor, optimization of heat exchangers, optimization of non- linear functions, fuzzy decision-making problems of fuel ethanol production, and multi sensor fusion. (Routroy & Kodali, 2005). Moreover, Falcone, Lopes, and Coelho (2008) have highly recommended DE for the SC optimization problem as well as for other logistics-related problems due to its robustness. Advantages of DE such as a simple structure, ease of use, a great robustness, and a fast convergence lead to promote its reputation in many fields (Damak et al., 2009; Routroy & Kodali, 2005). A brief description of the developed DE strategy is elaborated as followings:

4.1.1. Population structure

DE starts with randomly generated X^G_i ($i = 1, ..., NP$) vector to create the initial population, where the index i represents the *ith* solution of population, and G denotes the current generation. Supposing the index j indicates the *jth* element in the solution vector which contains D elements. In addition, NP indicates the population size. The initial population and the vectors that belong to it are, respectively, represented as follows:

$$P^G = X_i^G \quad \text{with} \quad i = 1, ..., NP \tag{21}$$

```
Begin
  First step:
    Initialise parameters value of DE algorithm, population size, control parameter
        of the evolution rate, crossover rate,
    Initialise parameters value of PSO algorithm, inertial weight, the weight of the
        stochastic acceleration terms (individual and plural learning), a spread of
        searching space and maximum iteration;
  Second step: production-inventory planning
  Initialise the population
  P^G = X_i^G = X_{i,j}^G   with  i = 1,...,NP;   j = 1,...,D;  and  G = 1,...,G_max;
    Repeat{
            Until (the DE's population size is not met; NP)
            Randomly select r1, r2, r3 ∈ {1,...,NP} such that r1 ≠ r2 ≠ r3≠ i;
              V_i^G = X_{r1}^G + F × (X_{r2}^G − X_{r3}^G);
              Repeat{
                    Until (the problem dimension is not met; D)
                    Randomly select k ∈ {1, ..., D};
                    Create rand_j ∈ [0,1];
                        If rand_j < CR   or  j = k
                            U_{i,j}^G = V_{i,j}^G;
                        Otherwise
                            U_{i,j}^G = X_{i,j}^G;
                        End
                              If f(U_i^G) ≤ f(X_i^G)
                                  X_i^{G+1} = U_i^G;
                              Else
                                  PSO activated
                                  Create a new particle applying (26), (27),
                                  namely TX_i;
                                      If f(TX_i) ≤ f(X_i^G)
                                          X_i^{G+1} = TX_i
                                      Else
                                          X_i^{G+1} = X_i^G
                                      End
                              End
              }
    }
  Third step: output optimization results
End
```

$$X_i^G = X_{i,j}^G \quad \text{with} \quad j = 1, ..., D; \quad \text{and} \quad G = 1, ..., G_{max} \tag{22}$$

Figure 9. Pseudo-code of the applied hybrid optimization algorithm.

As shown in Figure 10, the initial population of the DE algorithm can be represented as a string of element as people.

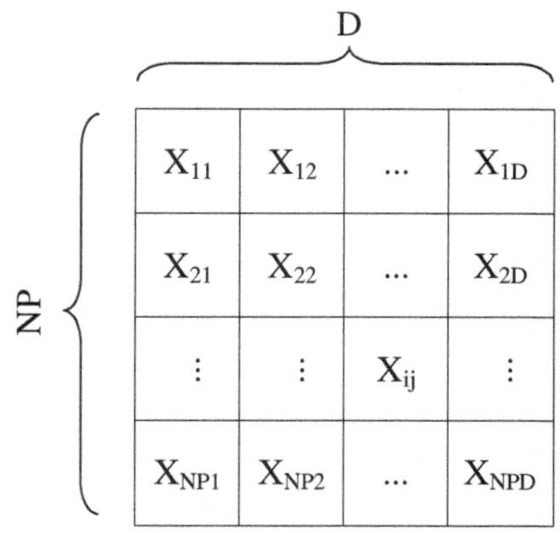

Figure 10. The general schematic of the population in DE.

The population size as a main factor of the DE has a significant role in the algorithm convergence speed. Whereas too small size leads to insufficient exploration, on contrary, too large size leads to increase the number of function evolutions upon which the algorithm convergence speed is negatively affected.

4.1.2. Mutation

After initialization of the population, mutation operator starts with selecting three vectors $\{X_{r1}, X_{r2}, X_{r3}\}$ randomly in the population set of *NP* elements such that $r1 \neq r2 \neq r3 \neq i$. The selected vectors are combined to create a mutant vector as depicted in Equation 23 in which *F* is the control parameter of the evolution rate such that $F \epsilon [0, 1]$.

$$V_i^G = X_{r1}^G + F \times \left(X_{r2}^G - X_{r3}^G\right); \quad i = 1, \ldots, NP, \quad r1, r2, r3 \in \{1, \ldots, NP\} \tag{23}$$

4.1.3. Crossover

Once the mutant vector is generated, crossover operator attempts to create a trial vector applying the following equation:

$$U_{i,j}^G = \begin{cases} V_{i,j}^G & \text{if} \quad rand_j \leq; CR \quad j = k \\ X_{i,j}^G & \text{otherwise} \end{cases} \tag{24}$$

where $j = 1, \ldots, D$, $k \in \{1,\ldots,D\}$ indicates the random parameter's index, chosen once for each *i*. $CR \in [0, 1]$ is the crossover factor upon which the crossover of various elements is executed. And, *r* and *j* are uniform random numbers which are compared with value of the crossover rate to determine the element of the trial vector. If the value of the random number is less or equal to crossover rate, the trial element is copied from the mutant vector; otherwise, it is selected from the current vector.

4.1.4. Selection

The selection step is performed by comparing the current population with its corresponding trial vector based on their objective function values. As shown in Equation 25, if the objective function value of the trial vector is less than the current population's one, it takes place in the next generation; otherwise, the PSO algorithm is activated.

$$X_i^{G+1} = \begin{cases} U_i^G & \text{if} \quad f(U_i^G) \leq f(X_i^G) \\ \text{PSO activated} & \text{otherwise} \end{cases} \tag{25}$$

As a result, all solutions of the next generation are as good as or better than the current population. Once the current generation is updated, mutation, crossover, and selection operators are repeated until the optimum, or the predefined termination criterion is met.

4.2. Particle swarm optimization

The presented PSO by Kennedy and Eberhart (1995) is an evolutionary calculations technique. The mechanism of PSO has been driven from the simulation of social behavior of flying birds which exchange their information to solve optimization problems. The PSO's solution is a set of particle in the position of search space. Each particle maintains information on itself experience as well as other particles' ones upon which moves toward the best point in the search space. To do so, two basic equations including velocity vector and position vector have been introduced as follows:

$$V_i(t+1) = w\,V_i(t) + c_1\,r_1\left(x_{ibest}(t) - x_i(t)\right) + c_2\,r_2\left(x_{gbest}(t) - x_i(t)\right) \tag{26}$$

$$x_i(t+1) = x_i(t) + V_i(t+1) \tag{27}$$

The velocity vector comprises of three parts; the first part refers to the inertia of the previous velocity. The second part, namely, cognition component, presents the personal thinking of the particle. And, the third one, namely, social component, presents the cooperation among the particles. $V_i(t)$ and $x_i(t)$indicate the velocity and the position of *ith* particle in *tth* iteration, respectively. x_{ibest} and x_{gbest} represent the best previous position of each particle, and the best position among all the particles in the swarm, respectively. w as the inertial weight has a role of balancing the global exploration and the local exploitation abilities of the swarm. $c1$ and $c2$ represent the weight of the stochastic acceleration terms, and $r1$ and $r2$ as the random functions in order to search the space in a better manner.

5. Numerical experiments

This section is developed to show validity of the proposed methodology tackling complexity of the proposed model. To do so, several instances in different sizes are extended. Table 2 corresponds to the generated problems in which different numbers of final products, raw materials, retailers, manufacturing firms, and suppliers are considered. The numerical data of the final products and raw materials at each level of the SC are a modified version from the ones given in Pasandideh, Niaki, and Nia (2011). The related data of problem 4 are only presented in Tables 3–5 for the sake of brevity. The initial values of the DE parameters including the crossover factor (*CR*), the control parameter of the evolution rate (*F*), and the population size (*NP*) are chosen to be 0.6, 0.7, and 20 by testing the

Table 2. The relevant data of the instances size

Index Problem No.	i	i'	j	k	z
1	2	3	2	2	2
2	2	3	4	3	3
3	3	5	4	3	3
4	5	5	4	3	5
5	5	5	4	4	5

Table 3. Relevant date of final products at the retailer's level

Final prod.	Retailer No.																			
	1					2					3					4				
	D^r_{ij}	Co^r_{ij}	Ch^r_{ij}	Cb^r_{ij}	Cl^r_{ij}	D^r_{ij}	Co^r_{ij}	Ch^r_{ij}	Cb^r_{ij}	Cl^r_{ij}	D^r_{ij}	Co^r_{ij}	Ch^r_{ij}	Cb^r_{ij}	Cl^r_{ij}	D^r_{ij}	Co^r_{ij}	Ch^r_{ij}	Cb^r_{ij}	Cl^r_{ij}
1	420	1	5	3	10	240	2	4	4	8	480	3	6	8	5	510	5	2	3	6
2	160	2	9	4	8	380	1	8	5	9	390	2	5	3	9	430	4	3	8	10
3	340	4	7	2	6	120	4	9	3	5	530	1	3	9	2	280	3	6	4	6
4	180	1	8	6	4	410	3	5	4	6	230	5	7	5	4	390	1	4	6	7
5	230	6	3	1	8	180	2	6	9	7	480	4	8	2	3	320	2	8	5	8

Table 4. Relevant data of final product at the manufacturing plant's level

Final product	Manufacturing No.											
	1				2				3			
	P^m_{ik}	Co^m_{ik}	Ch^m_{ik}	Cb^m_{ik}	P^m_{ik}	Co^m_{ik}	Ch^m_{ik}	Cb^m_{ik}	P^m_{ik}	Co^m_{ik}	Ch^m_{ik}	Cb^m_{ik}
1	2300	3	5	2	2000	2	3	6	1950	5	9	4
2	1900	4	8	4	2050	1	8	4	1850	3	4	3
3	2050	2	1	5	2300	4	7	5	2040	3	5	8
4	1590	5	4	6	1600	3	4	3	1650	4	3	4
5	2250	2	6	1	2100	2	2	6	2300	1	6	7

Table 5. Relevant data of raw material at the manufacturing plant's level.

Raw material No.	Manufacturing plant No.												Final product No.				
	1				2				3				1	2	3	4	5
	$Co^m_{i'k}$	$Ch^m_{i'k}$	$Cb^m_{i'k}$	$f_{i'k}$	$Co^m_{i'k}$	$Ch^m_{i'k}$	$Cb^m_{i'k}$	$f_{i'k}$	$Co^m_{i'k}$	$Ch^m_{i'k}$	$Cb^m_{i'k}$	$f_{i'k}$	$cc_{ii'}$	$cc_{ii'}$	$cc_{ii'}$	$cc_{ii'}$	$cc_{ii'}$
	2	5	5	2	1	8	5	2	5	3	7	2	2	3	1	2	4
	4	7	7	2	5	8	6	2	3	5	3	2	3	5	0	4	2
	6	8	9	4	3	1	3	4	3	6	4	4		0	3	1	2
	1	5	4	3	2	2	5	3	4	9	2	3	4	1	1	2	0
	2	6	2	1	5	3	2	1	2	7	5	1	0	0	3	0	2

Notes: f_k=14000; for all manufacturing plants, $f_{i'1} = f_{i'2} = \ldots = f_{i'k} = f_{i'}, cc_{ii'1} = cc_{ii'2} = \ldots = cc_{ii'k} = cc_{ii'}$.

problem, respectively. It is noted that the capability and the convergence speed of the proposed methodology lead to coverage the solution to the near optimal point after 30 iterations even for high iteration than 30. Then, in order to reduce the time of problem-solving, iteration factor is considered 30 in report.

And, the initial values of PSO parameters are selected as follows: the inertial weight is calculated according to $w = \chi = 2\Big/\left(\varphi - 2 + \sqrt{\varphi^2 - 4\varphi}\right)$ in the first iteration, where $\varphi = \varphi_1 + \varphi_2$, φ_1 and $\varphi_2$are similarly selected as 2.05; and gradually reduced by Equation 28, in which *wdamp* is supposed to be 0.99. Moreover, the weight of the stochastic acceleration terms are obtained applying $c_1 = 0.001\varphi_1\chi$, and $c_2 = 0.001\varphi_2\chi$. It is noted that the algorithm is stopped after 30 iterations.

$$w = w \times wdamp \tag{28}$$

5.1. Numerical sample of the proposed fuzzy-ANP framework

As mentioned in Section 3, the weights of order quantity for raw materials at manufacturing plants from each supplier are determined applying the proposed fuzzy-ANP framework. In this section, we implement the proposed structure only for problem 5 to demonstrate the application of the proposed ANP framework. In this regard, five manufacturing companies/suppliers are considered as alternatives. For instance, Tables 6 and 7 demonstrate the relevant judgments about alternatives with respect to criterion *Product flexibility* and the clusters with respect to cluster *Technology/Process-related*, respectively; the weights of comparison matrices are obtained as: 0.244, 0.166, 0.093, 0.063, and 0.435 for alternatives 1, 2, 3, 4, and 5, respectively. With respect to data in Table 7, the resulted weights are 0.362, 0.062, 0.044, 0.150, 0.106, and 0.276 for technology/process related, product related, market related, employee related, and information/organization management related, respectively. After performing the developed model and with respect to the results obtained in limit supermatrix, final weights $W = (0.0537, 0.0240, 0.0094, 0.0074, 0.0631) \times 10^{-3}$ are calculated.

Table 6. Judgment about alternatives with respect to criteria product flexibility

	A1	A2	A3	A4	A5
A1	(1, 1, 1)	(1, 5/2, 4)	(2, 4, 6)	(3, 5, 7)	(1/4, 5/8, 1)
A2	(1/4, 5/8, 1)	(1, 1, 1)	(1, 5/2, 4)	(2, 4, 6)	(1/6, 1/3, 1/2)
A3	(1/6, 1/3, 1/2)	(1/4, 5/8, 1)	(1, 1, 1)	(1, 5/2, 4)	(1/7, 5/21, 1/3)
A4	(1/7, 5/21, 1/3)	(1/6, 1/3, 1/2)	(1/4, 5/8, 1)	(1, 1, 1)	(1/8, 3/16, 1/4)
A5	(1, 5/2, 4)	(2, 4, 6)	(3, 5, 7)	(4, 6, 8)	(1, 1, 1)

Table 7. Judgment about clusters with respect to cluster technology/process related

	T	P	M	E	I	A
T	(1, 1, 1)	(4, 6, 8)	(5, 7, 9)	(2, 4, 6)	(3, 5, 7)	(1, 5/2, 4)
P	(1/8, 3/16, 1/4)	(1, 1, 1)	(1, 5/2, 4)	(1/6, 1/3, 1/2)	(1/4, 5/8, 1)	(1/7, 5/21, 1/3)
M	(1/9, 7/45, 1/5)	(1/4, 5/8, 1)	(1, 1, 1)	(1/7, 5/21, 1/3)	(1/6, 1/3, 1/2)	(1/8, 3/16, 1/4)
E	(1/6, 1/3, 1/2)	(2, 4, 6)	(3, 5, 7)	(1, 1, 1)	(1, 5/2, 4)	(1/4, 5/8, 1)
I	(1/7, 5/21, 1/3)	(1, 5/2, 4)	(2, 4, 6)	(1/4, 5/8, 1)	(1, 1, 1)	(1/6, 1/3, 1/2)
A	(1/4, 5/8, 1)	(3, 5, 7)	(4, 6, 8)	(1, 5/2, 4)	(2, 4, 6)	(1, 1, 1)

5.2. Computational results

The applied DE-PSO algorithm is coded in Matlab 7.8.0 (R2009a) software and run on a PC with 4.00 GB RAM and 2.00 Ghz processor. The results of objective function and elapsed time for the problem instances are summarized in Table 8. Each test problem has been run 15 times and the minimum objective function value along with its corresponding solutions is reported.

It must be noted that the supposed problem has also been evaluated applying the classic DE algorithm. And then, the proposed solution is applied to enhance the solution methodology performance. The results of the objective function for the problem instances which are solved by the DE algorithm are shown in Table 9, and compared with the results of Table 8. As shown in Table 9, in the most problems, the objective function values resulted upon the DE-PSO algorithm is yield less than the one, which are presented applying the classic DE algorithm.

Moreover, we evaluate the convergence rate of the developed algorithm in order to obtain a solution for all five considered problems. In this regard, the convergence trend of these solutions for each problem 1-5 is demonstrated in Figures 11-15, respectively. These figures illustrate capability of the DE-PSO algorithm in reaching a good solution in its early iterations as well as the reduction in the convergence speed by increasing the problem size.

Table 8. The value of objective functions and CPU time

	Problem No.				
	1	2	3	4	5
OFV	2602.4523	8607.4667	18378.9515	35538.3804	45438.0066
CPU time	2.60	7.45	9.33	14.72	19.75

Table 9. The comparison results of objective function value applying the DE and DE-PSO algorithms

	Problem No.				
	1	2	3	4	5
OFV-DE	2891.4194	9347.0349	17978.6103	39942.5499	50617.3324
OFV-DE-PSO	2602.4523	8607.4667	18378.9515	35538.3804	45438.0066

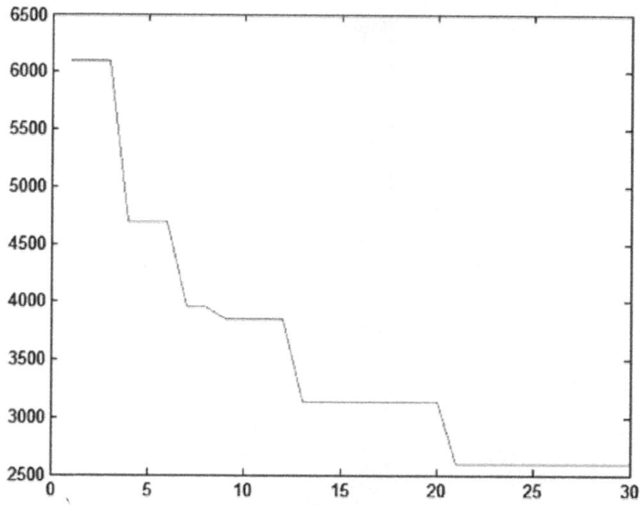

Figure 11. The convergence path (problem 1).

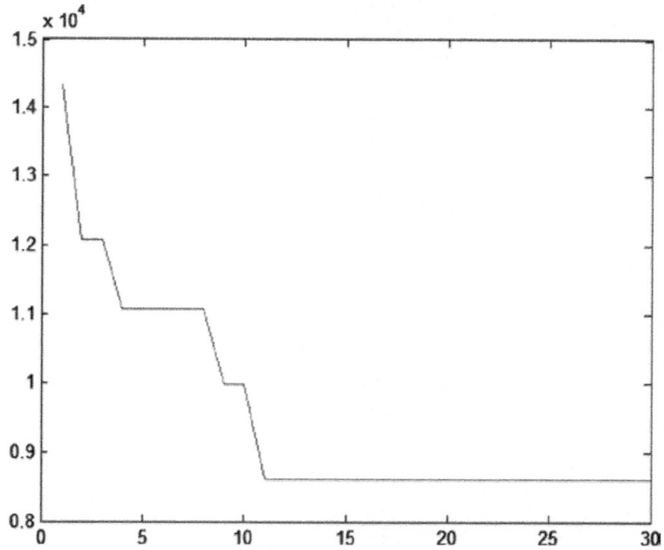

Figure 12. The convergence path (problem 2).

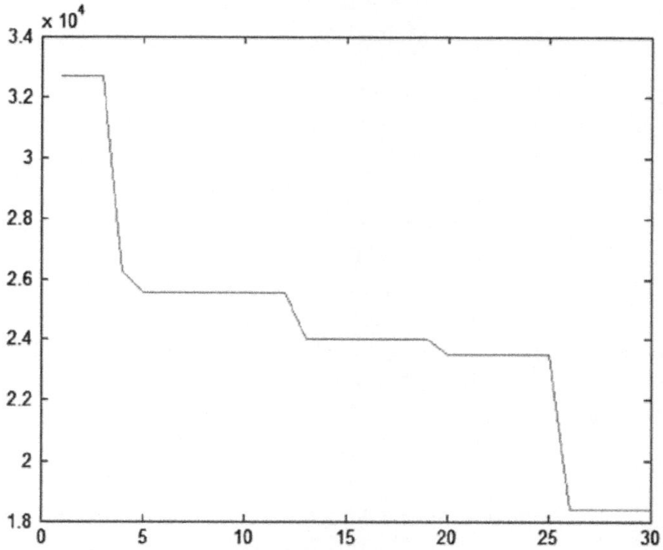

Figure 13. The convergence path (problem 3).

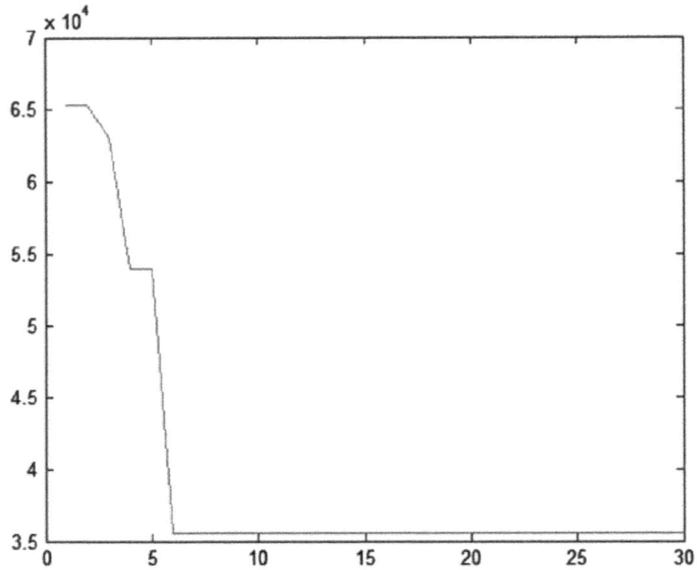

Figure 14. The convergence path (problem 4).

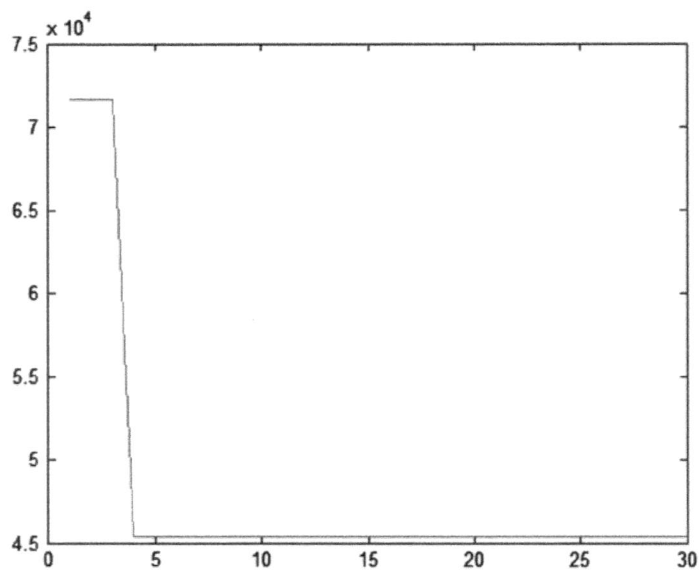

Figure 15. The convergence path (problem 5).

6. Conclusion and future research directions

This paper addressed a novel inventory management framework for a three-echelon multi-item SC having multiple suppliers, multiple manufacturing plants, and multiple retailers. In this regard, several inventory control policies corresponding to various levels have been proposed in an integrated point of view, leading to achieving the optimization of global SC. Moreover, in order to increase network flexibility and agility, different actors at each echelon were supposed to provide the lower levels' demands. To do so, a fuzzy-ANP model was developed to assess suppliers and elicit the suitable ones upon their agility level. Additionally, a mixed-integer non-linear mathematical model was proposed in order to coped with the inventory management problem. The proposed model was solved applying a hybrid DE–PSO algorithm to tackle the computational complexity of the problem.

To extend current direction of this paper, some suggestions are provided. First, other assumptions could be considered in SC to make it more practical, such as other kinds of inventory control policy, e.g. fixed/linear lead time, time costs. Second, modeling SC uncertainty could make it more realistic. Third, it is highly recommended to define other decision criteria in agile SCs, and other types of fuzzy numbers and preference methods, such as the eigenvalue method, the logarithmic least squares method, and the weighted least squares method, to determine the related weights of elements of pairwise comparison matrices and analyze them to elicit the best one. Lastly, developing other metaheuristics and heuristics to tackle complexity of the integrated inventory model is suggested.

Funding
The authors received no direct funding for this research.

Author details
Maryam Alimardani[1]
E-mail: m.alimardani@ut.ac.ir
Hamed Rafiei[1]
E-mail: hrafiei@ut.ac.ir
Masoud Rabbani[1]
E-mail: mrabani@ut.ac.ir
[1] School of Industrial Engineering, College of Engineering, University of Tehran, Tehran, Iran.

References

Banerjee, A., Kim, S.-L., & Burton, J. (2007). Supply chain coordination through effective multi-stage inventory linkages in a JIT environment. *International Journal of Production Economics, 108*, 271–280. http://dx.doi.org/10.1016/j.ijpe.2006.12.015

Ben-Daya, M., As'ad, R., & Seliaman, M. (2010). An integrated production inventory model with raw material replenishment considerations in a three layer supply chain. *International Journal of Production Economics.* doi:10.1016/j.ijpe.2010.10.024

Ben-Daya, M., & Al-Nassar, A. (2008). An integrated inventory production system in a three-layer supply chain. *Production Planning and Control, 19*, 97–104. http://dx.doi.org/10.1080/09537280801895930

Büyüközkan, G., & Çifçi, G. (2011). A novel fuzzy multi-criteria decision framework for sustainable supplier selection with incomplete information. *Computers in Industry, 62*, 164–174. http://dx.doi.org/10.1016/j.compind.2010.10.009

Clark, A. J., & Scarf, H. (1960). Optimal policies for a multi-echelon inventory problem. *Management Science, 6*, 475–490. http://dx.doi.org/10.1287/mnsc.6.4.475

Chan, C. K., & Kingsman, B. G. (2007). Coordination in a single-vendor multi-buyer supply chain by synchronizing delivery and production cycles. *Transportation Research Part E: Logistics and Transportation Review, 43*, 90–111. http://dx.doi.org/10.1016/j.tre.2005.07.008

Chen, Z. X., & Sarker, B. R. (2010). Multi-vendor integrated procurement-production system under shared transportation and just-in-time delivery system. *Journal of the Operational Research Society, 61*, 1654–1666. http://dx.doi.org/10.1057/jors.2009.115

Caggiano, K. E., Jackson, P. L., Muckstadt, J. A., & Rappold, J. A. (2009). Efficient computation of time-based customer service levels in a multi-item, multi-echelon supply chain: A practical approach for inventory optimization. *European Journal of Operational Research, 199*, 744–749. http://dx.doi.org/10.1016/j.ejor.2008.08.002

Christian, I., Ismail, H., Mooney, J., Snowden, S., Toward, M., & Zhang, D. (2001). Agile manufacturing transitional strategies. *Manufacturing information systems.* In *Proceedings of the Fourth SMESME International Conference,* Aalborg, Denmark.

Damak, N., Jarboui, B., Siarry, P., & Loukil, T. (2009). Differential evolution for solving multi-mode resource-constrained project scheduling problems. *Computers & Operations Research, 36*, 2653–2659.

Falcone, M. A., Lopes, H. S., & Coelho, L. D. S. (2008). Supply chain optimisation using evolutionary algorithms. *International Journal of Computer Applications in Technology, 31*, 158–167. http://dx.doi.org/10.1504/IJCAT.2008.018154

Goyal, S. K. (1976). An integrated inventory model for a single supplier-single customer problem. *International Journal of Production Research, 15*, 107–111.

Ganeshan, R., & Harrison, T. P. (1995). An introduction to supply chain management, Pennsylvania State University. *Supply Chain Management.* Retrieved from http://mason.wm.edu/faculty/ganeshan_r/documents/intro_supply_chain.pdf

Glock, C. H. (2012). The joint economic lot size problem: A review. *International Journal of Production Economics., 135*, 671–686. http://dx.doi.org/10.1016/j.ijpe.2011.10.026

Giannoccaro, I., Pontrandolfo, P., & Scozzi, B. (2003). A fuzzy echelon approach for inventory management in supply chains. *European Journal of Operational Research, 149*, 185–196. http://dx.doi.org/10.1016/S0377-2217(02)00441-1

Gümüs, A. T., & Güneri, A. F. (2007). Multi-echelon inventory management in supply chains with uncertain demand and lead times: literature review from an operational research perspective. *Proceedings of the Institution of Mechanical Engineers, Part B: Journal of Engineering Manufacture, 221*, 1553–1570. http://dx.doi.org/10.1243/09544054JEM889

Hill, R. M., Seifbarghy, M., & Smith, D. K. (2007). A two-echelon inventory model with lost sales. *European Journal of Operational Research, 181*, 753–766. http://dx.doi.org/10.1016/j.ejor.2006.08.017

Hoque, M. A. (2008). Synchronization in the single-manufacturer multi-buyer integrated inventory supply chain. *European Journal of Operational Research, 188*, 811–825. http://dx.doi.org/10.1016/j.ejor.2007.05.019

Joglekar, P. N., & Tharthare, S. (1990). The individually responsible and rational decision approach to economic lot sizes for one vendor and many purchasers. *Decision Sciences, 21*, 492–506. http://dx.doi.org/10.1111/deci.1990.21.issue-3

Kennedy, J. & Eberhart, R. (1995). Particle swarm optimization. *Proceeding of the 1995 IEEE International Conference on Neural Networks, 4*, 1942–1948.

Khouja, M. (2003). Optimizing inventory decisions in a multi-stage multi-customer supply chain. *Transportation Research Part E: Logistics and Transportation Review, 39*, 193–208. http://dx.doi.org/10.1016/S1366-5545(02)00036-4

Li, X., & Wang, Q. (2007). Coordination mechanisms of supply chain systems. *European Journal of Operational Research, 179*, 1–16. http://dx.doi.org/10.1016/j.ejor.2006.06.023

Lin, C-T., Chiu, H., & Tseng, Y. H. (2006). Agility evaluation using fuzzy logic. *International Journal of Production Economics., 101*, 353–368. http://dx.doi.org/10.1016/j.ijpe.2005.01.011

Mikhailov, L. (2002). Fuzzy analytical approach to partnership selection in formation of virtual enterprises. *Omega, 30*, 393–401. http://dx.doi.org/10.1016/S0305-0483(02)00052-X

Mitra, S. (2009). Analysis of a two-echelon inventory system with returns. *Omega, 37*, 106–115. http://dx.doi.org/10.1016/j.omega.2006.10.002

Pasandideh, S. H. R., Niaki, S. T. A., & Nia, A. R. (2011). A genetic algorithm for vendor managed inventory control system of multi-product multi-constraint economic order quantity model. *Expert Systems with Applications, 38*, 2708–2716. http://dx.doi.org/10.1016/j.eswa.2010.08.060

Pentico, D. W., & Drake, M. J. (2009). The deterministic EOQ with partial backordering: A new approach. *European Journal of Operational Research, 194*, 102–113. http://dx.doi.org/10.1016/j.ejor.2007.12.004

Pentico, D. W., Drake, M. J., & Toews, C. (2009). The deterministic EPQ with partial backordering: A new approach. *Omega, 37*, 624–636. http://dx.doi.org/10.1016/j.omega.2008.03.002

Rafiei, H., & Rabbani, M. (2011). Order partitioning and order penetration point location in hybrid make-to-stock/make-to-order production contexts. *Computers & Industrial Engineering, 61*, 550–560. http://dx.doi.org/10.1016/j.cie.2011.04.010

Rau, H., Wu, M.-Y., & Wee, H.-M. (2003). Integrated inventory model for deteriorating items under a multi-echelon supply chain environment. *International Journal of Production Economics, 86*, 155–168. http://dx.doi.org/10.1016/S0925-5273(03)00048-3

Routroy, S., & Kodali, R. (2005). Differential evolution algorithm for supply chain inventory planning. *Journal of Manufacturing Technology Management, 16*, 7–17.

Sajadieh, M. S., & Akbari Jokar, M. R. (2009). Optimizing shipment, ordering and pricing policies in a two-stage supply chain with price-sensitive demand. *Transportation Research Part E: Logistics and Transportation Review, 45*, 564–571. http://dx.doi.org/10.1016/j.tre.2008.12.002

Sarkis, J., Talluri, S., & Gunasekaran, A. (2007). A strategic model for agile virtual enterprise partner selection. *International Journal of Operations & Production Management, 27*, 1213–1234.

Sawik, T. (2009). Coordinated supply chain scheduling. *International Journal of Production Economics, 120*, 437–451. http://dx.doi.org/10.1016/j.ijpe.2008.08.059

Sharifi, H., & Zhang, Z. (1999). A methodology for achieving agility in manufacturing organisations: An introduction. *International Journal of Production Economics, 62*, 7–22. http://dx.doi.org/10.1016/S0925-5273(98)00217-5

Sherehiy, B., Karwowski, W., & Layer, J. K. (2007). A review of enterprise agility: Concepts, frameworks, and attributes. *International Journal of Industrial Ergonomics, 37*, 445–460. http://dx.doi.org/10.1016/j.ergon.2007.01.007

Tam, M. C. Y., & Tummala, V. M. R. (2001). An application of the AHP in vendor selection of a telecommunications system. *Omega, 29*, 171–182. http://dx.doi.org/10.1016/S0305-0483(00)00039-6

Torabi, S. A., & Hassini, E. (2009). Multi-site production planning integrating procurement and distribution plans in multi-echelon supply chains: An interactive fuzzy goal programming approach. *International Journal of Production Research, 47*, 5475–5499. http://dx.doi.org/10.1080/00207540801905460

Tsourveloudis, N. C., & Valavanis, K. P. (2002). On the measurement of enterprise agility. *Journal of Intelligent and Robotic Systems, 33*, 329–342. http://dx.doi.org/10.1023/A:1015096909316

Wu, C., & Barnes, D. (2011). A literature review of decision-making models and approaches for partner selection in agile supply chains. *Journal of Purchasing & Supply Management, 17*, 256–274.

Yang, P. C., & Wee, H. M. (2000). Economic ordering policy of deteriorated item for vendor and buyer: An integrated approach. *Production Planning and Control, 11*, 474–480. http://dx.doi.org/10.1080/09537280050051979

Yang, M. F., & Lin, Y. (2010). Applying the linear particle swarm optimization to a serial multi-echelon inventory model. *Expert Systems with Applications, 37*, 2599–2608. http://dx.doi.org/10.1016/j.eswa.2009.08.021

Zhou, W.-Q., Chen, L., & Ge, H.-M. (2012). A multi-product multi-echelon inventory control model with joint replenishment strategy. *Applied Mathematical Modelling*. Retrieved from http://dx.doi.org/10.1016/j.apm.2012.04.054

Appendix A

Developing Equation 4 from Equation 2

$$\max \quad \Pi\left(T_{ij}^{r}, F_{ijk}^{r}, q_{ijk}, K_{ijk}^{r}\right) = \left[\sum_{i}\sum_{j}\sum_{k} sr_{ij}\, D_{ij}^{r}\left[F_{ijk}^{r} + \beta_{ijk}^{r}\left(1 - F_{ijk}^{r}\right)\right]\right] -$$

$$\left[\begin{array}{l} \sum_{i}\sum_{j}\sum_{k} \frac{Co_{ij}^{r}}{q_{ijk}\, T_{ij}^{r}} + \sum_{i}\sum_{j}\sum_{k} Cpu_{ij}^{r} q_{ijk}\, D_{ij}^{r}\left[F_{ijk}^{r} + \beta_{ijk}^{r}\left(1 - F_{ijk}^{r}\right)\right] + \\ \sum_{i}\sum_{j}\sum_{k} \frac{Ch_{ij}^{r}\, D_{ij}^{r}\, q_{ijk}\, T_{ij}^{r}\, F_{ijk}^{r^{2}}}{2} + \sum_{i}\sum_{j}\sum_{k} \frac{\beta_{ijk}^{r}\, Cb_{ij}^{r}\, D_{ij}^{r}\, q_{ijk}\, T_{ij}^{r}\left(1-F_{ijk}^{r}\right)^{2}}{2} + \sum_{i}\sum_{j}\sum_{k} Cg_{ij}^{r}\, D_{ij}^{r}\left(1 - \beta_{ijk}^{r}\right)\left(1 - F_{ijk}^{r}\right) \end{array}\right]$$

$$= \sum_{i}\sum_{j}\sum_{k} D_{ij}^{r}\left(sr_{ij} - Cpu_{ij}^{r}\, q_{ijk}\right)\left[F_{ijk}^{r} + \beta_{ijk}^{r}\left(1 - F_{ijk}^{r}\right)\right] -$$

$$\left[\begin{array}{l} \sum_{i}\sum_{j}\sum_{k} \frac{Co_{ij}^{r}}{q_{ijk}\, T_{ij}^{r}} + \sum_{i}\sum_{j}\sum_{k} \frac{Ch_{ij}^{r}\, D_{ij}^{r}\, q_{ijk}\, T_{ij}^{r}\, F_{ijk}^{r^{2}}}{2} + \sum_{i}\sum_{j}\sum_{k} \frac{\beta_{ijk}^{r}\, Cb_{ij}^{r}\, D_{ij}^{r}\, q_{ijk}\, T_{ij}^{r}\left(1-F_{ijk}^{r}\right)^{2}}{2} \\ \qquad + \sum_{i}\sum_{j}\sum_{k} Cg_{ij}^{r}\, D_{ij}^{r}\left(1 - \beta_{ijk}^{r}\right)\left(1 - F_{ijk}^{r}\right) \end{array}\right]$$

$$= \sum_{i}\sum_{j}\sum_{k} D_{ij}^{r}\left(sr_{ij} - Cpu_{ij}^{r}\, q_{ijk}\right)\left[1 - \left(1 - \beta_{ijk}^{r}\right)\left(1 - F_{ijk}^{r}\right)\right] -$$

$$\left[\begin{array}{l} \sum_{i}\sum_{j}\sum_{k} \frac{Co_{ij}^{r}}{q_{ijk}\, T_{ij}^{r}} + \sum_{i}\sum_{j}\sum_{k} \frac{Ch_{ij}^{r}\, D_{ij}^{r}\, q_{ijk}\, T_{ij}^{r}\, F_{ijk}^{r^{2}}}{2} + \sum_{i}\sum_{j}\sum_{k} \frac{\beta_{ijk}^{r}\, Cb_{ij}^{r}\, D_{ij}^{r}\, q_{ijk}\, T_{ij}^{r}\left(1-F_{ijk}^{r}\right)^{2}}{2} \\ \qquad + \sum_{i}\sum_{j}\sum_{k} Cg_{ij}^{r}\, D_{ij}^{r}\left(1 - \beta_{ijk}^{r}\right)\left(1 - F_{ijk}^{r}\right) \end{array}\right]$$

$$
\begin{aligned}
&= \sum_i \sum_j \sum_k D^r_{ij}\left(sr_{ij} - Cpu^r_{ij}\, q_{ijk}\right) - \\
&\left[\begin{array}{l}
\sum_i \sum_j \sum_k \frac{Co^r_{ij}}{q_{ijk} T^r_{ij}} + \sum_i \sum_j \sum_k \frac{Ch^r_{ij} D^r_{ij} q_{ijk} T^r_{ij} F^{r^2}_{ijk}}{2} + \sum_i \sum_j \sum_k \frac{\beta^r_{ijk} Cb^r_{ij} D^r_{ij} q_{ijk} T^r_{ij}(1-F^r_{ijk})^2}{2} \\
+ \sum_i \sum_j \sum_k Cg^r_{ij} D^r_{ij}\left(1-\beta^r_{ijk}\right)\left(1-F^r_{ijk}\right) + \sum_i \sum_j \sum_k D^r_{ij}\left(sr_{ij} - Cpu^r_{ij}\, q_{ijk}\right)\left(1-\beta^r_{ijk}\right)\left(1-F^r_{ijk}\right)
\end{array}\right]
\end{aligned}
$$

$$
\begin{aligned}
&= \sum_i \sum_j \sum_k D^r_{ij}\left(sr_{ij} - Cpu^r_{ij}\, q_{ijk}\right) - \\
&\left[\begin{array}{l}
\sum_i \sum_j \sum_k \frac{Co^r_{ij}}{q_{ijk} T^r_{ij}} + \sum_i \sum_j \sum_k \frac{Ch^r_{ij} D^r_{ij} q_{ijk} T^r_{ij} F^{r^2}_{ijk}}{2} + \sum_i \sum_j \sum_k \frac{\beta^r_{ijk} Cb^r_{ij} D^r_{ij} q_{ijk} T^r_{ij}(1-F^r_{ijk})^2}{2} \\
+ \sum_i \sum_j \sum_k D^r_{ij}\left(sr_{ij} - Cpu^r_{ij}\, q_{ijk} + Cg^r_{ij}\right)\left(1-\beta^r_{ijk}\right)\left(1-F^r_{ijk}\right)
\end{array}\right]
\end{aligned}
$$

$$
\begin{aligned}
&= \sum_i \sum_j \sum_k D^r_{ij}\left(sr_{ij} - Cpu^r_{ij}\, q_{ijk}\right) - \\
&\left[\begin{array}{l}
\sum_i \sum_j \sum_k \frac{Co^r_{ij}}{q_{ijk} T^r_{ij}} + \sum_i \sum_j \sum_k \frac{Ch^r_{ij} D^r_{ij} q_{ijk} T^r_{ij} F^{r^2}_{ijk}}{2} + \sum_i \sum_j \sum_k \frac{\beta^r_{ijk} Cb^r_{ij} D^r_{ij} q_{ijk} T^r_{ij}(1-F^r_{ijk})^2}{2} \\
+ \sum_i \sum_j \sum_k D^r_{ij}\, Cl^r_{ij}\left(1-\beta^r_{ijk}\right)\left(1-F^r_{ijk}\right)
\end{array}\right]
\end{aligned}
$$

Based upon the definition in the notation section, $sr_{ij} - Cpu^r_{ij}\, q_{ijk} + Cg^r_{ij} = Cl^r_{ij}$, which gives

$$
\begin{aligned}
\text{Min} \quad & TC^r_{EOQ}(T^r_{ij}, F^r_{ijk}, q_{ijk}, K^r_{ijk}) = \sum_i \sum_j \Big(\sum_k \frac{Co^r_{ij}}{q_{ijk} T^r_{ij}} + \frac{Ch^r_{ij} D^r_{ij} q_{ijk} T^r_{ij} F^{r^2}_{ijk}}{2} \\
& + \frac{\beta^r_{ijk} Cb^r_{ij} D^r_{ij} q_{ijk} T^r_{ij}(1-F^r_{ijk})^2}{2} + Cl^r_{ij}\, D^r_{ij}(1-\beta^r_{ijk})(1-F^r_{ijk})\Big)
\end{aligned}
$$

Regarding to our decision variables, $T^r_{ij}, F^r_{ijk}, q_{ijk}$, and K^r_{ijk}, $\sum_i \sum_j \sum_k D^r_{ij}\left(sr_{ij} - Cpu^r_{ij}\, q_{ijk}\right)$ have constant value, therefore, $\Pi\left(T^r_{ij}, F^r_{ijk}, q_{ijk}, K^r_{ijk}\right)$ is maximized by minimizing the value of the expression inside the square bracket, which is $TC^r_{EOQ}(T^r_{ij}, F^r_{ijk}, q_{ijk}, K^r_{ijk})$ from Equation 4.

Permissions

All chapters in this book were first published in CE, by Cogent OA; hereby published with permission under the Creative Commons Attribution License or equivalent. Every chapter published in this book has been scrutinized by our experts. Their significance has been extensively debated. The topics covered herein carry significant findings which will fuel the growth of the discipline. They may even be implemented as practical applications or may be referred to as a beginning point for another development.

The contributors of this book come from diverse backgrounds, making this book a truly international effort. This book will bring forth new frontiers with its revolutionizing research information and detailed analysis of the nascent developments around the world.

We would like to thank all the contributing authors for lending their expertise to make the book truly unique. They have played a crucial role in the development of this book. Without their invaluable contributions this book wouldn't have been possible. They have made vital efforts to compile up to date information on the varied aspects of this subject to make this book a valuable addition to the collection of many professionals and students.

This book was conceptualized with the vision of imparting up-to-date information and advanced data in this field. To ensure the same, a matchless editorial board was set up. Every individual on the board went through rigorous rounds of assessment to prove their worth. After which they invested a large part of their time researching and compiling the most relevant data for our readers.

The editorial board has been involved in producing this book since its inception. They have spent rigorous hours researching and exploring the diverse topics which have resulted in the successful publishing of this book. They have passed on their knowledge of decades through this book. To expedite this challenging task, the publisher supported the team at every step. A small team of assistant editors was also appointed to further simplify the editing procedure and attain best results for the readers.

Apart from the editorial board, the designing team has also invested a significant amount of their time in understanding the subject and creating the most relevant covers. They scrutinized every image to scout for the most suitable representation of the subject and create an appropriate cover for the book.

The publishing team has been an ardent support to the editorial, designing and production team. Their endless efforts to recruit the best for this project, has resulted in the accomplishment of this book. They are a veteran in the field of academics and their pool of knowledge is as vast as their experience in printing. Their expertise and guidance has proved useful at every step. Their uncompromising quality standards have made this book an exceptional effort. Their encouragement from time to time has been an inspiration for everyone.

The publisher and the editorial board hope that this book will prove to be a valuable piece of knowledge for researchers, students, practitioners and scholars across the globe.

List of Contributors

Mhand Hifi and Labib Yousef
Laboratory EPROAD - EA 4669 (ROAD Team), UFR des Sciences, University of Picardie Jules Verne, 7 rue du Moulin Neuf, 80000 Amiens, France

Pairat Jiamruangjarus and Thanakorn Naenna
Faculty of Engineering, Department of Industrial Engineering, Mahidol University, Nakhonpathom 73170, Thailand

Azhar Khan and C.S. Saini
Department of Food Engineering and Technology, Sant Longowal Institute of Engineering & Technology, Longowal 148106, Punjab, India

Hoda Mahmoudi and Hamed Fazlollahtabar
Faculty of Industrial Engineering, Iran University of Science and Technology, Tehran, Iran

Mehmet Nuri Bahşiş
Mining Engineer, Garp Lignite Corporation, Kütahya, Turkey

Yaşar Kasap
Faculty of Engineering, Department of Mining Engineering, Dumlupınar University, Kütahya 43100, Turkey

Nita H. Shah
Department of Mathematics, Gujarat University, Ahmedabad 380 009, Gujarat, India

Joshua Prakash
Faculty of Engineering and Green Technology, Universiti Tunku Abdul Rahman, Jalan Universiti, Bandar Baru Barat, Kampar, Perak 31900, Malaysia

Jeng Feng Chin
School of Mechanical Engineering, Universiti Sains Malaysia (USM), Engineering Campus, 14300 Nibong Tebal, Seberang Perai Selatan, Penang, Malaysia

Sanjay Kumar and Manish Duhan
Department of Mechanical Engineering, International Institute of Technology and Management, Sonepat, Haryana, India

Abid Haleem
Faculty of Engineering and Technology, Department of Mechanical Engineering, Jamia Millia Islamia, New Delhi, India

Sunil Kumar and Nidhi Handa
Department of Mathematics, Gurukula Kangari Vishwavidyalaya, Hairdwar, Uttarakhand, India

S.R. Singh
Department of Mathematics, D.N. (P.G.) College, Meerut, Uttar Pradesh, India

Dharmendra Yadav
Department of Mathematics, Vardhaman (P.G.) College, MJP Rohilkhand University, Bijnor, Uttar Pradesh, India

Masoud Rabbani and Reyhaneh Siadatian
School of Industrial Engineering, College of Engineering, University of Tehran, P.O. Box 11155-4563 Tehran, Iran

Hamed Farrokhi-Asl
School of Industrial Engineering, Iran University of Science & Technology, Tehran, Iran

Neda Manavizadeh
Department of Industrial Engineering, KHATAM University, Tehran, Iran

Cun Rong Li and Jiadong Cheng
Department of Industrial Engineering, School of Mechanical & Electronics Engineering, Wuhan University of Technology, 122 Luoshi Road, Hongshan District, Wuhan City 430070, P.R. China

Kim Jansson and Iris Karvonen
Business Ecosystems, Value Chains and Foresight, VTT Technical Research Centre of Finland, Tekniikantie 2, P.O. Box 1000, Espoo FI-02044 VTT, Finland

Seyed Hossein Nikokalam-Mozafar1 and, Behzad Ashjari1,
Department of Industrial Engineering, Tafresh University, Tafresh, Iran

Reza Tavakkoli-Moghaddam
School of Industrial Engineering, College of Engineering, University of Tehran, Tehran, Iran

Aida Omidvar
School of Industrial Engineering, South Tehran Branch, Islamic Azad University, Tehran, Iran

Elwyn Rebello, Frazer Vaz, Avino Colaco, Wellon Rodrigues and Geethalakshmi K
Department of Mechanical Engineering, P. C. College of Engineering, Verna, Goa, India

Debasis Sen and S. Mazumder
Solid State Physics Division, Bhabha Atomic Research Centre, Mumbai, India

A.O. Surendranathan
Department of Metallurgical & Materials Engineering, National Institute of Technology Karnataka, Surathkal, India

Maryam Alimardani, Hamed Rafiei and Masoud Rabbani
School of Industrial Engineering, College of Engineering, University of Tehran, Tehran, Iran

Index